Differential Geometry and Global Analysis

In Honor of Tadashi Nagano

CONTEMPORARY MATHEMATICS

777

Differential Geometry and Global Analysis

In Honor of Tadashi Nagano

AMS Special Session
Differential Geometry and Global Analysis,
Honoring the Memory of Tadashi Nagano (1930-2017)
January 16, 2020
Denver, Colorado

Bang-Yen Chen
Nicholas D. Brubaker
Takashi Sakai
Bogdan D. Suceavă
Makiko Sumi Tanaka
Hiroshi Tamaru
Mihaela B. Vajiac
Editors

AMS AMERICAN MATHEMATICAL SOCIETY

2020 *Mathematics Subject Classification.* Primary 53B25, 53C05, 53C15, 53C21, 53C35, 53C40, 53C42, 53C43, 58A05, 58C40.

Library of Congress Cataloging-in-Publication Data

Names: Chen, Bang-Yen, editor.
Title: Differential geometry and global analysis : in honor of Tadashi Nagano / Bang-Yen Chen, Nicholas D. Brubaker, Takashi Sakai, Bogdan D. Suceavă, Makiko Sumi Tanaka, Hiroshi Tamaru, Mihaela B. Vâjiac, editors.
Description: Providence, Rhode Island : American Mathematical Society, [2022] | Series: Contemporary mathematics, 0271-4132 ; 777 | "AMS Special Session on Differential Geometry and Global Analysis Honoring the Memory of Tadashi Nagano (1930-2017), January 16, 2020, Denver, Colorado." | Includes bibliographical references.
Identifiers: LCCN 2021041271 | ISBN 9781470460150 (paperback) | ISBN 9781470468743 (ebook)
Subjects: LCSH: Geometry, Differential – Congresses. | Global analysis (Mathematics) – Congresses. | AMS: Differential geometry – Local differential geometry – Local submanifolds | Differential geometry – Global differential geometry | Global analysis, analysis on manifolds – General theory of differentiable manifolds | Global analysis, analysis on manifolds – Calculus on manifolds; nonlinear operators
Classification: LCC QA641 .D3846 2022 | DDC 516.3/6–dc23
LC record available at https://lccn.loc.gov/2021041271

DOI: https://doi.org/10.1090/suceava3/777

Color graphic policy. Any graphics created in color will be rendered in grayscale for the printed version unless color printing is authorized by the Publisher. In general, color graphics will appear in color in the online version.

∞ The paper used in this book is acid-free and falls within the guidelines
established to ensure permanence and durability.
Visit the AMS home page at `https://www.ams.org/`

10 9 8 7 6 5 4 3 2 1 27 26 25 24 23 22

This volume represents the celebration of the mathematical legacy of Dr. Tadashi Nagano (January 9, 1930 – February 1, 2017), Professor of Mathematics, University of Tokyo (1959-1967), University of Notre Dame (1967-1986), and Sophia University (1986-2000). Recipient of the Geometry Prize from the Mathematical Society of Japan, 1994.

Contents

Preface

As one of the great Japanese differential geometers, Professor Tadashi Nagano's (1930–2017) career embodied scholarly excellence. He was an inspiring mentor, a dedicated educator, and a creative, one-of-a-kind researcher whose insights about geometry will undoubtedly be felt far into the future. To commemorate his life and work, which impacted the worldwide mathematical community over many decades, the editors of the present volume organized an AMS Special Session at the 2020 Joint Mathematical Meetings (in Denver, Colorado) dedicated to his memory. The celebration took place on January 16, 2020.

FIGURE 1. Professor Makiko Sumi Tanaka at the Joint Mathematical Meetings 2020, in Denver, Colorado, during the AMS Special Session dedicated to the memory of Professor Tadashi Nagano. The picture is taken during a discussion session planned by the organizers, in which the former students and collaborators of T. Nagano were invited to share their memories.

This volume for the *Contemporary Mathematics* series documents the content of that Special Session, exemplifying the mathematical influence of Professor Nagano and providing historical information that is crucial to the development of differential geometry in the second half of the 20th century. Professor Michel L. Lapidus, AMS

Associate Secretary for the Western Section, and Professor Eriko Hironaka, from the AMS editorial team, deserve special recognition for their roles in promoting its completion. The editors also extend the utmost gratitude to the contributors who submitted articles; to the referees, who provided high-quality feedback that immeasurably improved the content and exposition of the material contained in these pages; and to the presenters and participants of the January 16, 2020 celebration of Professor Nagano.

From the inception of this work, the editors planned to invite Professors Bang-Yen Chen, Takushiro Ochiai, and Makiko Sumi Tanaka to write an essay describing Professor Tadashi Nagano's biography and work. This paper opens the present volume. It incorporates testimonials from Professors Richard H. Escobales, Michael Clancy, Jih-Hsin Cheng, and John Burns about their academic work and interactions with Professor Tadashi Nagano.

FIGURE 2. Professor Bang-Yen Chen at the Joint Mathematical Meetings 2020, in Denver, Colorado, during the AMS Special Session dedicated to the memory of Professor Tadashi Nagano. Bang-Yen Chen described his work under the guidance of Tadashi Nagano in his essay *My education in differential geometry and my indebtness*, published in the volume titled *Geometry of Submanifolds*, No. 756 in the *Contemporary Mathematics* series.

Additionally, the editors invited contributions from experts who could shed new light on some topics approached in the work of Professor Nagano, including recent developments and generalizations in the geometry of symmetric spaces; minimal surfaces and minimal submanifolds; totally geodesic submanifolds and their classification; Riemannian, affine, projective, and conformal connections; the (M_+, M_-) method and its applications; maximal antipodal subsets; biharmonic and biconservative hypersurfaces in space forms; the geometry of Laplace operator on Riemannian manifolds; Chen-Ricci inequalities for Riemannian maps; and many other topics

FIGURE 3. Professor Hiroshi Tamaru at the Joint Mathematical Meetings 2020, in Denver, Colorado, during the AMS Special Session dedicated to the memory of Professor Tadashi Nagano. Hiroshi Tamaru defended his doctoral dissertation under Tadashi Nagano's supervision at Sophia University in 1998, under the title *The orbit types of symmetric spaces and their applications to generalized symmetric spaces*

that could attract the interest of any scholar working in differential geometry and global analysis on manifolds.

* * *

Tadashi Nagano was born in Taipei in 1930, when Taiwan was administered by Japan. He pursued his undergraduate studies at the University of Tokyo (from 1951 to 1954), and defended his doctoral thesis titled *On compact transformation groups with* $(n-1)-$*dimensional orbits* under Kentaro Yano's supervision at University of Tokyo in 1959. He worked at the University of Tokyo from April in 1959 to May 1967 as a lecturer (1959–1962) and as an assistant professor (1962–1967). Nagano moved to United States to pursue an academic career with the University of Notre Dame in 1967. Subsequently, he became a full professor of University of Notre Dame in 1969.

Tadashi Nagano was a visiting professor at University of California at Berkeley from 1962–1964, National Tsing Hua University in Taiwan twice, first in 1966 and then one more time in 1978. During his sabbatical leave in 1983, Professor Nagano conducted research at Osaka University. After a successful academic career with

FIGURE 4. From left to right: Tadashi Nagano, Kentaro Yano, and Katsumi Nomizu.

University of Notre Dame, Tadashi Nagano returned to Japan and became a professor with Sophia University in 1986. He retired from Sophia University at 70 years old in 2000.

Most notably, Tadashi Nagano co-authored 10 papers with Shoshichi Kobayashi in the interval 1966–1972, including *A theorem on filtered Lie algebras and its applications*, Bull. Amer. Math. Soc. 70 (1964), pp. 401–403.

Tadashi Nagano served an editor-in-chief of Tokyo Journal of Mathematics for several years since 1990. In 1994, Tadashi Nagano was presented with the Geometry Prize from Mathematical Society of Japan for his research achievements over a large field of the differential geometry, including a geometric construction of compact symmetric spaces.

The editors express their most profound gratitude for the discussions and the important pieces of historical information provided by Professors Takushiro Ochiai, Koichi Ogiue, and Yusuke Sakane, which helped them prepare the present material. The editors express their highest thanks to Ms. Reiko Nagano and Ms. Junko Nagano for their support and most valuable feedback during their work on this book.

FIGURE 5. Professor Tadashi Nagano and his spouse, Mrs. Shizuko Nagano.

FIGURE 6. Tadashi Nagano reading his message of condolence at the Meeting in Memory of Professor Yozo Matsushima, organized by Science School of Osaka University in 1983. Professor Nagano spent his sabbatical leave in 1983 at Osaka University.

FIGURE 7. Professor Richard Escobales at the Joint Mathematical Meetings 2020, in Denver, Colorado, during the AMS Special Session dedicated to the memory of Professor Tadashi Nagano.

Contemporary Mathematics
Volume **777**, 2022
https://doi.org/10.1090/conm/777/15634

To the memory of Professor Tadashi Nagano

Bang-Yen Chen, Takushiro Ochiai, and Makiko Sumi Tanaka

Professor Tadashi Nagano (January 9, 1930 – February 1, 2017) is one of the great Japanese differential geometers. All three authors were extremely fortunate to have Professor Tadashi Nagano as a supervisor during our graduate school years. Chen and Tanaka were also essential collaborators of Professor Nagano for his lifelong research on the geometry of symmetric spaces.

We are always interested in honoring a great mathematician with the history and substance of mathematics, who shaped research areas and created pathways that many mathematicians follow. The authors firmly believe in Professor Nagano being one such visionary, and we wanted a way to honor his legacy.

We asked the editors to let us write a biographical essay of Professor Nagano himself and his mathematical achievements (with a complete list of publications), to honor his memory. To our great pleasure, the editors granted our request with an encouraging message for us along with the decision to include the essay in the front matter of this book.

Considering his mathematical career, we have divided it into three periods, according to the place of his professional affiliation. Period I (1930–1967) is from his birth to the time when he was on the faculty of the Department of Mathematics of the University of Tokyo in Japan. Please note that Period I includes his biography from his birth (1930), leading to his earning a doctorate (1959), in addition to his professional affiliation with the University of Tokyo. Period II (1967–1986) includes his work while on the faculty of the mathematics department of the University of Notre Dame in the USA. Period III (1986–2000) shows his work during his time in the mathematics department of Sophia University in Japan. Professor Nagano was the supervisor for Ochiai in Period I, for Chen in Period II, and for Tanaka in Period III. It was only natural for Ochiai to cover the Period I part, for Chen, the Period II part, and for Tanaka, the Period III part.

It is our intention and our hope with this essay, the result of our combined efforts, to share our knowledge of and appreciation of Professor Nagano and his legacy with the readers of this work.

1. Period I (1930–1967) by Takushiro Ochiai

We start with Professor Nagano's experience of encountering mathematics up to his graduation from the university. This part is mostly based on his essay, "My

Encounter with Mathematics," in the Japanese mathematics magazine, "Mathematics Seminar Readings," published in 1988.

Tadashi Nagano was born in Taipei in 1930 when Taiwan was under Japanese rule. He studied at elementary school for six years and junior high school for five years under the prewar education system. His mother never told him to study, so that he had no study habits and lived leisurely. However, she had her unique way of raising her child. One day when he was a fifth- or sixth-grader, she showed him the proof of "the vertical angles being equal," which impressed him very much. On another occasion, she bought him a bulk of books from a paperback series such as "Japanese Children," covering various cultural topics, and "Our Science." She said nothing to him, but Professor Nagano enjoyed them a lot. Fibonacci sequence in the mathematics history book piqued his interest, but "Atomic Story" was an excellent book, and the book convinced him that he would become an atomic physicist one day.

During his junior high school days, he had free time for himself, except at the height of World War II (WWII), when he had to provide labor in the mountain areas for the military. When time allowed, he taught himself calculus and differential equations. He liked their systematic methods, and his calculation ability had improved considerably. However, he confesses that the power series expansion did not come across clearly for him even then.

A few years after the end of WWII, he returned to Japan from Taiwan with his family. Having passed the difficult entrance examination, he attended the First Higher School, Japan, the preparatory boarding school of Tokyo Imperial University, the most elite university in Japan. He started his boarding school life, where the most gifted students in Japan gathered. In the mathematics-related dormitory, he lived with many young brilliant students who eventually led post-war Japanese mathematics (Nagayoshi Iwahori, Ichiro Satake, Goro Shimura, Akio Hattori, Shoshichi Kobayashi ...). Here, Professor Nagano's interest began to shift from atomic physics to mathematics. Just like everyone else in the dormitory, he self-taught "Modern Algebra" by Van der Waerden and "Die Idee der Riemannschen Fläche" by H. Wyle. He was also attracted to the Γ-function and was devoted to the calculations. He made some "discoveries," and continued pursuing it until he realized that it was the field known as difference equations.

Thus, in 1951, it came naturally to Professor Nagano to go on to the Department of Mathematics in the Faculty of Science, University of Tokyo, which had been reorganized from Tokyo Imperial University and renamed under the new School Education Law. He recalls that voluntary seminars organized only by students at the mathematics department were crucial for his encounter with mathematics. Heated discussions by many brilliant students, including his high-school dormitory friends, took place in such seminars. Despite difficulty obtaining the necessary texts (especially from foreign countries) due to Japan being still an under-developed country, the quality and quantity of the voluntary seminars were exceptional. The environment was very stimulating and got even easy-going Professor Nagano highly motivated.

Professor Nagano ends his essay with the following paragraph. "Did I encounter mathematics? I really don't know. I wish I could have learned more from the elders, seniors, teachers, and other outstanding scholars. I think I may have missed valuable opportunities, but I never had an unfortunate experience in the classroom that made me dislike mathematics. Rather, I feel that I have received an education that allows me to encounter mathematics."

I recall that he once told me that to become a good researcher in mathematics, one needs an environment where he meets excellent mentors and worthy rivals. He continued that he was fortunate to be in such an environment at crucial times. I believe he meant in the above paragraph that the encounter with mathematics did not happen at a particular time in his life, but that it was the environment that nurtured him to be a mathematician that was in itself his encounter with mathematics.

Graduated from the Department of Mathematics in the Faculty of Science in 1954, Professor Nagano continued to the master's program and studied under Professor Kentaro Yano, one of the most influential leaders in the field of geometry in Japan. He defended his doctoral thesis [**6**] under Professor Yano. He started to work at the University of Tokyo from April 1959 to May 1967, first as a lecturer (1959–1962), and then as an associate professor (1962–1967).

A few decades back, Professor Yano visited Paris in 1936–1938 and studied under Élie Cartan, who is one of the great geometers in the history of mathematics together with Gauss and Riemann. He absorbed the ideas of Élie Cartan and continued in the 1940s and 50s to study conformal geometry, holonomy groups, and then transformation groups. Research on transformation groups concerns the following two problems.

(i) We know the possible maximum dimension of the automorphism groups of finite geometrical structures such as Riemann, affine, projective, and conformal connections. We also know the geometric structure when the dimension of the automorphism group matches the possible maximum dimension. The problem is to categorize the geometric structures when the dimension of the automorphism groups is lower than the possible maximum.

(ii) When a geometric structure such as Riemann metric is given to a manifold, one can think of its automorphism group, G. When a geometric structure more general than the original one, like the associated Levi-Civita connection and the associated conformal structure, is given to the manifold, the latter structure determines its automorphism group, L. Naturally, L contains G. The problem is to find the conditions when L and G are equal.

The problems (i) and (ii) are aligned with F. Klein's "Erlangen Program," which is to characterize geometries based on group theory.

Professor Nagano tackled the above problems under Professor Yano. His papers in his early professional life (1959–1967) with the University of Tokyo were products of close collaboration with Professor Yano, and almost all research results are on the transformation groups connected to the above problems. The dissertation [**6**], supervised by Professor Yano, is concerned with the problem (i). His best achievement during his early professional life is the paper [**27**], which addresses the problem (ii). The main result of the paper [**27**] is the complete classification of connected compact symmetric spaces with Lie transformation groups that contain the largest connected groups of isometries. Professor Nagano used to say to me that to study transformation groups is to advance F. Klein's "Erlangen Program."

Professor Nagano, who was an associate professor at the University of Tokyo, took a leave of absence from the university to spend two years from 1962 to 1964 at the University of California, Berkeley in the USA. He conducted joint research with Professor Shoshichi Kobayashi, his classmate during high school and university undergraduate years.

They presented their results in co-authored papers, [**24**]–[**26**], [**28**]–[**32**],[**35**], whose central theme is on the filtered Lie algebra theory. As an interesting application of the theory, they studied in [**28**] some special cases of irreducible symmetric spaces in [**27**] purely algebraically to prove those symmetric spaces are so-called symmetric R-spaces.

I would like to mention here how I met with Professor Nagano. When I was a freshman at the University of Tokyo in 1961, my modern calculus course as one of the mandated liberal arts subjects to all the freshmen was held by Professor Nagano. Of course, the subject was presented through the so-called epsilon-delta definition of limits. He was young and baby-faced, handsome. I was so charmed with him that I managed to survive through the totally new experience of the epsilon-delta definition of limits. Unfortunately, it was not the case for almost all the other classmates, even though they had had no problem with high school mathematics and passed the hardest entrance examination. During the summer semester of my sophomore year (1962), all the sophomores had to choose their majors. My father was against my preference to be a professional mathematician in my future. He was afraid that I would live an economically poor life like most of the scholars in those days. Professor Nagano took charge of the guidance for sophomores interested in choosing their major in the department of mathematics. After the meeting was over, I asked him how difficult it was to live with the scholars' anticipated low income in the academic institutes. He replied with a charming smile, "No problem! You just take a job in the USA." Believing in him, I chose mathematics as my major against my father's wish. My first course of ultra-modern linear algebra in the mathematics department in the fall semester of my sophomore year was fortunately taught by Professor Nagano, who had quite enchanted me by then. His lecture style was quite fresh to me. He kept going around from left to right in front of the blackboard while he delivered his lecture with a lit cigarette in his hand and sometimes kept silent, unmoved for one minute or so to try a better way of explaining some fresh idea.

When Professor Nagano returned to Japan from Berkeley in the fall of 1964, I was a fourth-year mathematics student, and he kindly became my supervisor. Professor Nagayoshi Iwahori, who was my supervisor until then, took a leave of absence to stay at the Institute for Advanced Study in Princeton, NJ in the USA. After I went to the master's program, Professor Nagano continued to be my supervisor for the master thesis. He took a leave of absence from the University of Tokyo to visit Tsinghua University in Taiwan for the fall semester of 1966. That makes it only two years that I received direct guidance from professor Nagano. However, while Professor Nagano was in his office, I went into his office rather freely and had fruitful mathematical discussions. Professor Nagano treated me politely without making an unpleasant face. He never denied the student's assertions and guided them towards collecting and expanding their ideas. Looking back now, I am afraid that I was taking away a lot of his precious time. I was quite fortunate to meet Professor Nagano at the time when it was decisively important to start my professional career as a mathematician.

His stay as a visiting scholar at the University of California, Berkeley, one of the most excellent universities in the world, was extremely productive. However, I learned later that he also felt helpless toward Japan's poor mathematics research

FIGURE 1. Left to right: An unidenfied person, Takushiro Ochiai, Tadashi Nagano and Morio Obata in Stanford, California, 1973

environment in the postwar economic recovery time. In addition to a poor research environment, most young mathematicians were barely making a living due to their low salary. Around that time, many excellent scientists, including mathematicians, had given up their permanent jobs in Japan and took the permanent job offers from the outstanding universities in Europe and the USA, partially to stabilize their daily lives economically. The media used to take up the phenomenon as a "brain drain," which was said to jeopardize the Japanese industrial and cultural future.

Professor Nagano firmly believed that it was of fundamental importance to avoid the brain drain in order to nurture the next generation of talented mathematicians. He also thought that it was necessary, at the very least, for the government to work hard to improve its treatment of scholars so that their daily living concerns would not interfere with their research.

Joined by his colleague, Professor Michio Kuga, at the mathematics department, Professor Nagano worked aggressively to urge many people in various fields to improve the universities' research and educational environment. NHK, Japan's national broadcaster, produced a TV program on the brain drain and welcomed Professor Kuga as a special guest commentator. It turned out that the public was not as generous to the scholars as they expected, since most of the nation was still poor. As far as I can remember, this is the first and the last time that Professor Nagano worked on things other than mathematical research. It was rare for mathematicians to work on social issues like the brain drain. Unfortunately, Professor Nagano and his collaborators' efforts did not work out well then, but I was deeply impressed by his interest in such matters. It is amazing to see that the research environment has improved so much more than they dreamed of almost half a century later. It is incomparable to what it used to be.

Professor Yozo Matsushima, highly respected as was Professor Yano and an earlier example in the situation of the brain drain, had become a Professor at Notre Dame University in the USA. Professor Matsushima invited Professor Nagano to the university, and Professor Nagano decided to quit the University of Tokyo in 1967 to continue his research at Notre Dame University, becoming another case of the brain drain.

By that time, I had completed the master's program under Professor Nagano's supervision and was an assistant in the mathematics department when Professor Matsushima also invited me to Notre Dame University. I decided to take a leave of absence to enroll in the doctoral program there as a Fulbright international student. I obtained a Ph.D. degree at Notre Dame with Professor Nagano from 1967 to 1969 as my thesis advisor.

In the fall of 1969, I left the nest of Professor Nagano's warm and sincere guidance. A strong recommendation letter from Professor Nagano helped me land a visiting lectureship at the University of California, Berkeley. Because of his long-time guidance and support, I could start my career path with plenty of hope in my future, and I owe him very much for what I am now. Since then, I have had almost no chance to meet him to discuss mathematics, but he has always stayed in my heart as my model of the ideal mathematician.

2. Period II (1967–1986) by Bang-Yen Chen

In the Fall of 1965, I enrolled as master student at the Institute of Mathematics of National Tsinghua University located in Hsinchu, Taiwan. Graduate schools of mathematics in Taiwan at that time had only master programs. Although National Tsinghua University had very few graduate students, the University had enough resources to invite famous professors from Japan and United States to teach graduate courses for one or two semesters. My first teacher of differential geometry at Tsinghua is Professor Tominosuke Otsuki from Tokyo Institute of Technology in the Fall semester of 1965. Among other topics, he lectured on total absolute curvature introduced by S. S. Chern and R. K. Lashof in [On the total curvature of immersed manifolds, Amer. J. Math. 79 (1957), 306–318].

According to the director of the Institute, Professor Otsuki suggested to invite Nagano for the Fall semester of 1966. Then Professor Nagano visited Tsinghua as a Guest Professor together with his family. Professor Nagano taught one course. Among other, he lectured on "Geometry of G-structures". During the semester, I worked on my master thesis under his supervision.

T. J. Willmore and B. A. Saleemi published their article [The total absolute curvature of immersed manifolds, J. London Math. Soc. 41 (1966), 153–160] to extend Chern-Lashof's results from compact Euclidean submanifolds to compact submanifolds in complete simply-connectecd Riemannian manifolds with non-positive sectional curvature. In his 1966 review of Willmore–Saleemi's article published in Mathematical Reviews [MR0185553 (32 #3019)], N. Kuiper pointed out that

"*The proof (of Willmore-Saleemi) unfortunately contains a mistake in the last two paragraphs of page 159. Consequently, the statements in Section 4 remain interesting conjectures but unproved.*"

In August of 1966, Professor Nagano showed Kuiper's review to me and suggested to me to settle this open problem as my thesis. Fortunately, I was able to solve it and received my M.S. degree in June of 1967.

Before Professor Nagano leaved Taiwan, he told me that he will be at University of Notre Dame starting from the Fall semester of 1967. Certainly, I would like to study my doctoral program at Notre Dame under his supervision as soon as possible. But during that time every physical able male student needed to serve one year in military after graduation. However, I was able to teach the course "Differential Geometry" to senior students at National Tsinghua University in the 1967–1968 academic year instead.

In August of 1968, I went to Notre Dame for my doctoral program under supervision of Nagano. At Notre Dame, I took the courses "Kaehler Manifolds" from Yozo Matsushima and "Several Complex Variables" from Yum-Tong Siu.

At the beginning of Fall semester in 1968, I went to Professor Nagano's office to seek his advice. Professor Nagano informed me that I can choose research problems by myself under the condition that I shall see him at least once a week to let him know my progress. Each week I reported to him, Professor Nagano always provided me his best suggestions to improve my results. It turned out that his advice gave me the best research training. In fact, under his advice I had complete freedom to seek my research problems including my doctoral thesis. On the other hand, every week I had to work very hard in order to report to him. Consequently, Professor Nagano had made extremely important life-long influence on my research.

Although I did not write any joint paper with Professor Nagano during my two years (1968–1970) at University of Notre Dame, we did write six joint papers [**51, 52, 57, 58, 60, 63**] on several subjects starting about five years after my graduation. In the paper [**60**], we introduced and studied the notions of harmonic metric and harmonic tensor as well as finding some relations between harmonic metrics, harmonic tensor, geodesic vector field, and Gauss map (for details in this topic, see my recent survey article "Differential geometry of identity maps: a survey" Mathematics 8 (2020), no. 8, Art. 1264, 33 pp).

In the next two subsections, I will explain briefly our motivations for writing the other joint papers [**51, 52, 57**] and [**58, 63**], respectively.

2.1. The motivation to introduce the (M_+, M_-)-method. If $M = G/H$ is a symmetric space and o is a given point in M, then the map $\sigma : G \to G$ defined by $\sigma(g) = s_o g s_o$ is an involutive automorphism, where s_o denotes the point symmetry at o. Let $\mathfrak{g}$ and $\mathfrak{h}$ be the Lie algebras of G and H, respectively. Then σ gives rise to an involutive automorphism of $\mathfrak{g}$, also denoted by σ. Denote by $\mathfrak{h}$ the eigenspace of σ with eigenvalue 1 and by $\mathfrak{m}$ the eigenspace of σ with eigenvalue -1 on $\mathfrak{g}$. Then we have the decomposition:

$$\mathfrak{g} = \mathfrak{h} + \mathfrak{m},$$

known as the Cartan decomposition. It is well-known that $\mathfrak{m}$ can be identified with the tangent space of M at o in a natural way. A linear subspace $\mathcal{L}$ of $\mathfrak{m}$ is said to form a *Lie triple system* if it satisfies $[[\mathcal{L}, \mathcal{L}], \mathcal{L}] \subset \mathcal{L}$. A well-known criterion of É. Cartan states that a subspace $\mathcal{L}$ of $\mathfrak{m}$ forms a Lie triple system if and only if $\mathcal{L}$ is the tangent space of a totally geodesic submanifold of M through o.

Huei-Shyong Lue received his Ph.D. degree from the University of Notre Dame in 1974 under the supervision of Professor Nagano. After that Lue joined Michigan State University as a postdoctor for one year. One of my joint papers with Lue done at MSU is "Differential geometry of $SO(n+1)/SO(2) \times SO(n)$ I, Geometriae Dedicata 4 (1975), 253–261", in which we classified totally geodesic surfaces of

FIGURE 2. Tsing-Houa Teng (left), Tadashi Nagano (center) and Bang-Yen Chen (right) at Tamkang University, Tamsui, Taiwan, 1966

$SO(n+1)/SO(2)\times SO(n)$ using Cartan's criterion via Lie triple systems. I presented this paper to Professor Nagano in 1976 while I visited him at Notre Dame, Professor Nagano expressed his interest on this paper. This is the starting point of a series of my joint papers with Professor Nagano.

Using Cartan's criterion via Lie triple systems, Professor Nagano and I were able to classify totally geodesic submanifolds of $SO(n+1)/SO(2) \times SO(n)$, which was published as part I of the series [**51, 52, 57**]. After that, our nature target is to classify totally geodesic submanifolds of other symmetric spaces of rank ≥ 2. However, after many attempts we realized that it is very difficult in general to classify totally geodesic submanifolds of compact symmetric spaces of higher rank via Cartan's criterion. Therefore, it became clear to us that we need to find a new method in order to study compact symmetric spaces in this respect.

On the other hand, it is well-known that fixed point sets play important roles in many branches of mathematics. For each point p in a compact symmetric space M, there is a natural and important notion of point symmetry s_p which fixes p and reverses geodesics through that point. Therefore the most natural candidate is to study fixed point sets of geodesic symmetries. This idea provided us the motivation to introduce the (M_+, M_-)-method via fixed point sets of geodesic symmetries in which Professor Nagano is the major contributor. The first results in the (M_+, M_-)-theory we obtained were then collected in [**52**] as part II of this series.

The third part is the article [**57**] which was written in 1980 joint with Pui-Fai Leung while both Leung and I visited Nagano at Notre Dame. Part III consists of two portions. The first portion studied the stability of Lagrangian submanifolds in Kaehler manifold. The main result of the first portion stated as

THEOREM. *Let $\phi : M \to \widetilde{M}$ be a compact minimal Lagrangian submanifold of a Kaehler manifold $\widetilde{M}$.*

(1) *If $\widetilde{M}$ has positive Ricci curvature, then the index of ϕ satisfies $i(\phi) \geq \beta_1(M)$, where $\beta_1(M)$ is the first Betti number of M. In particular, if the first cohomology group of M is nontrivial, i.e., $H^1(M;\mathbf{R}) \neq 0$, then M is alway unstable;*

(2) *If $\widetilde{M}$ has non-positive Ricci curvature, then M is alway stable.*

Although part III was never submitted for publication, this result was presented in several of my talks delivered in Japan while I was a visiting professor at Science University of Tokyo during 1980–1981 and it was also included in my book:

"*Geometry of submanifolds and its applications*"

published by Science University of Tokyo in 1981.

In the second portion of [**57**], we introduced an algorithm for determining the stability of complete totally geodesic submanifolds of compact symmetric spaces and some of its applications to the (M_+, M_-)-theory. This portion of [**57**] was later included as the last section of my 1990 book:

"*Geometry of slant submanifolds*"

published by Katholieke Universiteit Leuven in Belgium. This algorithm from [**57**] was later reformulated by Y. Ohnita to include the index and nullity in his paper:

"*On stability of minimal submanifolds in compact symmetric spaces*, Compositio Math. 64 (1987) 157–189".

Since 1980, our algorithm and also Ohnita's reformulation have been applied by many geometers in their studies of stabilities of totally geodesic submanifolds in compact symmetric spaces.

For more details on the (M_+, M_-)-theory, see my survey article:

"*The (M_+, M_-)-method on compact symmetric spaces and its applications*"

to be included in this special volume honoring the memory of Professor Tadashi Nagano.

2.2. Motivation to introduce maximal antipodal sets and 2-numbers. The two papers [**58, 63**] on 2-numbers can be regarded as a continuation of [**51, 52, 57**]. Our motivation to introduce the notion of 2-numbers came from following paper of A. Borel and J.-P. Serre:

"*Sur certains sousgroupes des groupes de Lie compacts.* Comm. Math. Helv. 27 (1953), 128–139."

I recalled that one day in the Fall semester of 1981 while I was a visiting professor of the University of Notre Dame, Professor Nagano showed me Borel and Serre's paper and asked me whether it is possible to study 2-rank using geometry? I answered him very intuitively that the 2-rank of a compact Lie group G shall relate to antipodal points on G. The next morning, Professor Nagano showed me his proof which provided a nice link between 2-number and 2-rank for a compact connected Lie group. After that we introduced and investigated the notions of maximal antipodal sets and 2-number for Riemannian manifolds. Our study on 2-numbers relied heavily on the (M_+, M_-)-theory which we did earlier in the 1970s. The first results on 2-numbers obtained in the Fall semester of 1981 was then summarized as the short note [**58**] published in 1982 by Comptes Rendus Paris. After that we spent a lot of times worked together either at South Bend, Indiana or at Okemos, Michigan, mostly during spring, summer and winter breaks in the period of 1982–1985. The detailed expanded version of our results on 2-numbers entitled

"*A Riemannian geometric invariant and its applications to a problem of Borel and Serre.*"

was submitted to Transactions of American Mathematical Society for publication in early 1985. The final version was published in 1988 as paper [**63**].

For applications of 2-numbers, see my survey article:

"*Two-numbers and their applications – a survey*, Bull. Belg. Math. Soc. Simon Stevin 25 (2018), 565–596."

2.3. Some remarks about Nagano's daily life. During the period in which Professor Nagano and I worked on (M_+, M_-)-theory and 2-numbers, very often either I drove to South Bend, Indiana and stayed at Nagano's house for a few days or he drove to Okemos, Michigan and stayed in my house for a few days. The distance between South Bend and Okemos is about 165 miles and it took about 2.5 hours by car. Often, Professor Nagano came to my house for a few days with his family. Similarly, often my family and I went to South Bend and stayed at his house for a few days. When my family stayed at his house, Mrs. Nagano will cook Japanese food for us. Similarly, when Nagano came, my wife will cook Taiwanese food for them. I recalled that once after we were done with mathematics, I drove both families to Cedar Point Theme Park in Ohio and spent a couple days there, while Nagano's two daughters were still very young.

Usually, during the visits we will spend many hours each day concentrated on our research projects. Professor Nagano loved to have a cup of hot black coffee with a piece of chocolate and sometime smoking and walking outside the house during his deep thinking. After finishing our discussion on mathematics, we will chat on many topics; very often he told me many subjects about literatures, arts, and also history in mathematics.

Before he went back to Japan, he gave me his book "*Hilbert*" written by C. Reid published by Springer in 1970. In this book, I found many underlines and remarks throughout the book, which clearly indicated that Professor Nagano had read this book very carefully and loved this book.

As a final remark, Professor Nagano is a great mathematician who had many very clever ideas and worked on mathematics very seriously. I recalled that once after our discussion on a research problem, he had to drive home from Okemos. But he drove on a wrong highway for about one hour before he realized that. Later, he told me that he made the mistake since he was still thinking on the mathematical problem we discussed earlier while he was driving home.

2.4. Some comments from his formal Ph.D. students at Notre Dame. Professor Nagano had supervised ten Ph.D. students at Notre Dame. Listed in chronological order of graduation years, they are Takushiro Ochiai (1969), Bang-Yen Chen (1970), Richard H. Escobales (1972), Heui-Shyong Lue (1974), Michael J. Clancy (1980), Frank Barnet (1981), David J. Jr, Welsh (1982), Jih-Hsin Cheng (1983), John T. Burns (1985), and Steohen P. Peterson (1985).

The following are comments sent to me from his Ph.D. students at Notre Dame.

1. From Richard H. Escobales. Professor Nagano was my advisor and what a wonderful advisor he was. He was also the thesis adviser for Dr. Frank Barnet and Dr. David Welsh, two of our graduates from Canisius College.

I failed the first oral exam for my Ph.D. I asked Professor Nagano to be my adviser and he agreed. He then decided that he would question me regularly on topics that I should have studied thoroughly. God bless him, Professor Nagano made me work! I remember one time he apologized for doing that. He didn't need to apologize, our adviser was right on target! Professor Nagano told me something like this. 'Mr. Escobales, you must forgive me if I am rude, but I do not know the nuances of your language'. After the workout that he gave me, I did well on my second oral exam.

Professor Nagano suggested that I work on Riemannian submersions and gave me his own paper on that subject as well as Barrett O'Neil's fundamental paper on Riemannian submersions.

Our adviser had wide intellectual interests, but he was first and foremost a mathematician.

When I finished my Ph.D., I asked Professor Nagano whether my thesis was junk. He said that it was definitely not junk. The fact that Professor Chern communicated my first paper was helpful for a new Ph.D.

2. From Michael Clancy. I first encountered Professor Nagano in October of 1975 having arrived at the University of Notre Dame, together with another student from Ireland, late into the first semester. This lateness was on account of our taking examinations during September at the University in Galway. We were directed to Professor Nagano's office as his lecture was about to start. He came across as a charming, kind and gentle man. It was in slight shock that we left his office as he had just then presented us with the homework assignment due for the next week. The handing up of homework was not the custom in Irish Universities at that time and I had not done such a thing since I left secondary school. I cannot recall what was the outcome of that homework.

Sometime during my second year at Notre Dame I asked Professor Nagano to become my thesis advisor. To my joy, he agreed to do so and thus began our weekly meetings in his office. These meetings could last from an hour to five hours or more and this arrangement continued until I completed the Ph.D. in 1980. Our meetings were devoted largely to mathematics but by no means exclusively so. Our personalities were not that dissimilar, despite the differences in our separate national characteristics. The one leading him to be taciturn and the other leading me towards the complete opposite. I soon became accustomed to those moments of long silence where the thinking should not be hindered. I suppose that he grew used to my talkativeness. His suggestion on one occasion that I should have been a poet may have been an effort to alert me to this tendency of mine or perhaps a gentle nudge that may be considered as career guidance. In any case, he was never but encouraging towards me and each meeting had the atmosphere of hope and expectation that now, on this occasion, some pearl or gem of mathematical wisdom would reveal itself. Sometimes it did.

On many occasions during those years when he was my advisor, Professor Nagano invited me to his home. It was a happy home and these visits I enjoyed not only for the warm company provided by all but also for the lovely food provided, as far as I can recall, exclusively by the charming Mrs. Nagano. On one such visit during the Summer months, I recall that Professor Nagano brought me to the back garden so that I could see the results of his horticultural endeavours. The farming was not extensive as his wry smile admitted but that same smile displayed the joy that he felt on seeing those plants grow and give fruit. My guess is that he was very happy with this success and that it was sufficient to satisfy his yearnings for any further activities in agriculture. Perhaps my guess is wrong.

During the same year that I left Notre Dame, a new student from Ireland, John Burns, a friend of mine, started on the graduate program in Mathematics there. John, seeing the same attractive qualities in Professor Nagano as had I, chose also to request to become a student of Professor Nagano. His request was accepted. This was particularly fortuitous for me because while in the past my friendship with John centered around the playing of Irish traditional music it now had a second and equally strong base, namely, a mutual interest in the same area of mathematical research. These two interests have continued to the present day.

In conclusion, I wish to express my highest regard for Professor Nagano. His interactions with me always displayed his care and concern for me and undoubtedly his wish that I should succeed. His influence on me has been very great and he lives warmly in my memory.

3. From Jih-Hsin Cheng. During 1979–1983, I had been learning differential geometry at Notre Dame under the guidance of Professor Tadashi Nagano. There was a special academic issue that I would like to mention. In the first year there I took Professor Andrea Sommese's course on algebraic geometry while studying Professor Koichi Ogiue's paper "Differential geometry of Kaehler submanifolds". It ended up that I could apply a result of Sommese or Van de Ven on algebraic geometry to solve a conjecture in submanifold geometry made by Ogiue for the case of algebraic submanifolds. The result was published in 1981. This is the first published paper of mine.

4. From John Burns. I fondly remember Prof. Nagano as an inspiring lecturer and a patient, encouraging Ph.D. supervisor. At the outset of my research Prof. Nagano was already supervising four other graduate students. This meant that a large part of every afternoon was spent helping students with their research. He was a man of unstinting generosity, both with his time and ideas. On my afternoons he would politely listen to my optimistic ideas on what I wished to be true, point to the board, smile with his warm sense of humour and say prove it! Of course, as he well knew, I almost never could. He derived great enjoyment from seeing his students develop. I recall him joyously declaring one day that I would be able to write a thesis, after his thorough checking of a proof (written on the board!) and he suggested that I show the proof to my fellow students. Another important support Prof. Nagano generously provided for graduate students was the careful guidance offered through the uncertainty of graduate work. My spirits would rise when told I know you don't believe me but that is an important step . That was encouragement to keep going. No recollection of Prof. Nagano would be complete without mentioning his modestness. Despite his hugely impressive body of research, I recall only one hint of self-acknowledgement when teaching a course on symmetric spaces he very early on introduced Polar sets and Meridians, stating that this approach was not standard yet. As well as sharing the hospitality of the family home on several occasions with students, I recall a particular treat when on one occasion after dinner Prof. Nagano offered to show us his toy. This turned out to be an electric synthesizer! and he entertained us with his playing, the enjoyment increasing for both him and us whenever a mistake was made. I was so surprised and delighted watching the Gods at play.

3. Period III (1986–2000) by Makiko Sumi Tanaka

After his return from the USA, Professor Nagano joined the department of Mathematics at Sophia University in 1986. Professors Soji Kaneyuki and Takeo Yokonuma who were knowledgeable in Lie algebra theory were there at the time. Professor Nagano continued his dedicated research on symmetric spaces, and he managed to improve his geometric theory of compact symmetric spaces, which he had started in collaboration with Prof. Bang-Yen Chen. I entered the Master's course of Sophia University, supervised by Professor Nagano, in 1987. Professor Nagano once told me that each compact symmetric space seemed to include individual geometric features, and he wanted to make it clear. This might have originated in [**27**]. His theory was developed in the category of symmetric spaces. He described the category of compact symmetric spaces in [**64**], whereas he defined the category of symmetric spaces in [**78**] by the following axioms:(i) The category of symmetric spaces is a subcategory of the category of smooth manifolds. (ii) For each point x of a symmetric space M there exists a point symmetry $s_x : M \to M$ which satisfies (a) $s_x \circ s_x = \mathrm{id}_M$, and (b) x is an isolated fixed point of s_x. (iii) A map $f : M \to N$ from a symmetric space M to another N is a morphism when it commutes with the geodesic symmetries, that is, $f \circ s_x = s_{f(x)} \circ f$ for any $x \in M$. (iv) Each point symmetry is a morphism (hence, an automorphism). We can show that there exists a unique affine connection on a connected symmetric space which is preserved by every automorphism; therefore, a point symmetry is a geodesic symmetry. When a symmetric space admits a Riemmanian metric g preserved by every automorphism, the invariant connection is the Levi-Civita connection of g, and hence the symmetric space is a Riemannian symmetric space. A subspace is a submanifold, the inclusion map of which is an injective morphism. Therefore, a subspace is totally geodesic and hence is a symmetric space.

Special subspaces called "a polar" and "a meridian" of a symmetric space play important roles in his theory, the so-called (M_+, M_-)-theory. The definition of a polar appeared in [**63**, Definition 1.8], and that of a meridian appeared in [**69**, 1.5 Definition], although these notions were first introduced in [**52**]. Let M be a symmetric space, and we take a point o in M. Each connected component of the fixed point set $F(s_o, M)$ of s_o is called a *polar* of o in M and denoted by M^+ or by $M^+(p)$ if it contains a point p with $s_o(p) = p$. There is a unique connected subspace whose tangent space at p is the orthogonal complement of the tangent space $T_pM^+(p)$ in T_pM, which is the connected component of $F(s_p \circ s_o, M)$ containing p. We call the subspace the *meridian* to $M^+(p)$ through p and denote it by M^- or $M^-(p)$. When M is a compact symmetric space, there exits a polar of a positive dimension except for when M is a product of spheres of a dimension greater than or equal to one. Polars and meridians of a compact symmetric space M are subspaces that inherit geometric structures from M. Each polar is a connected subspace of a lower dimension than the dimension of M, and each meridian is a connected subspace of the same rank as the rank of M. Therefore, an induction on polars or meridians can sometimes be used, and the arguments on M can sometimes be reduced to arguments on polars or meridians.

Professor Nagano proved the following fundamental theorem in [**69**, 1.15. Theorem]: A compact connected simple symmetric space M is determined by any one pair (M^+, M^-) of a polar M^+ and a meridian M^- to M^+. More precisely, another such symmetric space N is isomorphic with M if a pair (N^+, N^-) of a polar in N and a meridian to it is isomorphic to (M^+, M^-) in the following strong sense: M^+ is isomorphic with N^+, and M^- is isometric to N^- up to a constant multiple of the metric. Here, a symmetric space M is called semisimple if the identity component of the group of automorphisms of M is semisimple, and M is called simple if M is semisimple but not a local product of two symmetric spaces of positive dimensions ([**69**, 1.7. Definition]). The paper is one of a series of five papers entitled "The involutions of compact symmetric spaces." Part I [**68**] and Part II [**69**] constitute the main part of Professor Nagano's research in Period III. In Part I [**68**], Professor Nagano made a global study of polars, where their local study was done in [**52**], and he determined all the involutions t and the fixed point set $F(t, M)$ for every compact simple symmetric space M and a few others. He also described the shortest geodesics to a polar using root systems. He explained some interesting applications to illustrate the significance of the results and the use of the geometric method. In Part II [**69**], he proved that any compact connected simple symmetric space M is determined by any one pair (M^+, M^-) as mentioned before. He also gave a local characterization of M^-, that is, he proved that the root system of M^- is obtained from that of M with a simple rule. Part III [**74**], Part IV [**76**], and Part V [**77**] are joint papers with me. In Part III, we constructed Riemannian submersions of a certain type out of polars from which we got a double-tiered fibration using the Cartan embedding $M = G/K$ into G. We studied the correspondence between these double-tiered fibrations and the real simple graded Lie algebras of the second kind. In Part IV, we determined the signature of every compact oriented symmetric space M as well as the self-intersections of subspaces N in M. In Part V, we proved that the roots of a symmetric space defined with the curvature using the Jacobi equation, made a root system, and we determined their multiplicities. We established the known facts in a more geometric way.

In 1994, Professor Nagano was presented with the Geometry Prize from the Mathematical Society of Japan for his research achievements over a large field of differential geometry, including a geometric construction of theory of compact symmetric spaces. Professor Nagano's new geometric approach to symmetric space theory was highly evaluated because it was not just a reconstruction of known theory, and it provided new knowledge and various applications that could not be clarified using conventional methods.

Incidentally, when I was an undergraduate student, I was a student of Professor Shigeru Ishihara, who was one of the collaborators of Professor Kentaro Yano. When I told Professor Ishihara that I wanted to go on to graduate school, he recommended that I go to Sophia University and that I study under Professor Nagano's supervision. I met Professor Nagano for the first time when I took my oral examination. He asked me many questions with a gentle smile and took a sincere interest in my answers.

When I was a Master's student, Professor Nagano made time for discussing mathematics with me almost everyday. In the preface of his book written in Japanese "Mathematics of Surfaces," published in 1968, he wrote: "I think that we are slow to conduct original research activities these days in Japan. Basic steps need

FIGURE 3. Tadashi Nagano (left) and Makiko Sumi Tanaka (right) at Nagano's retirement party in Sophia University, 2000

to be learned as soon as possible so that research activities can be done earlier." Professor Nagano suggested that I start collaborating on research activities with him. I had to study very hard every day, but I was able to experience the joy of research on mathematics early in my career. The first research problem I tackled was classifying subspaces isomorphic to spheres in compact symmetric spaces [**67**]. Professor Nagano was very tolerant and patient in teaching me. He would always listen to my ideas carefully and let me finish my thoughts, even when he already clearly understood everything I wanted to say. He also would tell me his ideas, which usually astonished me and always deepened my understanding. Through the experience of discussing mathematics with Professor Nagano, I learned so many things, and I was able to grow as a researcher.

Professor Nagano loved chocolate. Every time we finished our discussion, we had some chocolate and a cup of coffee together. He sometimes justified eating chocolate by saying that we used our brains a lot, and the brain runs on glucose. These are my heartwarming memories of Professor Nagano.

Dr. Hiroshi Tamaru is another former Ph.D. student of Professor Nagano at Sophia University. Currently he is a professor at Osaka City University in Japan. When he was a freshman, he attended Professor Nagano's class of set theory. He was impressed that the layout of material on the blackboard was designed to convey the structure of the proof; that was also mentioned by Dr. Michael Clancy, a former Ph.D. student at the University of Notre Dame. Dr. Tamaru uses the same method when he teaches a class, he said. On the other hand, the content was quite high level. Professor Nagano gave a homework question to the class "Prove that the direct product of two connected topological spaces is connected." This question was given for the consecutive holidays from the end of April to the beginning of May, whereas the class itself started at the beginning of April. When he was a Master's student, Dr. Tamaru attended Professor Nagano's lecture on the introduction to symmetric spaces. After Professor Nagano gave the definition and various examples of symmetric spaces, he suddenly asked "Mr. Tamaru, do you want to define a morphism of symmetric spaces?" Dr. Tamaru said: Once one defines a notion, a morphism of it should be naturally be defined. It was not surprising, but it was what I was really aware of at the time. Dr. Tamaru remembered that Professor Nagano always said that if you had to learn everything you needed to know and if you had to have all the necessary background knowledge beforehand, you would never be able to start your research. Instead, you could learn what you needed to know as you went along, even after you started your research. Dr. Tamaru said that Professor Nagano's stance suited him very well, and he felt really lucky. Dr. Tamaru succeeded with Professor Nagano's stance, and some of his students got doctoral degrees as a result. He said it would be impossible without the influence of Professor Nagano.

FIGURE 4. Left to right: Hiroshi Tamaru, Makiko Sumi Tanaka, Tadashi Nagano and Bang-Yen Chen at Professor Nagano's retirement party Tokyo, February 23, 2000

Ms. Terumi Ogawa, who was a former Master's student of Professor Nagano at Sophia University, remembered Professor Nagano and said: Professor Nagano came to the graduate students' room almost every evening and said to me "Do you have a minute?", then we would start a discussion on mathematics. He guided me really warmly and cordially. I think that he took care of his graduate students as well as or better than anyone, and I felt truly privileged and pleased to have Professor Nagano's daily guidance.

Finally, in our joint paper published in 2012, Dr. Hiroyuki Tasaki and I proved that the intersection of any two real forms of a Hermitian symmetric space of compact type is an antipodal set of each real form if the intersection is discrete; moreover, it is a great antipodal set if two real forms are congruent. The result was amazing to me. We proved the latter part using an induction on polars. I wanted to tell Professor Nagano this result; unfortunately Professor Nagano was already in a nursing home at the time, and we were not able to let him know. I believe that if he could have known the result, he would have said "That's interesting!" with a big smile, like he used to have.

List of Professor Nagano's publications

Reseach papers

[1] Kentaro Yano and Tadashi Nagano, *Some theorems on projective and conformal transformations*, Nederl. Akad. Wetensch. Proc. Ser. A. 60 = Indag. Math. **19** (1957), 451–458. MR0110993

[2] Tadashi Nagano, *Sur des hypersurfaces et quelques groupes d'isométries d'un espace riemannien* (French), Tohoku Math. J. (2) **10** (1958), 242–252, DOI 10.2748/tmj/1178244662. MR101534

[3] Tadashi Nagano, *On conformal transformations of Riemannian spaces*, J. Math. Soc. Japan **10** (1958), 79–93, DOI 10.2969/jmsj/01010079. MR110991

[4] Kentaro Yano and Tadashi Nagano, *Einstein spaces admitting a one-parameter group of conformal transformations*, Ann. of Math. (2) **69** (1959), 451–461, DOI 10.2307/1970193. MR101535

[5] Tadashi Nagano, *Compact homogeneous spaces and the first Betti number*, J. Math. Soc. Japan **11** (1959), 4–9, DOI 10.2969/jmsj/01110004. MR103944

[6] Tadashi Nagano, *Transformation groups with $(n-1)$-dimensional orbits on non-compact manifolds*, Nagoya Math. J. **14** (1959), 25–38. MR104760

[7] Tadashi Nagano, *Isometries on complex-product spaces*, Tensor (N.S.) **9** (1959), 47–61. MR107877

[8] Tadashi Nagano, *Homogeneous sphere bundles and the isotropic Riemann manifolds*, Nagoya Math. J. **15** (1959), 29–55. MR108810

[9] Tadashi Nagano, *The projective transformation on a space with parallel Ricci tensor*, Kōdai Math. Sem. Rep. **11** (1959), 131–138. MR109330

[10] Tadashi Nagano, *On some compact transformation groups on spheres*, Sci. Papers College Gen. Ed. Univ. Tokyo **9** (1959), 213–218. MR113968

[11] Kentaro Yano and Tadashi Nagano, *The de Rham decomposition, isometries and affine transformations in Riemannian spaces*, Jpn. J. Math. **29** (1959), 173–184, DOI 10.4099/jjm1924.29.0_173. MR120591

[12] Tadashi Nagano, *The conformal transformation on a space with parallel Ricci tensor*, J. Math. Soc. Japan **11** (1959), 10–14, DOI 10.2969/jmsj/01110010. MR124010

[13] Tadashi Nagano, *On the space problems of Wang-Tits-Freudenthal. A characterization of the classical spaces by the congruence theorem for segments* (Japanese), Sūgaku **11** (1959/60), 205–217. MR137075

[14] Tadashi Nagano, *Almost complex structures* (Japanese), Sūgaku **11** (1959/60), 130–133. MR138076

[15] Tadashi Nagano and Tsunero Takahashi, *Homogeneous hypersurfaces in euclidean spaces*, J. Math. Soc. Japan **12** (1960), 1–7, DOI 10.2969/jmsj/01210001. MR114183

[16] Tadashi Nagano, *On fibred Riemann manifolds*, Sci. Papers College Gen. Ed. Univ. Tokyo **10** (1960), 17–27. MR157325

[17] Kentaro Yano and Tadashi Nagano, *On geodesic vector fields in a compact orientable Riemannian space*, Comment. Math. Helv. **35** (1961), 55–64, DOI 10.1007/BF02567005. MR124854

[18] Kentaro Yano and Tadashi Nagano, *Les champs des vecteurs géodésiques sur les espaces symétriques* (French), C. R. Acad. Sci. Paris **252** (1961), 504–505. MR124855

[19] Tadashi Nagano, *On the minimum eigenvalues of the Laplacians in Riemannian manifolds*, Sci. Papers College Gen. Ed. Univ. Tokyo **11** (1961), 177–182. MR144283

[20] T. Nagano, *Differential geometry and analytic group theory method in foundations of geometry*, Algebraical and Topological Foundations of Geometry (Proc. Colloq., Utrecht, 1959), Pergamon, Oxford, 1962, pp. 131–134. MR0141729

[21] Soji Kaneyuki and Tadashi Nagano, *On the first Betti numbers of compact quotient spaces of complex semi-simple Lie groups by discrete subgroups*, Sci. Papers College Gen. Ed. Univ. Tokyo **12** (1962), 1–11. MR145008

[22] Soji Kaneyuki and Tadashi Nagano, *On certain quadratic forms related to symmetric riemannian spaces*, Osaka Math. J. **14** (1962), 241–252. MR159347

[23] Akihiko Morimoto and Tadashi Nagano, *On pseudo-conformal transformations of hypersurfaces*, J. Math. Soc. Japan **15** (1963), 289–300, DOI 10.2969/jmsj/01530289. MR155341

[24] Shoshichi Kobayashi and Tadashi Nagano, *On projective connections*, J. Math. Mech. **13** (1964), 215–235. MR0159284

[25] Shoshichi Kobayashi and Tadashi Nagano, *A theorem on filtered Lie algebras and its applications*, Bull. Amer. Math. Soc. **70** (1964), 401–403, DOI 10.1090/S0002-9904-1964-11113-7. MR162892

[26] Shoshichi Kobayashi and Tadashi Nagano, *On filtered Lie algebras and geometric structures. I*, J. Math. Mech. **13** (1964), 875–907. MR0168704

[27] Tadashi Nagano, *Transformation groups on compact symmetric spaces*, Trans. Amer. Math. Soc. **118** (1965), 428–453, DOI 10.2307/1993971. MR182937

[28] Shoshichi Kobayashi and Tadashi Nagano, *On filtered Lie algebras and geometric structures. II*, J. Math. Mech. **14** (1965), 513–521. MR0185042

[29] Shoshichi Kobayashi and Tadashi Nagano, *On filtered Lie algebras and geometric structures. III*, J. Math. Mech. **14** (1965), 679–706. MR0188364

[30] Shoshichi Kobayashi and Tadashi Nagano, *On a fundamental theorem of Weyl-Cartan on G-structures*, J. Math. Soc. Japan **17** (1965), 84–101, DOI 10.2969/jmsj/01710084. MR192438

[31] Shoshichi Kobayashi and Tadashi Nagano, *On filtered Lie algebras and geometric structures. IV*, J. Math. Mech. **15** (1966), 163–175. MR0195993

[32] Shoshichi Kobayashi and Tadashi Nagano, *On filtered Lie algebras and geometric structures. V*, J. Math. Mech. **15** (1966), 315–328. MR0195994

[33] Tadashi Nagano, *Linear differential systems with singularities and an application to transitive Lie algebras*, J. Math. Soc. Japan **18** (1966), 398–404, DOI 10.2969/jmsj/01840398. MR199865

[34] T. Nagano, *On transitive infinite Lie algebras* (Japanese), Sūgaku **18** (1966), 65–74. MR212060

[35] Shoshichi Kobayashi and Tadashi Nagano, *A report on filtered Lie algebras*, Proc. U.S.-Japan Seminar in Differential Geometry (Kyoto, 1965), Nippon Hyoronsha, Tokyo, 1966, pp. 63–70. MR0215887

[36] Tadashi Nagano, *A problem on the existence of an Einstein metric*, J. Math. Soc. Japan **19** (1967), 30–31, DOI 10.2969/jmsj/01910030. MR205196

[37] Tadashi Nagano, *1-forms with the exterior derivative of maximal rank*, J. Differential Geometry **2** (1968), 253–264. MR239526

[38] Tadashi Nagano, *Homotopy invariants in differential geometry. I*, Trans. Amer. Math. Soc. **144** (1969), 441–455, DOI 10.2307/1995291. MR253246

[39] Tadashi Nagano, *Recent trends in differential geometry* (Japanese), Sūgaku **21** (1969), no. 1, 54–64. MR514028

[40] Tadashi Nagano, *Homotopy invariants in differential geometry*, Memoirs of the American Mathematical Society, No. 100, American Mathematical Society, Providence, R.I., 1970. MR0268906

[41] Shoshichi Kobayashi and Tadashi Nagano, *Riemannian manifolds with abundant isometries*, Differential geometry (in honor of Kentaro Yano), Kinokuniya, Tokyo, 1972, pp. 195–219. MR0358623

[42] T. Nagano and K. Yagi, *The affine structures on the real two-torus. I*, Bull. Amer. Math. Soc. **79** (1973), 1251–1253, DOI 10.1090/S0002-9904-1973-13400-7. MR326611

[43] Tadashi Nagano and Katsumi Yagi, *The affine structures on the real two-torus. I*, Osaka Math. J. **11** (1974), 181–210. MR377917

[44] K. deCesare and T. Nagano, *On compact foliations*, Differential geometry (Proc. Sympos. Pure Math., Vol. XXVII, Stanford Univ., Stanford, Calif., 1973), Amer. Math. Soc., Providence, R. I., 1975, pp. 277–281. MR0377924

[45] Tadashi Nagano and Brian Smyth, *Sur les surfaces minimales hyperelliptiques dans un tore* (French, with English summary), C. R. Acad. Sci. Paris Sér. A-B **280** (1975), no. 22, Aii, A1527–A1529. MR380598

[46] Tadashi Nagano and Brian Smyth, *Minimal varieties and harmonic maps in tori*, Comment. Math. Helv. **50** (1975), 249–265, DOI 10.1007/BF02565749. MR390974

[47] Tadashi Nagano and Brian Smyth, *Minimal varieties in tori*, Differential geometry (Proc. Sympos., Pure Math., Vol. XXVII, Part 1, Stanford Univ., Stanford, Calif., 1973), Amer. Math. Soc., Providence, R.I., 1975, pp. 189–190. MR0417993

[48] T. Nagano and B. Smyth, *Minimal surfaces in tori by Weyl groups*, Proc. Amer. Math. Soc. **61** (1976), no. 1, 102–104 (1977), DOI 10.2307/2041673. MR431047

[49] Tadashi Nagano, *On compact transformation groups*, Differential geometry (Proc. Conf., Michigan State Univ., East Lansing, Mich., 1976), Dept. Math., Michigan State Univ., East Lansing, Mich., 1976, pp. 75–84. MR0454996

[50] Frank Connolly and Tadashi Nagano, *The intersection pairing on a homogeneous Kähler manifold*, Michigan Math. J. **24** (1977), no. 1, 33–39. MR451273

[51] Bang-yen Chen and Tadashi Nagano, *Totally geodesic submanifolds of symmetric spaces. I*, Duke Math. J. **44** (1977), no. 4, 745–755. MR458340

[52] Bang-yen Chen and Tadashi Nagano, *Totally geodesic submanifolds of symmetric spaces. II*, Duke Math. J. **45** (1978), no. 2, 405–425. MR487902

[53] Tadashi Nagano and Brian Smyth, *Periodic minimal surfaces*, Comment. Math. Helv. **53** (1978), no. 1, 29–55, DOI 10.1007/BF02566064. MR487909

[54] Tadashi Nagano and Brian Smyth, *Minimal surfaces in tori by Weyl groups. II*, Minimal submanifolds and geodesics (Proc. Japan-United States Sem., Tokyo, 1977), North-Holland, Amsterdam-New York, 1979, pp. 117–120. MR574260

[55] T. Nagano and B. Smyth, *Periodic minimal surfaces and Weyl groups*, Acta Math. **145** (1980), no. 1-2, 1–27, DOI 10.1007/BF02414183. MR586592

[56] T. Nagano, *Stability of harmonic maps between symmetric spaces*, Harmonic maps (New Orleans, La., 1980), Lecture Notes in Math., vol. 949, Springer, Berlin-New York, 1982, pp. 130–137. MR673587

[57] Bang-Yen Chen, Pui-Fai Leung and Tadashi Nagano, *Totally geodesic submanifolds of symmetric spaces III*, preprint 1980; eprint: arXiv:1307.7325v2 [math.DG].

[58] Bang-yen Chen and Tadashi Nagano, *Un invariant géométrique riemannien* (French, with English summary), C. R. Acad. Sci. Paris Sér. I Math. **295** (1982), no. 5, 389–391. MR684733

[59] Tadashi Nagano and Masaru Takeuchi, *Signature of quaternionic Kaehler manifolds*, Proc. Japan Acad. Ser. A Math. Sci. **59** (1983), no. 8, 384–386. MR726530

[60] Bang-Yen Chen and Tadashi Nagano, *Harmonic metrics, harmonic tensors, and Gauss maps*, J. Math. Soc. Japan **36** (1984), no. 2, 295–313, DOI 10.2969/jmsj/03620295. MR740319

[61] Tadashi Nagano and Takushiro Ochiai, *On compact Riemannian manifolds admitting essential projective transformations*, J. Fac. Sci. Univ. Tokyo Sect. IA Math. **33** (1986), no. 2, 233–246. MR866391

[62] Tadashi Nagano and Masaru Takeuchi, *Cohomology of quaternionic Kaehler manifolds*, J. Fac. Sci. Univ. Tokyo Sect. IA Math. **34** (1987), no. 1, 57–63. MR882124

[63] Bang-Yen Chen and Tadashi Nagano, *A Riemannian geometric invariant and its applications to a problem of Borel and Serre*, Trans. Amer. Math. Soc. **308** (1988), no. 1, 273–297, DOI 10.2307/2000963. MR946443

[64] Tadashi Nagano, *The involutions of compact symmetric spaces*, Tokyo J. Math. **11** (1988), no. 1, 57–79, DOI 10.3836/tjm/1270134261. MR947946

[65] Tadashi Nagano, *Work of Shoshichi Kobayashi* (Japanese), Sūgaku **41** (1989), no. 1, 64–75. MR994402

[66] Tadashi Nagano and Sumi Makiko, *The structure of the symmetric space with applications*, Geometry of manifolds (Matsumoto, 1988), Perspect. Math., vol. 8, Academic Press, Boston, MA, 1989, pp. 111–128. MR1040520

[67] Tadashi Nagano and Sumi Makiko, *The spheres in symmetric spaces*, Hokkaido Math. J. **20** (1991), no. 2, 331–352, DOI 10.14492/hokmj/1381413846. MR1114410

[68] Tadashi Nagano and Jirō Sekiguchi, *Commuting involutions of semisimple groups*, Tokyo J. Math. **14** (1991), no. 2, 319–327, DOI 10.3836/tjm/1270130375. MR1138170

[69] Tadashi Nagano, *The involutions of compact symmetric spaces. II*, Tokyo J. Math. **15** (1992), no. 1, 39–82, DOI 10.3836/tjm/1270130250. MR1164185

[70] Tadashi Nagano and Sumi Makiko, *Stability of p-harmonic maps*, Tokyo J. Math. **15** (1992), no. 2, 475–482, DOI 10.3836/tjm/1270129472. MR1197114

[71] Tadashi Nagano, *The development of the theory of symmetric spaces and differential geometry* (Japanese), Sūgaku **44** (1992), no. 3, 245–249. MR1204635

[72] Tadashi Nagano, *Morio Obata and his works*, Geometry and its applications (Yokohama, 1991), World Sci. Publ., River Edge, NJ, 1993, pp. 3–8. MR1343253

[73] Tadashi Nagano, *A hymn to the symmetric spaces*, Geometry and its applications (Yokohama, 1991), World Sci. Publ., River Edge, NJ, 1993, pp. 147–159. MR1343268

[74] Tadashi Nagano and Makiko Sumi Tanaka, *The involutions of compact symmetric spaces. III*, Tokyo J. Math. **18** (1995), no. 1, 193–212, DOI 10.3836/tjm/1270043621. MR1334718

[75] Tadashi Nagano, *Symmetric spaces and quaternionic structures*, Quaternionic structures in mathematics and physics (Trieste, 1994), Int. Sch. Adv. Stud. (SISSA), Trieste, 1998, pp. 203–218. MR1645781

[76] Tadashi Nagano and Makiko Sumi Tanaka, *The involutions of compact symmetric spaces. IV*, Tokyo J. Math. **22** (1999), no. 1, 193–211, DOI 10.3836/tjm/1270041622. MR1692030

[77] Tadashi Nagano and Makiko S. Tanaka, *The involutions of compact symmetric spaces. V*, Tokyo J. Math. **23** (2000), no. 2, 403–416, DOI 10.3836/tjm/1255958679. MR1806473

[78] Tadashi Nagano, *Geometric theory of symmetric spaces* (Japanese), Sūrikaisekikenkyūsho Kōkyūroku **1206** (2001), 55–82. Geometry of submanifolds (Japanese) (Kyoto, 2001). MR1853312

Books

[1] *Mathematics of Surfaces*, Baifukan Co., Ltd. 1968 (Japanese).

[2] *Global Methods for Calculus of Variations*, Kyoritsu Shuppan Co., 1971 (Japanese).

Professor emeritus, Michigan State University, Address: 2231 Tamarack Drive, Okemos, Michigan 48864

Email address: chenb@msu.edu

Professor emeritus, the University of Tokyo, Address: 4-11-7-107 Yushima, Bunkyou-Ku, Tokyo 113-0034, Japan

Email address: takushiroochiai@gmail.com

Department of Mathematics, Faculty of Science and Technology, Tokyo University of Science, Noda, Chiba, 278-8510, Japan

Email address: tanaka_makiko@ma.noda.tus.ac.jp

Contemporary Mathematics
Volume **777**, 2022
https://doi.org/10.1090/conm/777/15635

Bifurcations of minimal surfaces via index theory

Nicholas D. Brubaker

ABSTRACT. We present a method for constructing new families of minimal surfaces from an existing minimal surface by applying techniques for proving bifurcations from a simple eigenvalues to the minimal surface equation $H = 0$. Although there is no explicit free parameter in $H = 0$, we consider surfaces over compact domains and then vary the effective radius of the domain, which relates the method to the index of a complete minimal surface. We demonstrate the method for the standard catenoid and degree n Enneper surfaces, although it is certainly more widely applicable.

1. Introduction

The index of a minimal surface defines the number of independent, compactly-supported normal deformations that reduce the (local) area of the surface. By interpreting the area as an energy — in a manifest connection to soap films — the index generalizes the notion of physical stability [**9**]. If it is zero, then the surface is stable since there are no functional reducing deformations. If it is positive, then the surface is unstable, and the resulting integer hierarchically characterizes the degree of instability, relating it structurally to an infinite-dimensional saddle surface.

A straightforward method for calculating the index of a minimal surface is to add the dimensions of the eigenspaces associated with the self-adjoint differential operator J defined by the second variation of the area [**15**, p.102]. This operator is commonly called the *Jacobi* or *stability operator*. When the surface is non-compact (e.g., if the Σ is embedded in $\mathbb{R}^3$), resolving the eigenspaces involves a limiting process: Computations are performed on a patch of the surface formed by an intrinsic ball with fixed center, and then the radius of the ball is taken to infinity [**7**]. Consequently, the eigenvalues shift as the radius increases, which typically causes the value of the index to jump at least once.

In addition to generating conclusions about stability, changes in the index during the limiting process convey information about the local existence of multiple minimal surfaces. A minimal patch has mean curvature $H = 0$. Hence, its local uniqueness is directly connected (by the implicit function theorem) to the solution set of the Jacobi equation [**3**]; namely, non-trivial solutions imply that there are multiple minimal patches. The Jacobi equation, which is derived by differentiating $H = 0$ with respect to boundary preserving normal variations, is exactly null space condition for the stability operator. That is, it is identical to zero-eigenvalue problem of J. So at an index change [**16**], which produces a zero eigenvalue, uniqueness fails and solutions to minimal surface equation emerge.

For all smooth non-compact minimal surfaces besides the plane, there is undoubtedly an index change. The reasoning is similar to arguments for proving the fundamental theorem of algebra via topological methods [**10**]. First, all non-planar surfaces have a positive index [**12, 14**]. However, if they are restricted to a sufficiently small intrinsic ball, then the resulting patch approximates a plane, which have index 0. So the patch also has index 0. As the size of the ball increases, the planar correspondence breaks down and the index must transition to a positive number, which — given that the eigenvalues of the stability operator vary smoothly [**25**] — implies that there must be at least one intermediate patch with a zero eigenvalue. Taking the ball's radius, which we interpret as a bifurcation parameter, to be strictly increasing removes the possibility of having a saddle-node bifurcation, which from topological restrictions suggests that another family of surfaces must emerge, most likely via a transcritical or a pitchfork bifurcation. The index of the non-compact surface dictates how many zero eigenvalues crossings occur; hence multiple bifurcations may be present.

This paper demonstrates the overall procedure for finding these new families of minimal surfaces bifurcating from an index change. Establishing the existence of such a family relies on the following classical result by Crandall and Rabinowitz on bifurcations from simple eigenvalues.

THEOREM 1.1 ([**11**]). *Let X and Y be Banach spaces and $F\colon X\times\mathbb{R}\to Y$ be a twice continuously Fréchet-differentiable map of the variables (x,λ). Suppose that $F(0,\lambda)=0$ for λ in an open interval I of $\mathbb{R}$, and assume the following conditions hold for a specific λ^* in I:*

(i) $\dim \operatorname{null}(D_xF(0,\lambda^*)) = 1$;
(ii) $\operatorname{codim} \operatorname{range}(D_xF(0,\lambda^*)) = 1$;
(iii) $D_\lambda D_xF(0,\lambda^*)x_0 \notin \operatorname{range}(D_xF(0,\lambda^*))$ *for all non-zero* $x_0 \in \operatorname{null}(D_xF(0,\lambda^*))$.

Then there exists an additional continuously differentiable family $(x(t),\lambda(t))$ of solutions of $F=0$ for all t in an interval $(-\varepsilon,\varepsilon)$ such that $x(0)=0$ and $\lambda(0)=\lambda^*$. Furthermore, in a sufficiently small neighborhood around $(0,\lambda^*)$, the curves $(0,\lambda^*t)$ and $(x(t),\lambda(t))$ are the only solutions to $F(x,\lambda)=0$, and $x(t)$ must be of the form $x(t)=tx_0+t\xi(t)$ for ξ contained in a space complementary to $\operatorname{null}(D_xF(0,\lambda^*))$.

Section 2 illustrates the method for a catenoid. The catenoid provides an example where everything is explicitly calculable and fits into a well-understood context. It has only one index change, yielding a transcritical bifurcation that produces a secondary catenoid. Section 3 again demonstrates the method but instead uses Enneper surfaces, which provide exploratory examples. With n-dihedral symmetry, their index is $2n-1$ and changes n times [**5, 8**]. We prove that bifurcations occur at each of these index changes. Also, for $n=2$, we give a local reconstructions of emanating surfaces. The first bifurcation (i.e., the one with smallest radius) is a standard pitchfork. Namely, two new stable surfaces appear, which flatten the interior region, driving the Enneper surfaces towards a plane. The second bifurcation (i.e., the one with largest radius) flips between between being transcritical and supercritical (see [**21**]) depending on the value of the parameter $\theta\in[0,\pi)$, which prescribes the nodal lines of the perturbation creating the second family of surfaces. If the nodal and dihedral symmetry lines align exactly, then the new surfaces inherit the dihedral invariance and predictably produce a pitchfork. Otherwise, the invariance is broken, and the result is transcritical. For the standard Enneper surface

($n = 1$), the bifurcation results are well-known; see [**24**]. Of particular note is the existence of three minimal surfaces with the same boundary [**22**]. (As shown in [**2**], the bifurcation is part of a cusp catastrophe built from arbitrary deformations of the boundary of the intrinsic ball.) Our work generalizes these ideas to all the degree n Enneper surfaces. In particular, the results for $n = 2$ prove the existence of infinitely minimal surfaces with the same boundary.

2. Catenoid

A catenoid is a mathematical idealization of a soap film stretched between two identical circles of radius ρ contained in parallel planes separated by a distance 2ζ. In fact, varying ρ and ζ produces a two-parameter family of films. Two catenoids exist for $\zeta \leq (\alpha^* \operatorname{sech} \alpha^*)\rho$, where α^* is the the unique positive root of $1 - \alpha \tanh \alpha = 0$; none exist for $\zeta > (\alpha^* \operatorname{sech} \alpha^*)\rho$; and a fold bifurcation occurs along the line $\zeta = (\alpha^* \operatorname{sech} \alpha^*)\rho$ separating these two regions.

Figure 1 gives a concise representation of this family of surfaces by showing the catenoid's neck size η in terms of ρ and ζ. For fixed ρ and ζ, the catenoid with larger η has the smaller area of the two surfaces and is the stable version of the two configurations. Slicing the two-parameter family for $\eta = 1$ produces a one-parameter family representing finite portions of the standard catenoid whose boundary circles satisfy $\rho = \cosh \zeta$. For this boundary, there is also another distinct family of catenoids with varying neck-size. These to families intersect at $(\rho, \zeta, \eta) = (\cosh \alpha^*, \alpha^*, 1)$, giving a transcritical bifurcation where the stability of the surfaces is exchanged; see Figure 2. As expected, the standard catenoids are stable for $\zeta \approx 1$ and then transition to unstable for large ζ.[1] Also, the index of the standard catenoid is 1, which is exactly the number of additional surfaces with the same boundary and smaller area.

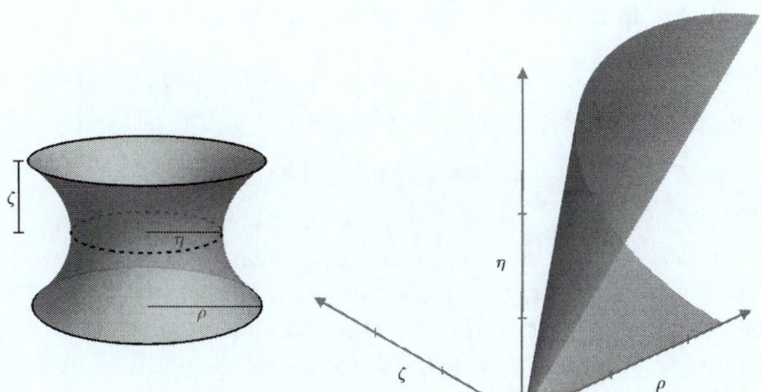

FIGURE 1. Left: Catenoid. Right: Stability surface

We next illustrate the local construction of Figure 2. Fully reconstructing exact bifurcation diagrams is a difficult task. It requires a rarely-available explicit representation of a multiple parameter family of surfaces; however, as is often the

[1]While it may seem contradictory that there is another stable family of catenoids, only the unstable family of fixed-η catenoids converges as $\zeta \to \infty$ (with $\rho = \cosh \zeta$). That is, the other stable family diverges since the neck-size blows up and all points of the limiting structure end up infinitely far from z-axis, effectively producing a cylinder of infinite radius.

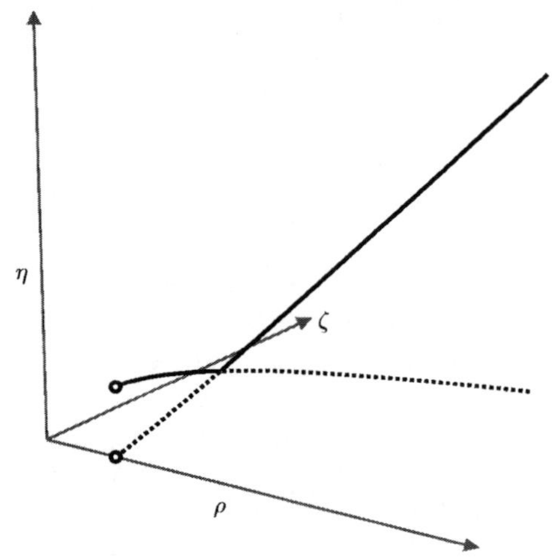

FIGURE 2. Bifurcation diagram for the catenoids

case, much of the global information contained in this plot is inferred from rectifying the local features of the branches of the surfaces. So one minimal surface suffices, and the family of surface is generated by choosing compact regions on the surface of increasing size.

Consider the standard catenoid Σ (of neck-size $\rho = 1$) that is centered at the origin and rotationally symmetric about the vertical axis. Let Σ_α be a portion of Σ created by intersecting it with extrinsic ball $B_R(0) = \{\boldsymbol{x} \in \mathbb{R}^3 : |\boldsymbol{x}| < R\}$ of radius $R = \sqrt{\alpha^2 + \cosh^2\alpha}$, where α is the height of Σ_α above the horizontal plane. The boundary $\partial\Sigma_\alpha$ of Σ_α is formed by two circles in the planes $z = \pm\alpha$ each of radius $\cosh\alpha$.

For any $\alpha > 0$, the index of Σ_α, denoted $\operatorname{ind}(\Sigma_\alpha)$, is defined to be the number of negative eigenvalues of the differential operator

$$L_\alpha := -\Delta_{\Sigma_\alpha} + 2K_{\Sigma_\alpha} \tag{2.1}$$

defined on $C_0^\infty(\Sigma_\alpha)$, i.e., the space of functions that are smooth on Σ_α and vanish on $\partial\Sigma_\alpha$ **[9]**. In (2.1), Δ_{Σ_α} and K_{Σ_α} are the intrinsic Laplacian and Gaussian curvature of Σ_α, and L_α is the negation of the standard Jacobi operator. The eigenvalues of L_α are well-ordered. Also, as functions of the parameter α, they varying continuously and are decreasing **[25]**. Hence, each eigenvalue may change sign only once by going from positive to negative. At these changes, the index increases by the dimension of the corresponding eigenspace[2] [**15**, p.109]; namely,

$$\operatorname{ind}(\Sigma_\alpha) = \sum_{t<\alpha} \dim \operatorname{null}(L_t). \tag{2.2}$$

Calculating the index from (2.2) is a straightforward endeavor: Compute the solutions of $L_\alpha(\phi) = 0$ for each α, and then count only the ones that can be made to satisfy homogeneous Dirichlet boundary conditions. The map

$$X_\alpha(u, v) = (\cosh(\alpha u)\cos v, \cosh(\alpha u)\sin v, \alpha u), \tag{2.3}$$

[2]These results are only valid for fixed surfaces, and therefore do not apply to the other branch of catenoids, which deform as the boundary $\partial\Sigma_\alpha$ varies in α.

for $(u,v) \in \Omega = [-1,1] \times [0,2\pi)$, parameterizes the catenoid patch Σ_α. Hence, the null space equation $L_\alpha(\phi) = 0$ explicitly becomes

$$\frac{1}{\alpha^2 \cosh^2(\alpha u)} \left(\phi_{uu} + \alpha^2 \phi_{vv} + \frac{2\alpha^2}{\cosh^2(\alpha u)} \phi \right) = 0 \tag{2.4}$$

where ϕ is an unknown function that is 2π-periodic in v. Decomposing ϕ into a standard Fourier series, namely,

$$\phi(u,v) = \sum_{k \in \mathbb{Z}} A_k(u) e^{ikv},$$

reduces (2.4) to a sequence of ordinary differential equations for each amplitude A_k:

$$A_k'' + \alpha^2(2\,\mathrm{sech}^2(\alpha u) - k^2) A_k = 0. \tag{2.5}$$

If $k = 0$, then the general solution is $A_0(u) = a_0 \tanh(\alpha u) + b_0(1 - \alpha u \tanh(\alpha u))$, which is a linear combination of an odd and an even functions. Applying the homogeneous Dirichlet boundary conditions at $u = \pm 1$ implies that $a_0 = 0$ and that $1 - \alpha \tanh \alpha = 0$. The latter equation has one unique positive root,

$$\alpha^* = 1.19968\ldots, \tag{2.6}$$

and accordingly, $A_0(u) = b_0(1 - \alpha^* u \tanh(\alpha^* u))$. For $k \neq 0$, the analysis is very similar; however, the solutions are all trivial [**15**, p.109].

Consequently, the dimension of $\mathrm{null}(L_{\alpha^*})$ is 1; the dimension of $\mathrm{null}(L_\alpha)$ is 0 for $\alpha \neq \alpha^*$; and (2.2) implies that

$$\mathrm{ind}(\Sigma_\alpha) = \begin{cases} 0, & 0 < \alpha \leq \alpha^* \\ 1, & \alpha > \alpha^* \end{cases}. \tag{2.7}$$

Expression (2.7) gives both the stability of Σ_α and reveals the value of α where new minimal surfaces appear. Explicitly, Σ_α is stable for $\alpha < \alpha^*$ and unstable for $\alpha > \alpha^*$. At the transition point $\alpha = \alpha^*$, there is a normal variation of Σ_α that preserves its surface area to first order, suggesting the appearance of a new branch of distinct surfaces [**4**, **18**, **19**].

We next establish the existence of a bifurcation for $\alpha = \alpha^*$ using the mean curvature equation $H = 0$. Consider the minimal immersion $X_\alpha \colon \Sigma_\alpha \to \mathbb{R}^3$ given in (2.3) with Gauss map N_α. Let $\hat{X}_\alpha$ be an arbitrary normal variation of X_α defined by $\hat{X}_\alpha = X_\alpha + \varphi N_\alpha$ for φ in an open neighborhood U of $C_0^\infty(\Sigma_\alpha)$ containing 0. Here U is chosen to be sufficiently small so that $\hat{X}_\alpha$ is an immersion. By denoting the mean curvature of $\hat{X}_\alpha$ as $H(\hat{X}_\alpha)$ and defining the function $F(\varphi, \alpha) = 2H(\hat{X}_\alpha)$, which maps $U \times \mathbb{R}^+$ to $C^\infty(\Sigma_\alpha)$, then the minimality of the varied surface $\hat{X}_\alpha$ is equivalent to

$$F(\varphi, \alpha) = 0. \tag{2.8}$$

By construction, $(0, \alpha) \in U \times \mathbb{R}^+$ is a solution of (2.8) for all $\alpha \in \mathbb{R}^+$.

For fixed $\alpha \neq \alpha^*$, the solution $(0,\alpha)$ is locally unique, meaning there are no other minimal surfaces near Σ_α. This result follows directly from the implicit function theorem: If the Fréchet derivative $D_\varphi F(0,\alpha) \colon C_0^\infty(\Sigma_\alpha) \to C^\infty(\Sigma_\alpha)$ is bijective, then $\varphi = 0$, for a given α, is the only solution of $F = 0$ in a sufficiently small open neighborhood of $(0,\alpha)$. Here the derivative $D_\varphi F(0,\alpha) = -L_\alpha$. So when $\alpha \neq \alpha^*$, the index analysis gives that $\dim \mathrm{null}(D_\varphi F(0,\alpha)) = 0$, implying injectivity.

Also, surjectivity for functions in $L^2(\Sigma_\alpha)$ holds by standard arguments [**13**], and local uniqueness of $(0,\alpha) \in U \times \mathbb{R}^+$ applies for all $\alpha \neq \alpha^*$.

However, for $\alpha = \alpha^*$, local uniqueness of $(0,\alpha)$ fails since $D_\varphi F(0,\alpha^*)$ is not injective. In fact, another distinct one-parameter family of solutions to (2.8) bifurcates from $\varphi = 0$.

We prove the existence of a second branch of solutions by applying Theorem 1.1. The first condition of the theorem follows directly from our index analysis. In particular, zero is an eigenvalue of $D_\varphi F(0,\alpha^*)$, and the function

$$\varphi_0(u,v) = 1 - \alpha^* u \tanh \alpha^* u \tag{2.9}$$

spans the corresponding eigenspace. Hence, $\dim \operatorname{null}(D_\varphi F(0,\alpha^*)) = 1$. Proofs of the other two conditions of the theorem capitalize on the Riesz-Schauder alternative theorem [**17**]: A function $f \in L^2(\Sigma_\alpha)$ is in the range of $D_\varphi F(0,\alpha^*)$ if and only if

$$\int_{\Sigma_\alpha} \varphi_0 f \, d\sigma = 0.$$

This equivalence directly implies condition (ii) since the L^2-orthogonality between $\operatorname{range}(D_\varphi F(0,\alpha^*))$ and φ_0 enforces that $\operatorname{range}(D_\varphi F(0,\alpha^*))^\perp = \operatorname{span}\{\varphi_0\}$, i.e., the codimension of the range of $D_\varphi F(0,\alpha^*)$ equals 1. For condition (iii), note that the derivative operator $D_\varphi F(0,\alpha) = -L_\alpha$ is the negation of the expression on the left side of (2.4). By differentiating this expression with respect to α, we deduce

$$D_\alpha D_\varphi H(0,\alpha)\psi = -2\left(\frac{(1+\alpha u \tanh \alpha u)}{\alpha^3 \cosh^2 \alpha u}\psi_{uu} + \frac{u \tanh \alpha u}{\cosh^2 \alpha u}\psi_{vv} + \frac{4u \tanh \alpha u}{\cosh^4 \alpha u}\psi\right).$$

lieu So for $\psi = \varphi_0(u,v)$, with the help of $\partial_{uu}\varphi_0 = -2\alpha^* \operatorname{sech}^2(\alpha^*)\varphi_0$ and $\partial_{uu}\varphi_0 = 0$, we find that

$$D_\alpha D_\varphi F(0,\alpha^*)\varphi_0 = -\frac{4}{\alpha^*} K_{\Sigma_{\alpha^*}} {\varphi_0}^2,$$

where the Gaussian curvature $K_{\Sigma_{\alpha^*}} = -\operatorname{sech}^4 \alpha^* u$. Thus,

$$\begin{aligned} \iint_{\Sigma_{\alpha^*}} \varphi_0 (D_\alpha D_\varphi F(0,\alpha^*)\varphi_0)\, dS &= 4\int_0^{2\pi}\int_{-1}^{1} \frac{(1-\alpha^* u \tanh \alpha^* u)^3}{\cosh^2 \alpha^* u}\, du\, dv \\ &= 4\pi(\alpha^*)^2, \end{aligned} \tag{2.10}$$

which by the alternative theorem implies that $D_\alpha D_\varphi F(0,\alpha^*)\varphi_0 \notin \operatorname{range}(D_\varphi F(0,\alpha^*))$.

Therefore, conditions (i)–(iii) of Theorem 1.1 hold for (2.8) at $(\varphi,\alpha) = (0,\alpha^*)$, inducing a bifurcation and proving the existence of a second family of minimal surfaces with the same boundary as Σ_α.

THEOREM 2.1. *Consider two circles of radius R in parallel planes separated by a distance 2α for $\alpha > 0$. Set $R = \sqrt{\alpha^2 + \cosh^2 \alpha}$, and let Σ_α be the catenoid that is stretched between the circles and whose mid-plane cross-section is a circle of radius 1. For varying α, another distinct family of minimal surfaces $\hat{\Sigma}_\alpha$ with the same boundary (i.e., $\partial\hat{\Sigma}_\alpha = \partial\Sigma_\alpha$) bifurcates from Σ_α if and only if $\alpha = \alpha^*$, where α^* is the unique positive solution of $1 - \alpha \tanh \alpha = 0$.*

Determining the type of bifurcation requires that we compute the local expansion of the new family of solutions. Let $(\varphi(\,\cdot\,;t), \alpha(t))$ represent the family for $t \in (-\varepsilon,\varepsilon)$, which must satisfy $(\varphi(\cdot;0),\alpha(0)) = (0,\alpha^*)$. By definition, this pair solves

$$F(\varphi(\,\cdot\,;t),\alpha(t)) = 0, \quad t \in (-\varepsilon,\varepsilon), \tag{2.11}$$

and Taylor series coefficients found from expanding (2.11) around $t = 0$ fix the form of the local solution.

Implied by the proof of existence of the bifurcation, the function

$$\dot{\varphi}(\cdot\,;0) = \varphi_0 \equiv 1 - \alpha^* u \tanh \alpha^* u \tag{2.12}$$

is the first term in the φ-series. This choice makes $\dot{F}|_{t=0} = 0$ identically.

The second terms $\dot{\alpha}(0)$ and $\ddot{\varphi}(\cdot\,;0) \equiv \varphi_1 \in C_0^\infty(\Sigma_{\alpha^*})$ are governed by the second derivative[3] of (2.11) evaluated at $t = 0$. Upon using $D_\alpha F(0, \alpha^*) = D_{\alpha\alpha} F(0, \alpha^*) = 0$, this derivative is equivalent to the non-homogeneous, linear partial differential equation

$$-D_\varphi F(0, \alpha^*)\varphi_1 = D_{\varphi\varphi} F(0, \alpha^*)(\varphi_0, \varphi_0) + 2\dot{\alpha}(0)\, D_\alpha D_\varphi F(0, \alpha^*)\varphi_0, \tag{2.13}$$

for φ_1. By self-adjointness of $-D_\varphi F(0, \alpha^*)$ over functions with homogeneous Dirichlet boundary conditions, a solution of (2.13) exists only if the right side of equation (2.13) is orthogonal to φ_0 in $L^2(\Sigma_{\alpha^*})$; namely,

$$\iint_{\Sigma_{\alpha^*}} \varphi_0 (D_{\varphi\varphi} F(0, \alpha^*)(\varphi_0, \varphi_0))\, dS + 2\dot{\alpha}(0) \iint_{\Sigma_{\alpha^*}} \varphi_0\, D_\alpha D_\varphi F(0, \alpha^*)\varphi_0\, dS = 0.$$

The second integral equals $4\pi(\alpha^*)^2$; see (2.10). A similar computation shows that the first integral equals $-4\pi(\alpha^*)^3$. Hence,

$$\dot{\alpha}(0) = \frac{\alpha^*}{2}. \tag{2.14}$$

Although, it is not needed for the determination of the type of bifurcation, the resulting solution of (2.13), found by separating variables, is

$$\varphi_1(u, v) = -\frac{\varphi_0(u, v)^2}{\cosh^2(\alpha^* u)}. \tag{2.15}$$

With quantities given in (2.12) and (2.14), this work classifies the exact type of bifurcation observed in Theorem 2.1.

COROLLARY 2.2. *The second family of minimal surfaces $\hat{\Sigma}_\alpha$ emerging from the standard catenoid Σ_α in* (2.3) *is locally parameterized by*

$$\hat{X}_\alpha(u, v) = X_\alpha(u, v) + (t\,(1 - \alpha^* u \tanh \alpha^* u) + \mathcal{O}(t^2))\, N_\alpha(u, v) \tag{2.16a}$$

with

$$\alpha = \alpha^* + \frac{\alpha^*}{2} t + \mathcal{O}(t^2) \tag{2.16b}$$

where t is restricted to a sufficiently small neighborhood of 0 *and N_α denotes the normal vector field of X_α. This implies that the bifurcation at $\alpha = \alpha^*$ is transcritical. Also, for values of α sufficiently close to α^*, the surface $\hat{\Sigma}_\alpha$ is unstable for $\alpha < \alpha^*$ and stable for $\alpha > \alpha^*$, meaning its index transitions from* 1 *to* 0 *across $\alpha = \alpha^*$.*

The stability of the new branch follow from the local expansion. At $\alpha = \alpha^*$, the new branch has the same eigenvalues of the known branch of catenoid, i.e., one zero eigenvalue, with all the others being positive. Stability of the new branch

[3]Note that $\dot{\alpha}(0)$ does not appear in the $\dot{F}|_{t=0} = 0$ due to the fact that $F(0, \cdot) \equiv 0$ for any argument.

is then determined by the sign of the minimal eigenvalue λ_0 of $-\Delta_{\hat{\Sigma}_\alpha} + 2K_{\hat{\Sigma}_\alpha}$. Given (2.16), the local expansion of this minimal eigenvalue

$$\lambda_0 = \frac{12}{3\cosh(2\alpha^*) - 7}\, t + \mathcal{O}(t^2).$$

which shows that the new branch is stable for $t > 0$ and unstable for $t < 0$.

Overall the derived results in Theorem 2.1 and Corollary 2.2 reproduce the local picture shown in Figure 2: A transcritical bifurcation at $\alpha = \alpha^*$ with the new family of surfaces transitioning from unstable to stable for increasing α.

3. Dihedral Enneper surfaces

We next apply the overall methodology of the previous section to prove the existence of bifurcations from each degree n Enneper surface Σ^n [**1**, **5**]. Here Σ^1 is the standard Enneper surface, and the others (i.e., Σ^n for $n = 2, 3, \ldots$) are similar but with increasing orders of dihedral symmetry. Modulo rotations, translations and dilations in $\mathbb{R}^3$, the general map

$$\sigma^n(u,v) = \frac{u}{2}\left(\cos v - \frac{u^{2n}\cos(2n+1)v}{2n+1}, \sin v + \frac{u^{2n}\sin(2n+1)v}{2n+1}, \frac{2u^n\cos(n+1)v}{n+1}\right),$$

for polar coordinates $(u,v) \in \mathbb{R}^+ \times [0, 2\pi)$, parameterizes each Σ^n. Accordingly,

$$X_\alpha^n(u,v) = \sigma^n(\alpha\, u, v), \quad (u,v) \in \Omega = [0,1] \times [0, 2\pi) \tag{3.1}$$

is a surface patch, Σ_α^n, of Σ^n that is centered at $(0,0,0)$ with an effective radius $\alpha > 0$. As in section 2, bifurcations appear at values of α where $\operatorname{ind}(\Sigma_\alpha^n)$ has jump discontinuities.

The index of the degree n Enneper surface Σ_α^n depends on both n and α. To determine its exact value, consider the differential operator

$$L_\alpha^{(n)} := -\Delta_{\Sigma_\alpha^n} + 2K_{\Sigma_\alpha^n}$$

defined on the space $C_0^\infty(\Sigma_\alpha^n)$ of smooth functions vanishing on $\partial\Sigma_\alpha$. Here $\Delta_{\Sigma_\alpha^n}$ and $K_{\Sigma_\alpha^n}$ denote the intrisic Laplacian and Gauss curvature of X_α^n. In coordinate form,

$$L_\alpha^{(n)} := -\frac{4}{\alpha^2(1+(\alpha u)^{2n})^2}\left(\frac{\partial^2}{\partial u^2} + \frac{1}{u}\frac{\partial}{\partial u} + \frac{1}{u^2}\frac{\partial^2}{\partial v^2} + \frac{8n^2(\alpha u)^{2n}}{u^2(1+(\alpha u)^{2n})^2}\right),$$

which implies that its null space, for each pair (n, α), is the set of solutions to

$$\frac{\partial^2\phi}{\partial u^2} + \frac{1}{u}\frac{\partial\phi}{\partial u} + \frac{1}{u^2}\frac{\partial^2\phi}{\partial v^2} + \frac{8n^2(\alpha u)^{2n}}{u^2(1+(\alpha u)^{2n})^2}\phi = 0, \quad (u,v) \in (0,1) \times (0, 2\pi); \tag{3.2a}$$

$$\phi_u(0,\cdot) = \phi(1,\cdot) = 0, \quad \phi(\cdot,0) = \phi(0,2\pi), \quad \frac{\partial\phi}{\partial v}(\cdot,0) = \frac{\partial\phi}{\partial v}(\cdot,2\pi). \tag{3.2b}$$

This is a separable boundary value problem. A general Fourier series expansion that is 2π-periodic in v, i.e.,

$$\phi(u,v) = A_0(u) + \sum_{k=1}^{\infty}(A_k(u)\cos kv + B_k(u)\sin kv),$$

reduces it to a sequence of boundary value problems indexed by k; namely, the amplitudes A_k and B_k both solve

$$f_k'' + \frac{f_k'}{u} + \frac{1}{u^2}\left(\frac{8n^2(\alpha u)^{2n}}{(1+(\alpha u)^{2n})^2} - k^2\right) f_k = 0; \quad f_k(1) = 0, |f_k(0)| < \infty, \tag{3.3}$$

for the specified subscript values. The function

$$\frac{u^k(n+k-(n-k)\alpha^{2n}u^{2n})}{1+\alpha^{2n}u^{2n}} \tag{3.4}$$

is a solution of (3.3). Reduction of order supplies a second solution; however, it is unbounded at $u=0$, regardless of the value of k. So the general solution of (3.3) must be a scalar multiple of (3.4). Applying $f_k(1)=0$ to this general solution leads to the algebraic equation $n+k=(n-k)\alpha^{2n}$, which has real roots

$$\alpha^*_{n,k} = \left(\frac{n+k}{n-k}\right)^{1/(2n)}, \tag{3.5}$$

if and only if the integer k satisfies $0 \le k < n$. Notice that $\alpha^*_{n,0}=1$ is always root. In summary, the boundary value problem (3.3) for a given $k \in \mathbb{N}_0$ has the general solution

$$f_k(u) = c_k \frac{n\,u^k(1-u^{2n})}{n-k+(n+k)u^{2n}}, \tag{3.6}$$

with c_k is an arbitrary constant, if $\alpha=\alpha^*_{n,k}$ for $k\in[0,n)$. Otherwise, if $\alpha \ne \alpha^*_{n,k}$, which includes the condition where $k \ge n$, the solutions of (3.3) are zero. By relating these results back to the null space problem in (3.2), we deduce that $\dim\operatorname{null}(L_{\alpha^*_{n,0}})=1$, $\dim\operatorname{null}(L_{\alpha^*_{n,k}})=2$ (for $k\in\{1,2,\dots,n-1\}$) and $\dim\operatorname{null}(L_\alpha)$ $=0$ for all other values of α. With an adapted version of formula (2.2), the explicit expression for $\operatorname{ind}(\Sigma^n_\alpha)$ is

$$\operatorname{ind}(\Sigma^n_\alpha) = \begin{cases} 0, & \alpha\in(0,1] \\ 1, & \alpha\in(1,\alpha^*_{n,1}] \\ 3, & \alpha\in(\alpha^*_{n,1},\alpha^*_{n,2}] \\ 5, & \alpha\in(\alpha^*_{n,2},\alpha^*_{n,3}] \\ \vdots & \vdots \\ 2n-1, & \alpha\in(\alpha^*_{n,n-1},\infty). \end{cases} \tag{3.7}$$

As a function of α, this index is discontinuous at n points. In other words, there will be n bifurcations from Σ^n_α, one at each discontinuity. Also, the values of the jumps in the index give direct insight into the codimension of the bifurcation.

The method for proving there are bifurcations at the discontinuities of $\operatorname{ind}(\Sigma^n_\alpha)$ is the same as before: Apply Theorem 1.1 to the minimal surface equation $2H_n=0$ for local perturbations of the Enneper surface patch Σ^n_α. In particular, given the immersion $X^n_\alpha\colon \Sigma^n_\alpha \to \mathbb{R}^3$, define a normal variation $\hat{X}^n_\alpha = X^n_\alpha + \varphi N^n_\alpha$ for an arbitrary function φ in an open neighborhood U of $C^\infty_0(\Sigma^n_\alpha)$, which contains 0 and is sufficiently small so that $\hat{X}^n_\alpha$ remains immersed. Then the variation $\hat{X}^n_\alpha$ is a minimal surface if and only if

$$F_n(\varphi,\alpha) := 2H_n(\varphi,\alpha) = 0, \tag{3.8}$$

where $H_n\colon U\times\mathbb{R}^+ \to C^\infty(\Sigma^n_\alpha)$ denotes the mean curvature of $\hat{X}^n_\alpha$. By definition, the pair $(0,\alpha)\in U\times\mathbb{R}^+$ is a solution of (3.8), and Theorem 1.1 gives conditions on the derivatives

(3.9)

$$D_\varphi F_n(0,\alpha) = \frac{4}{\alpha^2(1+(\alpha u)^{2n})^2}\left(\frac{\partial^2}{\partial u^2}+\frac{1}{u}\frac{\partial}{\partial u}+\frac{1}{u^2}\frac{\partial^2}{\partial v^2}+\frac{8n^2(\alpha u)^{2n}}{u^2(1+(\alpha u)^{2n})^2}\right)$$

and

$$
\begin{aligned}
D_\alpha D_\varphi F_n(0,\alpha) &= -\frac{2(1+(2n+1)(\alpha u)^{2n})}{\alpha(1+(\alpha u)^{2n})} D_\varphi F_n(0,\alpha) \\
&\quad + \frac{64n^3(\alpha u)^{2(n-1)}\left(1-(\alpha u)^{2n}\right)}{\alpha(1+(\alpha u)^{2n})^5} I
\end{aligned} \tag{3.10}
$$

for guaranteeing that additional solutions of (3.8) emanate from $(0,\alpha)$ with $\varphi \neq 0$.

However, complications arise in attempting to directly apply Theorem 1.1 after the first bifurcation point. So we separately consider the cases when $\alpha = 1$ and when $\alpha = \alpha^*_{n,k}$ for $k \in \mathbb{N} \cap [1,n)$.

When $\alpha = 1$, the conditions that guarantee a bifurcation are straightforward to verify. The index analysis in (3.2)–(3.6) shows that the function

$$
\varphi_0(u,v) = \frac{1-u^{2n}}{1+u^{2n}}. \tag{3.11}
$$

spans the nullspace of $D_\varphi F_n(0,1)$. That is, $\dim \operatorname{null}(D_\varphi F_n(0,1)) = 1$. Also, $\operatorname{codim}\operatorname{range}(D_\varphi F_n(0,1)) = 1$, since the self-adjointness of $D_\varphi F_n(0,1)$ with respect to the $L^2(\Sigma_1^n)$ inner product implies that $\operatorname{range}(D_\varphi F_n(0,1))^\perp = \operatorname{null}(D_\varphi F_n(0,1))$. To verify the last condition of the theorem, first observe that

$$
D_\alpha D_\varphi F_n(0,1)\varphi_0 = -4nK_{\Sigma_1^n}\varphi_0^2,
$$

where $K_{\Sigma_1^n} = -16n^2u^{2(n-1)}/(1+u^{2n})^4$ is the Gaussian curvature of Σ_α^n for $\alpha = 1$. Then

$$
\iint_{\Sigma_1^n} \varphi_0(D_\alpha D_\varphi F_n(0,1)\varphi_0)\, dS = -4n \iint_{\Sigma_1^n} K_{\Sigma_1^n}\varphi_0^3\, dS = 2\pi n^2, \tag{3.12}
$$

which is nonzero and implies that $D_\alpha D_\varphi F_n(0,1)\varphi_0 \notin \operatorname{range}(D_\varphi F_n(0,1))$. So $(\varphi,\alpha) = (0,1)$ is a bifurcation point of (3.8).

When $\alpha = \alpha^*_{n,k}$ (for $k \geq 1$), condition (i) of Theorem 1.1 fails. At these values, the operator $D_\varphi F_n(0,\alpha)\colon C_0^\infty(\Sigma_\alpha^n) \to C^\infty(\Sigma_\alpha^n)$ has a two-dimensional null space induced by a continuous translational symmetry in the angular v-direction. There are bifurcations at each $\alpha = \alpha^*_{n,k}$ for $k \geq 1$ but mirroring the steps of the $\alpha = 1$ case requires us to place extra assumptions on the set of admissible variations.

Thus, we change the setup leading to (3.8) by imposing that the varied family of surfaces $\hat{X}_\alpha^n$ preserve the radial curves $X_\alpha^n(\cdot,\theta)$ and $X_\alpha^n(\cdot,\theta+\pi)$ for an arbitrary $\theta \in [0,\pi)$. This requirement eliminates the symmetry of the angular coordinate v and makes it sufficient to consider (3.8) on the half sector, Ξ_α^n, of the surface Σ_α^n bounded between these curves; namely, the function F_n now maps $U \times \mathbb{R}^+$ to $C^\infty(\Xi_\alpha^n)$, where U is a sufficiently small neighborhood of 0 in the space $C_0^\infty(\Xi_\alpha^n)$ of smooth functions vanishing on $\partial\Xi_\alpha^n$.

Given this modification, both the dimension of $\operatorname{null}(D_\varphi F_n(0,\alpha^*_{n,k}))$ and the codimension of $\operatorname{range}(D_\varphi F_n(0,\alpha^*_{n,k}))$ are 1. Switching from the full immersion Σ_α^n to the sector Ξ_α^n only changes the associated domain and boundary conditions of $D_\varphi F_n(0,\alpha)$. Problem (3.2) again determines its null space but with $\varphi = 0$ at $v = \theta$ and $v = \theta+\pi$ replacing the 2π-periodic conditions in (3.2b), which alters the Fourier basis to $\{\sin k(v-\theta)\}_{k=1}^\infty$. All the other steps leading to (3.7) remain the same, implying that (for $k \geq 1$)

$$
\operatorname{null}(D_\varphi F_n(0,\alpha^*_{n,k})) = \{c_0\,\varphi_0 : c_0 \in \mathbb{R}\} \tag{3.13a}
$$

for

$$\varphi_0 \equiv \frac{n\,u^k(1-u^{2n})}{n-k+(n+k)u^{2n}} \sin k(v-\theta) \tag{3.13b}$$

Therefore, $\dim \operatorname{null}(D_\varphi F_n(0,\alpha^*_{n,k}))=1$ and $\operatorname{codim} \operatorname{range}(D_\varphi F_n(0,\alpha^*_{n,k}))=1$, where the latter equality follows from (3.13) coupled with the operator being self-adjoint on Ξ^n_α with homogeneous Dirichlet boundary conditions.

Also, for each $k \geq 1$, the function $D_\alpha D_\varphi F_n(0,\alpha^*_{n,k})\varphi_0 \notin \operatorname{range}(D_\varphi F_n(0,\alpha^*_{n,k}))$, where φ_0 is defined in (3.13b) and spans $\operatorname{null}(D_\varphi F_n(0,\alpha^*_{n,k}))$. Formula (3.10) implies that

$$D_\alpha D_\varphi F_n(0,\alpha^*_{n,k})\varphi_0 = \frac{64n^3(\alpha^*_{n,k}u)^{2(n-1)}(1-(\alpha^*_{n,k}u)^{2n})}{\alpha^*_{n,k}(1+(\alpha^*_{n,k}u)^{2n})^5}\varphi_0. \tag{3.14}$$

This function is in the $\operatorname{range}(D_\varphi F_n(0,\alpha^*_{n,k}))$ only if it is orthogonal to all the elements in the null space of $D_\varphi F_n(0,\alpha^*_{n,k})$ using the L^2-inner product. However, evaluating the following inner product proves that such a conclusion is not possible:

$$\iint_{\Xi^n_\alpha} \varphi_0(D_\alpha D_\varphi F_n(0,\alpha^*_{n,k})\varphi_0)\,dS = \frac{n^2\pi}{2\,\alpha^*_{n,k}} > 0, \qquad \text{for } \alpha = \alpha^*_{n,k}. \tag{3.15}$$

Theorem 1.1 is thus valid for the restricted problem, and by combining with the result for $k=0$, we deduce that a new branch of solutions of (3.8) bifurcates from $(0,\alpha)$ at $\alpha = \alpha^*_{n,k}$ for each integer $k \in [0,n)$. For all other values of α, the solution $\varphi = 0$ is locally unique, which follows from the bijectivity of $D_\varphi F_n(0,\alpha)$ for $\alpha \neq \alpha^*_{n,k}$, as demonstrated in the catenoid situation.

These results prove that n new families of minimal surfaces emerge from Σ^n_α with the same boundary.

THEOREM 3.1. *For $\alpha > 0$, let Σ^n_α be a patch of Enneper surface of degree $n \geq 1$ that is parameterized by (3.1). Another distinct family of minimal surfaces $\hat{\Sigma}^n_\alpha$ with the same boundary bifurcates from Σ_α if and only if*

$$\alpha = \alpha^*_{n,k} \equiv \Big(\frac{n+k}{n-k}\Big)^{1/(2n)} \tag{3.16}$$

for any integer $k = 0,1,2,\ldots,n-1$.

Interestingly, each bifurcation, aside from the one at $\alpha = 1$, produces a two parameter family of new minimal surfaces for varying α and θ, where θ is the angle used to partition the Enneper surface into two components and represents a nodal line of the perturbation that leaves the original Enneper patch unchanged. Thus, for any degree n Enneper surface Σ^n_α with $n \geq 2$, there are infinitely many minimal surfaces with the same boundary.

Also, all of the bifurcations occur for α in small interval contained in $[1,2]$. When $n=1$, there is only one bifurcation at $\alpha = \alpha_{1,0} = 1$. When $n > 1$, bifurcations are at $\alpha = \alpha_{2,0} = 1$ and at $\alpha = \alpha_{n,n-1} = (2n-1)^{1/(2n)}$, and the rest occur for values of α in between. At $n=2$, the last bifurcation $\alpha_{n,n-1}$ obtains its max value of $\sqrt[4]{3}$. Then it monotonically decreases with the asymptotic relationship $\alpha_{n,n-1} = 1 + (1/2)n^{-1}\log n + \mathcal{O}(n^{-1})$, implying that the interval containing all the bifurcations converges degenerately to the point $\alpha = 1$ as $n \to \infty$. For example, if $n = 100$, then 100 families of minimal surfaces appear from Σ^{100}_α for α in $[1, 1.02682\ldots]$.

3.1. Classification of the bifurcations for $n = 2$. We next locally construct the two minimal surface families bifurcating Σ^2_α at $\alpha = \alpha^*_{2,0} = 1$ and $\alpha = \alpha^*_{2,1} = \sqrt[4]{3}$. We use Σ^2_α because the degree 1 case is known, yielding one pitchfork bifurcation at $\alpha = 1$ [**23**], and the arbitrary degree situation is difficult to fully classify due to the sheer number of bifurcations for large n. The degree 2 surface provides a preliminary step to generalizing for arbitrary n in a specific case that involves both the one-parameter and two-parameter families of minimal surfaces.

Local constructions of the surfaces proceed from the local expansions of the new solution branches to the mean curvature equation (3.8). As dictated by the bifurcation theorem, these solutions (φ, α) are of the form

$$\varphi(\cdot\,;t) = t\,\varphi_0 + \frac{t^2}{2}\varphi_1 + \cdots, \qquad \alpha(t) = a_0 + t\,a_1 + \frac{t^2}{2}\,a_2 + \cdots \tag{3.17}$$

for sufficiently small t, say in $(-\varepsilon, \varepsilon)$. Accordingly, they solve

$$F_2(\varphi(\cdot\,;t), \alpha(t)) = 0, \quad t \in (-\varepsilon, \varepsilon). \tag{3.18}$$

The leading order terms of (3.17) for the new solutions appearing at $\alpha = 1$ and $\alpha = \sqrt[4]{3}$ are

$$\varphi_0 = \frac{1-u^4}{1+u^4}, \qquad a_0 = 1, \tag{3.19}$$

and

$$\varphi_0 = \frac{2\,u(1-u^4)}{1+3u^4}\sin(v-\theta), \qquad a_0 = \sqrt[4]{3}, \tag{3.20}$$

respectively. Such choices ensure that the minimal surface equation (3.18) and its first derivative with respective to t vanish at $t = 0$.

The next non-zero terms of the α expansion — specifically, a_1 or a_2 — dictate the type of bifurcation, and the second and third derivatives of (3.18) fix their values. A direct computation, coupled with the observation that $F_2(0, \cdot) = 0$, implies that the second derivative of (3.18) at $t = 0$ produces the partial differential equation

$$-D_\varphi F_2(0, a_0)\varphi_1 = D_{\varphi\varphi} F_2(0, a_0)(\varphi_0, \varphi_0) + 2a_1(D_\alpha D_\varphi F_2(0, a_0))\varphi_0. \tag{3.21}$$

In accordance with space of admissible variations, φ_1 vanishes on $\partial\Sigma^n_{a_0}$ when $a_0 = 1$ or on $\partial\Xi^n_{a_0}$ when $a_0 = \sqrt[4]{3}$. Both boundary conditions make the differential operator $-D_\varphi F_2(0, a_0)$ self-adjoint; hence, a solution of (3.21) exists only if the function of the right side of the equation is orthogonal in $L^2(\Sigma^n_{a_0})$ to φ_0.

For the leading order terms given in (3.19), the constant $a_1 = 0$. In this case, the derivative

$$D_{\varphi\varphi} F_2(0,1)(\varphi_0, \varphi_0) = -\frac{2^{11}u^3(1-10u^4+7u^8)}{(1+u^4)^8}\cos 3v.$$

This function is orthogonal to φ_0 given the presence of the trigonometric function $\cos 3v$. So the solvability condition of (3.21) dictates that a_1 must vanish, and the classification of the bifurcation follows from the determination of a_2 from the third derivative equations. To this end, we need the corresponding solution of (3.21) modulo the null space of the linear operator, i.e.,

$$\varphi_1 = u^3 \left(\frac{32u^2}{(1+u^4)^4} - \frac{2(5+u^4)}{3(1+u^4)} \right) \cos 3v.$$

For the leading order terms in (3.20), the constant a_1 is not longer strictly zero. Instead, it depends on the angle θ dictating the sector of the surface. Also, the derivative changes to

$$\begin{aligned} D_{\varphi\varphi}F_2(0, \sqrt[4]{3})(\varphi_0, \varphi_0) = (R_1 \sin(v-\theta) + R_3 \sin 3(v-\theta) + R_5 \sin 5(v-\theta)) \sin 3\theta \\ - (R_1 \cos(v-\theta) + R_3 \cos 3(v-\theta) \\ + R_5 \cos 5(v-\theta)) \cos 3\theta, \end{aligned}$$

where R_1,R_3 and R_5 are rational functions in u:

$$R_1(u) = \frac{2^8 u \left(63u^{16} + 84u^{12} - 294u^8 + 84u^4 - 1\right)}{\sqrt[4]{3}(1+3u^4)^8}$$

$$R_3(u) = -\frac{2^{11} u^5 \left(9u^{12} + 51u^8 - 89u^4 + 13\right)}{\sqrt[4]{3}(1+3u^4)^8}, \quad R_5(u) = \frac{2^{12} u^5 \left(21u^8 - 26u^4 + 1\right)}{\sqrt[4]{3}(1+3u^4)^8}.$$

From the orthogonality of trigonometric functions, the inner product of φ_0 (in (3.20)) and $D_{\varphi\varphi}F_2(0, \sqrt[4]{3})(\varphi_0, \varphi_0)$ is set by the $\sin(v-\theta)$ term; namely,

$$\begin{aligned} &\iint_S \varphi_0 \, D_{\varphi\varphi}F_2(0, \sqrt[4]{3})(\varphi_0, \varphi_0) \, dS \\ &\qquad = \frac{\sqrt{3}\sin 3\theta}{2} \int_0^1 \int_\theta^{\theta+\pi} u^2(1-u^4)(1+3u^4) R_1(u) \sin^2(v-\theta) \, dv \, du \\ &\qquad = \frac{3\sqrt[4]{3}\pi}{2} \sin 3\theta, \end{aligned}$$

where the first integral is over the surface $S = \Sigma^2_{a_0}$ for $a_0 = \sqrt[4]{3}$. With this computation, solvability of (3.21) is equivalent to

$$-\frac{3\sqrt[4]{3}\pi}{2} \sin 3\theta + 2a_1 \frac{2\pi}{\sqrt[4]{3}} = 0,$$

which rearranges to

$$a_1(\theta) = -\frac{3\sqrt{3}}{8} \sin 3\theta. \tag{3.22}$$

At the nonzero values of a_1, the expansion, which is fixed out to order $\mathcal{O}(t)$, proves that the corresponding bifurcation is transcritical.

When $a_1 = 0$, which occurs for (3.19) or for (3.20) with $\theta \in \{0, \pi/3, 2\pi/3\}$, there is no linear correction for $\alpha(t)$ in (3.17), implying that higher order nonlinear terms, namely a_2, classify the bifurcation. Steps identical to the preceding ones for a_1, but for the third derivative instead of the second, show that the second order term φ_2 satisfies the partial differential equation

$$-D_\varphi F_2(0, a_0)\varphi_2 = \eta(\varphi_0, \varphi_1) + 3a_2(D_\alpha D_\varphi F_2(0, a_0)\varphi_0), \tag{3.23}$$

which is coupled with homogeneous Dirichlet boundary conditions on $\partial\Sigma^n_{a_0}$ when $a_0 = 1$ or on $\partial\Xi^n_{a_0}$ when $a_0 = \sqrt[4]{3}$. The function η is defined by

$$\eta(\varphi_0, \varphi_1) = 3D_{\varphi\varphi}F_2(0, a_0)(\varphi_0, \varphi_1) + D_{\varphi\varphi\varphi}F_2(0, a_0)(\varphi_0, \varphi_0, \varphi_0),$$

where $D_{\varphi\varphi}F_2$ and $D_{\varphi\varphi\varphi}F_2$ denote the second and third Frèchet derivatives of F_2 with respect to the first argument. By orthogonality arguments, (3.23) has a solution if and only if

$$0 = \iint_{S(a_0)} \varphi_0 \eta(\varphi_0, \varphi_1)\, dS + 3a_2 \iint_{S(a_0)} \varphi_0 (D_\alpha D_\varphi F_2(0, a_0)\varphi_0)\, dS$$

where the surface $S(a_0)$ is equal to $\Sigma^2_{a_0}$ or the sector $\Xi^2_{a_0}$ for $a_0 = 1$ and $a_0 = \sqrt[4]{3}$, respectively. Positivity of the second integral, following from the proof of Theorem 3.1, ensures that we can solve for a_2 and that (3.23) has a solution.

Then the next term a_2 is exactly

$$a_2 = \frac{-\iint_{S(a_0)} \varphi_0 \eta(\varphi_0, \varphi_1)\, dS}{3\iint_{S(a_0)} \varphi_0 (D_\alpha D_\varphi F_2(0, a_0)\varphi_0)\, dS} \tag{3.24}$$

when $a_1 = 0$. With φ_0 and a_0 from (3.19), the denominator equals 24π; see (3.12). Also,

$$\eta(\varphi_0, \varphi_1) = \frac{2^{10}u^6 p(u)}{(1+u^4)^{11}} \cos 6v - \frac{2^9 u^2 (3u^{16} + 16u^{12} + 62u^8 - 80u^4 + 15)}{(1+u^4)^8},$$

where p is a degree 20 polynomial of only even powers, implying that the numerator splits into two terms. The first term vanishes after integrating over v, while the second term is exactly -56π; hence, the coefficient $a_2 = 7/3$ at the first bifurcation. At the second bifurcation, i.e., choosing the leading order terms in (3.19), the denominator of the a_2 expression in (3.24) reduces to $2\pi 3^{3/4}$; see (3.15). However, calculating the numerator is much more involved. It requires the solution of (3.21), which can be found with a Fourier expansion in v using the basis $\{\sin k(v-\theta)\}_{k=1}^{\infty}$ for θ being 0, $\pi/3$ or $2\pi/3$. Since the forcing function of the differential equation is

$$D_{\varphi\varphi}F_2(0, \sqrt[4]{3})(\varphi_0, \varphi_0) = -(R_1 \cos(v-\theta) + R_3 \cos 3(v-\theta) + R_5 \cos 5(v-\theta)) \cos 3\theta,$$

then the Fourier expansion of the solution contains only the sines with even k. After finding the solution, its expression goes into the numerator of (3.24) and the result is integrated term-by-term to give a non-zero number. So for $\theta = 0, \pi/3$, and $2\pi/3$, we have that $a_2 \neq 0$.

With these calculations, we have that the two new branches of solutions of $F_2 = 0$ intersecting with the known branch $(\varphi, \alpha) \in \{0\} \times \mathbb{R}^+$ at the bifurcation points $(0,1)$ and $(0, \sqrt[4]{3})$ are locally given by

$$\begin{aligned} \varphi(u, v; t) &= t\frac{1-u^4}{1+u^4} + \frac{t^2}{2}u^3 \left(\frac{32u^2}{(1+u^4)^4} - \frac{2(5+u^4)}{3(1+u^4)} \right) \cos 3v + \cdots, \\ \alpha(t) &= 1 + t^2 \frac{7}{6} + \cdots, \end{aligned} \tag{3.25}$$

and

$$\begin{aligned} \varphi(u, v; t) &= t\frac{2u(1-u^4)}{1+3u^4} \sin(v-\theta) + \cdots, \\ \alpha(t) &= \sqrt[4]{3} - t\frac{3\sqrt{3}}{8} \sin 3\theta + \cdots, \end{aligned} \tag{3.26}$$

for sufficiently small t. These expansions inserted into the perturbation $\hat{X}^2_\alpha = X^2_\alpha + \varphi N^2_\alpha$ of the degree 2 Enneper immersion prove the following corollary to Theorem 3.1.

COROLLARY 3.2. *The families of minimal surfaces $\hat{\Sigma}^2_\alpha$ coalescing with the degree 2 Enneper patch Σ^2_α (given in (3.1)) at $\alpha = 1$ and $\alpha = \sqrt[4]{3}$, respectively, have local parameterizations that are given by*

$$\hat{X}^2_\alpha = X^2_\alpha + t\Big(\frac{1-u^4}{1+u^4} + t\left(\frac{16u^5}{(1+u^4)^4} - \frac{u^3(5+u^4)}{3(1+u^4)}\right)\cos 3v + \mathcal{O}(t^2)\Big)N_\alpha \tag{3.27a}$$

with

$$\alpha = 1 + \frac{7}{6}t^2 + \mathcal{O}(t^3), \tag{3.27b}$$

and

$$\hat{X}^2_\alpha = X^2_\alpha + t\Big(\frac{2u(1-u^4)}{1+3u^4}\sin(v-\theta) + \mathcal{O}(t)\Big)N_\alpha \tag{3.28a}$$

with

$$\alpha = \sqrt[4]{3} - t\frac{3\sqrt{3}}{8}\sin 3\theta + \mathcal{O}(t^2). \tag{3.28b}$$

Here $|t|$ must be sufficiently small, and N_α denotes the normal vector field of the patch X^2_α.

Expansion (3.27) dictates that the first bifurcation at $\alpha = 1$ is a supercritical pitchfork; see Figure 3. That is, for $\alpha \in (1, 1+\varepsilon)$ with ε sufficiently small, there are three minimal surfaces with the same boundary. One surface is the patch Σ^2_α, given in (3.1), of the degree 2 Enneper surface. The other two are deformations of Σ^2_α achieved by vertically shifting the horizontal tangent plane of its central point.

Family (3.28), as previously mentioned, depends on two parameters, α and $\theta \in [0, \pi)$. For each fixed θ not equal to 0, $\pi/3$ or $2\pi/3$, the bifurcation at $\alpha = \sqrt[4]{3}$ is transcritical, i.e., there are two distinct minimal surfaces for each α in an small neighborhood around $\sqrt[4]{3}$. The new surface deforms the interior of the Enneper surface by tilting the horizontal tangent plane of central point along a horizontal

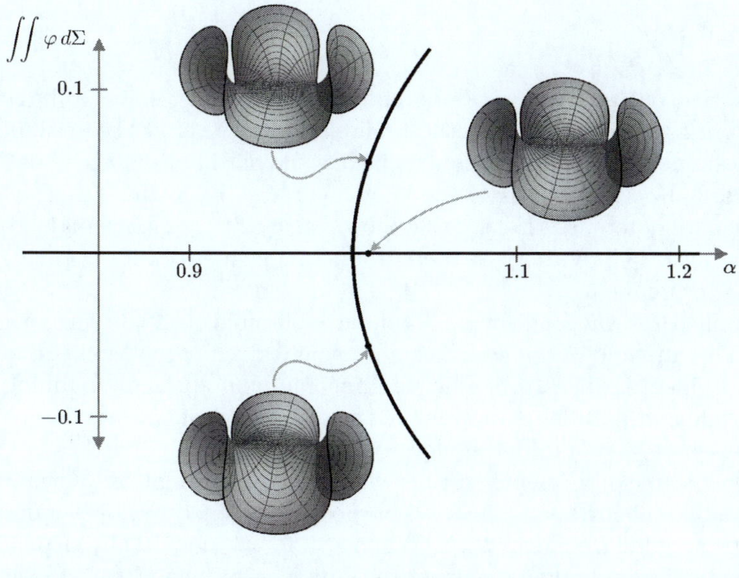

FIGURE 3. Bifurcation diagram of $2H_2(\varphi, \alpha) = 0$ near $\alpha = 1$.

axis of rotation. In the deformation (i.e., (3.28a)), t controls the degree of the tilt, while θ fixes the orientation of the axis of rotations; see Figure 4. When $\theta = 0$, $\pi/3$ or $2\pi/3$, the deformed surface again forms by tilting, but now the axis rotation exactly aligns with one of the main folds the Enneper surfaces. This alignment induces a mirror symmetry (across the vertical plane through the axis of rotation) between the oppositely tilted surfaces. That symmetry manifests as a pitchfork bifurcation in the solution set.

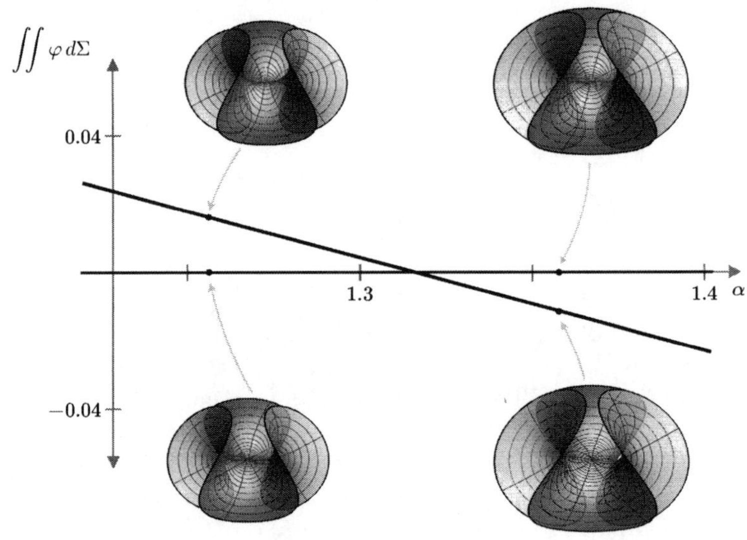

FIGURE 4. Bifurcation diagram of $2H_2(\varphi, \alpha) = 0$ near $\alpha = \sqrt[4]{3}$ and for $\theta = \pi/4$.

4. Discussion

This article outlined a general method for constructing new families of minimal surfaces from a given one. The method relies on results about the existence bifurcations from simple eigenvalues to find solutions the minimal surface equation $H = 0$ that are normal perturbations of the known surface. Even though $H = 0$ contains no explicit parameters, restricting the known surface to compact regions defines an effective radius $\alpha > 0$. Varying α then produces a framework for applying standard bifurcation techniques.

We applied the method for the standard catenoid and the degree n Enneper surfaces. The catenoid provided a base case where all the properties are well-known, since the problem reduces to solving finding minimal surfaces spanning circles of the same radius in parallel planes, while the degree n Enneper surfaces allowed for a situation where we could derive novel results.

For the degree n Enneper surface there are n bifurcations, meaning n new families of minimal surfaces appear, each separately generated by a k-mode perturbation with angular frequency $2\pi k$. When $k = 0$, the new family depends on only one parameter, α, and emanates from the Enneper surface at $\alpha = 1$. All the other families depend on two parameters, α and θ, where $\theta \in [0, \pi)$ prescribes a nodal line of the mode on the Enneper surface. The corresponding the bifurcation points of these two-parameter families are at

$$\alpha^*_{n,k} = \left(\frac{n+k}{n-k}\right)^{1/(2n)}, \qquad \theta \in [0, \pi)$$

for each $k = 1, 2, \ldots n-1$. Also, as n increases the values of $\alpha^*_{n,k}$ get compressed to an interval $[1, 1+\varepsilon(n)]$, where $\varepsilon(n) \to 0$ as $n \to \infty$.

For the $n = 2$, the first bifurcation at $\alpha = 1$ is a pitchfork, similar the degree $n = 1$ case elucidated [**24**], and the second bifurcation hinges on the value of θ. When $\theta = 0, \pi/3$, or $2\pi/3$, the nodal lines of the deformation exactly align with the dihedral symmetries of the Enneper surface, yielding pitchforks. Other values of θ break the symmetry, making the bifurcations transcritical. We conjecture that local structure connecting the varying behavior for θ involves sliding another saddle-node bifurcation whose fold exactly aligns main branch of surfaces at $\theta = 0, \pi/3$, or $2\pi/3$; see the stability diagram in Figure 5. This structure is likely created from a slice of a higher-dimensional catastrophe [**26**]. A potential method for uncovering the catastrophe is to remove the symmetry of the surface patch by adding waves to the boundary [**23**].

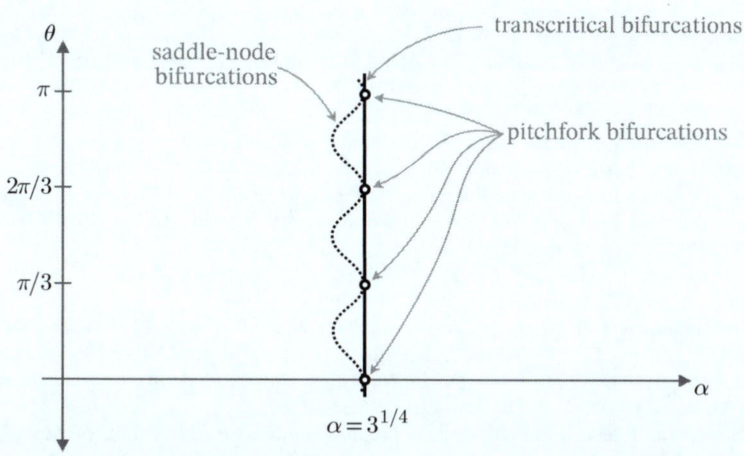

FIGURE 5. Possible bifurcation structure in the parameter space near the bifurcation line $\alpha = \sqrt[4]{3}$.

We expect that the degree n case generalizes the degree 2 situation. That is, the first bifurcation should be a pitchfork, while all the other bifurcations are transcritical, except for the pitchforks that appear when the nodal lines of the deformations align with the dihedral symmetries of the base surface. It would also be very interesting to determine if any of the families arising from the different bifurcations connect. Such a thought is especially puzzling in the situation for large n where the bifurcation are compressed in a small interval. In either situation, results of Troomba and Beeson establishing the existence of three minimal surfaces with the same boundary extend larger finite numbers after appropriately choosing the location of the nodal lines.

Additionally, the outlined method is widely applicable. All non-planar minimal surfaces have positive index. Thus, the value of the index, often determined from compact normal deformations on surface disks of varying size, predicts the number of new families of minimal surfaces. Of course, these new surfaces have a boundary but extending them to complete surfaces follows from a Björling representation [**6**, **20**]. This idea suggests that all non-planar, two-sided complete minimal surfaces, besides catenoids, are connected to another homothetically-distinct minimal surface. Catenoids provide the exception since they are index 1 and have a continuous rotational symmetry that aligns with the normal 0-mode deformation. The degree 1 Enneper surface also has index 1 but doesn't have the rotational symmetry. All other minimal surfaces have index greater than 1, inducing an asymmetric mode. It would be interesting to investigate if symmetry methods explain the continuous transformations between the surfaces. Also, the described method naturally extends to multiple coverings of a surface and, therefore, can be used to create additional minimal immersions that are not embedded. Preliminary calculations show that a double covering of a catenoid can create double barreled catenoid.

References

[1] E. Abbena, S. Salamon, and A. Gray, *Modern differential geometry of curves and surfaces with mathematica*, CRC press, 2017.

[2] M. J. Beeson and A. J. Tromba, *The cusp catastrophe of Thom in the bifurcation of minimal surfaces*, Manuscripta Math. **46** (1984), no. 1-3, 273–308, DOI 10.1007/BF01185204. MR735523

[3] Renato G. Bettiol and Paolo Piccione, *Instability and bifurcation*, Notices Amer. Math. Soc. **67** (2020), no. 11, 1679–1691, DOI 10.1090/noti. MR4201907

[4] Nicholas D. Brubaker, *A continuation method for computing constant mean curvature surfaces with boundary*, SIAM J. Sci. Comput. **40** (2018), no. 4, A2568–A2583, DOI 10.1137/17M1143228. MR3845275

[5] N. D. Brubaker, T. Murphy, and K. O. Negron, *Numerically destabilizing minimal discs*, Experimental Mathematics (2019), 1–12.

[6] Antonio Bueno, *The Björling problem for prescribed mean curvature surfaces in* $\mathbb{R}^3$, Ann. Global Anal. Geom. **56** (2019), no. 1, 87–96, DOI 10.1007/s10455-019-09657-w. MR3962027

[7] Otis Chodosh and Davi Maximo, *On the topology and index of minimal surfaces*, J. Differential Geom. **104** (2016), no. 3, 399–418. MR3568626

[8] Otis Chodosh and Davi Maximo , *On the topology and index of minimal surfaces II*, Preprint, `arXiv:1808.06572`, 2018.

[9] Tobias Holck Colding and William P. Minicozzi II, *A course in minimal surfaces*, Graduate Studies in Mathematics, vol. 121, American Mathematical Society, Providence, RI, 2011, DOI 10.1090/gsm/121. MR2780140

[10] Richard Courant and Herbert Robbins, *What is mathematics?: An elementary approach to ideas and methods*, Oxford University Press, New York, 1979. MR552669

[11] Michael G. Crandall and Paul H. Rabinowitz, *Bifurcation from simple eigenvalues*, J. Functional Analysis **8** (1971), 321–340, DOI 10.1016/0022-1236(71)90015-2. MR0288640

[12] M. do Carmo and C. K. Peng, *Stable complete minimal surfaces in* $\mathbf{R}^3$ *are planes*, Bull. Amer. Math. Soc. (N.S.) **1** (1979), no. 6, 903–906, DOI 10.1090/S0273-0979-1979-14689-5. MR546314

[13] Lawrence C. Evans, *Partial differential equations*, Graduate Studies in Mathematics, vol. 19, American Mathematical Society, Providence, RI, 1998. MR1625845

[14] Doris Fischer-Colbrie and Richard Schoen, *The structure of complete stable minimal surfaces in 3-manifolds of nonnegative scalar curvature*, Comm. Pure Appl. Math. **33** (1980), no. 2, 199–211, DOI 10.1002/cpa.3160330206. MR562550

[15] A. T. Fomenko and A. A. Tuzhilin, *Elements of the geometry and topology of minimal surfaces in three-dimensional space*, Translations of Mathematical Monographs, vol. 93, American Mathematical Society, Providence, RI, 1991. Translated from the Russian by E. J. F. Primrose, DOI 10.1007/s10958-011-0212-2. MR1134130

[16] Hansjörg Kielhöfer, *Bifurcation theory: An introduction with applications to partial differential equations*, 2nd ed., Applied Mathematical Sciences, vol. 156, Springer, New York, 2012, DOI 10.1007/978-1-4614-0502-3. MR2859263

[17] Miyuki Koiso, *Deformation and stability of surfaces with constant mean curvature*, Tohoku Math. J. (2) **54** (2002), no. 1, 145–159. MR1878932

[18] Rafael López, *Bifurcation of cylinders for wetting and dewetting models with striped geometry*, SIAM J. Math. Anal. **44** (2012), no. 2, 946–965, DOI 10.1137/11082484X. MR2914256

[19] Rafael López, *Stability and bifurcation of a capillary surface on a cylinder*, SIAM J. Appl. Math. **77** (2017), no. 1, 108–127, DOI 10.1137/16M107311X. MR3597165

[20] Rafael López and Matthias Weber, *Explicit Björling surfaces with prescribed geometry*, Michigan Math. J. **67** (2018), no. 3, 561–584, DOI 10.1307/mmj/1531447375. MR3835564

[21] Jerrold E. Marsden, *Qualitative methods in bifurcation theory*, Bull. Amer. Math. Soc. **84** (1978), no. 6, 1125–1148, DOI 10.1090/S0002-9904-1978-14549-2. MR508450

[22] Johannes C. C. Nitsche, *Contours bounding at least three solutions of Plateau's problem*, Arch. Rational Mech. Anal. **30** (1968), 1–11, DOI 10.1007/BF00253243. MR226515

[23] Johannes C. C. Nitsche, *Non-uniqueness for Plateau's problem. A bifurcation process*, Ann. Acad. Sci. Fenn. Ser. A I Math. **2** (1976), 361–373. MR0467552

[24] Johannes C. C. Nitsche, *Lectures on minimal surfaces: vol. 1*, Cambridge university press (1989).

[25] S. Smale, *On the Morse index theorem*, J. Math. Mech. **14** (1965), 1049–1055, DOI 10.1111/j.1467-9876.1965.tb00656.x. MR0182027

[26] Henry C. Wente, *A surprising bubble catastrophe*, Pacific J. Math. **189** (1999), no. 2, 339–376, DOI 10.2140/pjm.1999.189.339. MR1696127

Department of Mathematics, California State University, Fullerton, Fullerton California, 92831

Email address: nbrubaker@fullerton.edu

Contemporary Mathematics
Volume **777**, 2022
https://doi.org/10.1090/conm/777/15624

The (M_+, M_-)-method on compact symmetric spaces and its applications

Bang-Yen Chen

ABSTRACT. The (M_+, M_-)-method on compact symmetric spaces was introduced by the author and Tadashi Nagano in 1978. Since then, this theory have been studied and applied to several important areas in mathematics by many authors. The main purpose of this article is to provide a comprehensive survey on the (M_+, M_-)-method, including its applications.

1. Basics of compact symmetric spaces

For a circle S^1 in the Euclidean plane $\mathbb{R}^2$, the *antipodal point* q of a point $p \in S^1$ is the point in S^1 which is diametrically opposite to p. A geodesic on a Riemannian manifold M is locally the shortest curves between any two nearby points. Since each closed geodesic in a Riemannian manifold M is isometric to a circle S^1, antipodal points can also be defined for any closed geodesic in M. A closed geodesic in a Riemannian manifold is simply called a *circle* for short.

Let R and ∇ denote the Riemannian curvature tensor and the Levi-Civita connection of a Riemannian manifold M. The class of Riemannian manifolds with parallel curvature tensor, i.e., $\nabla R = 0$, was introduced by P. A. Shirokov in 1925. This class is known today as the class of locally symmetric Riemannian spaces.

An isometry s of a Riemannian manifold is said to be *involutive* if $s^2 = id$. A Riemannian manifold M is called a *symmetric space* if for each point $p \in M$ there exists an involutive isometry s_p with p as an isolated fixed point. The involutive isometry $s_p \neq id$ is called the *geodesic symmetry at p*.

A symmetric space is a Riemannian manifold M such that for each point $p \in M$ the geodesic symmetry s_p at p exists. Èlie Cartan noticed in 1926 that irreducible symmetric spaces are separated into ten large classes each of which depends on one or two arbitrary integers, and in addition there exist twelve special classes corresponding to the exceptional simple groups. Based on this, É. Cartan initiated his theory of symmetric spaces in his famous two papers "Sur une classe remarquable d'espaces de Riemann, I and II" [**13, 14**] and vigorously developed by him in the late 1920s. Cartan achieved his classification of symmetric spaces by reducing the problem to the classification of simple Lie algebras over real field **R**, a problem which Cartan himself solved earlier in 1914.

2020 *Mathematics Subject Classification.* Primary 53C35; Secondary 22E40, 53-03, 53C40.

Key words and phrases. (M_+, M_-)-theory, 2-number, 2-rank, compact symmetric space, homotopy, homology, antipodal set, Betti number, Euler number, arithmetic distance, real form.

FIGURE 1. The author received his Ph.D. degree from the University of Notre Dame in May 1970 under the supervision of Professor Tadashi Nagano.

Symmetric spaces are the most beautiful and important Riemannian manifolds. Such spaces arise in a wide variety of situations in both mathematics and physics. This class of spaces contains many prominent examples which are of great importance for various branches of mathematics, like compact Lie groups, Grassmannians and bounded symmetric domains. Symmetric spaces are also important objects of study in representation theory, harmonic analysis and in differential geometry.

Denote by G_M the closure of the group of isometries on a symmetric space M generated by geodesic symmetries $\{s_p : p \in M\}$. If there is no confusion, we may denote G_M simply by G. The G_M has an analytic structure compatible with the compact-open topology in which G is a Lie transformation group of the symmetric space M. Therefore, the typical *isotropy subgroup* H, say at a point $o \in M$, is compact and $M = G/H$. Hence, the symmetric space M is homogeneous. From the point of view of Lie theory, a symmetric space M is the quotient G/H of Lie group G by a Lie subgroup H, where the Lie algebra $\mathfrak{h}$ of H is also required to be the $(+1)$-eigenspace of an involution of the Lie algebra $\mathfrak{g}$ of G.

In this article, we follow standard symbols as in Helgason's book [**40**] to denote symmetric spaces mostly. Here are a few minor exceptions. More specifically, we denote $AI(n)$ by $SU(n)/SO(n)$ and $AII(n)$ by $SU(2n)/Sp(n)$, etc. And let $G_d(\mathbb{R}^n)$, $G_d(\mathbb{C}^n)$ and $G_d(\mathbb{H}^n)$ denote the Grassmannians of d-dimensional subspaces in real, complex and quaternion vector spaces, respectively. The standard notations for the exceptional spaces such as $G_2, F_4, E_6, ..., GI, ..., EIX$ denote the simply-connected spaces, where we write GI for $G_2/SO(4)$. For a symmetric space M, we denote by M^* the bottom space, i.e., the adjoint space of M in Helgason's book.

A *Helgason sphere* in a compact symmetric space is a totally geodesic sphere of maximal dimension with maximal sectional curvature. For a symmetric space M, the dimension of a maximal flat totally geodesic submanifold of M is called the *rank* of M, denoted by $rk(M)$. Clearly, the rank of a symmetric space is at least one. It is well-known that the class of rank one compact symmetric spaces consists of n-sphere S^n, a projective space $\mathbb{F}P^n (\mathbb{F} = \mathbb{R}, \mathbb{C}, \mathbb{H})$, and the 16-dimensional Cayley plane $FII = \mathbb{O}P^2$ with $\mathbb{O}$ being the Cayley algebra.

Every complete totally geodesic submanifold of a symmetric space is also a symmetric space. If B is a complete totally geodesic submanifold B of a symmetric space M, the equation of Gauss implies $rk(B) \leq rk(M)$.

2. (M_+, M_-)-theory

In this section, we provide a brief survey of the (M_+, M_-)-theory for compact symmetric spaces introduced by the author and T. Nagano in [**26, 29**] (see [**18, 64**] for earlier surveys on (M_+, M_-)-theory). Our approach to compact symmetric spaces was based on antipodal points and fixed point sets of compact symmetric spaces. Consequently, our approach to compact symmetric spaces is different from other approaches done by É. Cartan and others which are mostly based on the classification of simple Lie algebras and root systems.

For the references of this section, we refer to [**18, 26, 29, 62–64, 67, 69**].

2.1. Polars $M_+^o(p)$.

The following results were proved in [**26**].

THEOREM 2.1. *Let $M = G/H$ be a compact symmetric space. Then*

(1) *If p is an antipodal point of o, then $s_o s_p = s_p s_o$.*
(2) *If p is an antipodal point of o, then $H(p)$ is a complete totally geodesic submanifold of M through p.*
(3) *If M is connected, then $H(p)$ is the connected component of the fixed point set $F(s_o, M)$ if and only if p is an antipodal point of o.*
(4) *If M is connected, then $p \in F(s_o, M) \setminus \{o\}$ if and only if p is an antipodal point of o.*

Let o be a given point of a compact symmetric space $M = G/H$. A connected component of the fixed point set $F(s_o, M) \setminus \{o\}$ of the symmetry s_o is called a *polar* of o. We denote it by M_+^o; or $M_+^o(p)$ if M_+^o contains a point p.

FIGURE 2. Tadashi Nagano (left) and Bang-Yen Chen (right) in Tokyo, 1998

THEOREM 2.2. *Let $M = G/H$ be a compact symmetric space. Then for each antipodal point p of $o \in M$, the isotropy subgroup H at o acts transitively on the polar $M_+^o(p)$. Further, we have $H(p) = M_+^o(p)$ and $H(p)$ is connected. Consequently, we have $M_+^o(p) = H/H_p$, where $H_p := \{h \in H : h(p) = p\}$.*

For polars of a compact symmetric space M, we have the following.

THEOREM 2.3. *Let $\{o, p\}$ be an antipodal pair in a compact symmetric space M. Then (1) $M_+^o(p)$ and $M_+^p(o)$ are isometric and (2) If M is connected and A is a maximal flat torus of M containing o, then A meets every polar of $o \in M$.*

THEOREM 2.4. *Let p and q be two antipodal points of o in a compact symmetric space M. Then $s_q(M_+^o(p)) = M_+^o(p)$ and $F(s_q, M_+^o(p)) \neq \emptyset$.*

If a polar $M_+^o(p)$ of a compact symmetric space is a singleton, it is called a *pole* of o. We have the following properties for poles.

THEOREM 2.5. *Let o be a point in a connected compact symmetric space M. If $\bar{o}$ is a pole of o in M, then we have:*
(1) *$\bar{o}$ lies in every M_-^o.*
(2) *$\bar{o}$ is an antipodal point of any antipodal point $q\,(\neq \bar{o})$ of o in M.*
(3) *If M is a compact flat symmetric space, the each polar of o is a pole of o.*

We have following characterizations of poles from [**29**, page 277].

THEOREM 2.6. *The following six conditions are equivalent to each other for two distinct points o, p of a connected compact symmetric space $M = G_M/K_G$.*
(i) *p is a pole of $o \in M$;*
(ii) *$s_p = s_o$;*
(iii) *$\{p\}$ is a polar of $o \in M$;*
(iv) *there is a double covering totally geodesic immersion $\pi = \pi_{\{o,p\}} : M \to M''$ with $\pi(p) = \pi(o)$;*
(v) *p is a point in the orbit $F(\sigma, G_M)(o)$ of the group $F(\sigma, G_M)$ through o, where $\sigma = \mathrm{ad}(s_o)$;*

(vi) *the isotropy subgroup of SG_M at p is that, SK_G (of SG_M at o), where SG_M is the group generated by G_M and the symmetries; SG_M/G_M is a group of order ≤ 2.*

2.2. Meridians $M_-^o(p)$. The notion of meridians is defined in [**26**].

THEOREM 2.7. *Under the hypothesis of Theorem 2.1, the normal space to $M_+^o(p)$ at $p \in M$ is the tangent space of a connected complete totally geodesic submanifold $M_-^o(p)$. Thus $\dim M_+^o(p) + \dim M_-^o(p) = \dim M$. Moreover, we have*

(1) $rk(M_-^o(p)) = rk(M)$.
(2) *$M_-^o(p)$ is a connected component of the fixed point set $F(s_p \circ s_o, M)$ of $s_p \circ s_o$ through p.*
(3) $M_-^o(p) = M_-^p(o)$.
(4) *Every M_-^o meets every M_+^o.*

The set $M_-^o(p)$ is called a *meridian* of o. Polars and meridians of a compact symmetric space are compact totally geodesic submanifolds. Thus they are compact symmetric spaces as well.

THEOREM 2.8. *Let o, p, q be mutually antipodal points in a compact symmetric space M. Then we have $s_q(M_-^o(p)) = M_-^o(p)$.*

From [**18**, page 31], we have the following.

THEOREM 2.9. *Let (o, p) be an antipodal pair in a compact symmetric space M. Then $s_o s_p = s_q$ for some point $q \in M$ if and only if the polar $M_+^o(p)$ and the meridian $M_-^o(p)$ are conjugate under the action of the isotropy subgroup at p.*

Moreover, under this case, we have:

(a) *o, p, q are mutually antipodal,*
(b) *$s_p = s_o s_q = s_q s_o$ and $s_o = s_p s_q = s_q s_p$.*
(c) *$\dim M$ is even.*
(d) $M_+^o(p) = M_+^o(q) = M_-^p(q) = M_-^q(p)$, $M_+^p(o) = M_+^p(q) = M_-^o(q) = M_+^q(o)$ *and* $M_+^q(o) = M_+^q(p) = M_-^o(p) = M_-^p(o)$.
(e) *$M_+^o(p), M_+^p(q)$ and $M_+^q(o)$ are conjugate under the actions of the isotropy subgroups at o, p and q.*
(f) *The polars $M_+^o(p), M_+^p(q)$ and $M_+^q(o)$ meet pairwise at o, p, and q as singletons.*
(g) *The polars $M_+^o(p), M_+^p(q)$ and $M_+^q(o)$ meet pairwise orthogonally at o, p, or q.*
(h) $d(o, p) = d(p, q) = d(q, o)$.

REMARK 2.1. Theorem 2.9 shows that if there is an antipodal pair (o, p) in a compact symmetric space M such that $s_o s_p = s_q$ for some point $q \in M$, then there exists a *right polar triangle* with vertices o, p and q and the three sides are given by the polars $M_+^o(p), M_+^p(q)$ and $M_+^q(o)$. The three sides are conjugate under the actions of the isotropy subgroups at o, p and q. The three sides meet pairwise at o, p and q as singletons and meet mutually orthogonal at the vertices. If we replace the sides by $M_-^o(p), M_-^p(q)$ and $M_-^q(o)$, then we have the same triangle. Conversely, Theorem 2.9 implies that the existence of such right polar triangle guarantees the existence of a point q satisfying $s_o s_p = s_q$. The projective planes RP^2, CP^2, HP^2 and the Cayley plane $\mathbb{O}P^2$ are examples of compact symmetric spaces which admits a right polar triangle. In fact, there are some other compact symmetric spaces which also admit right polar triangles (see [**18**, pages 31-32]).

REMARK 2.2. The classification of polars and meridians was globally completed by T. Nagano [**26**, **62**]. T. Nagano and M. Sumi (= M. S. Tanaka) proved in [**64**] that the root system $R(M_-)$ of a meridian $M_-(M_- \neq M)$ is obtained from the Dynkin diagram of the root system $R(M)$ for a compact symmetric space M. By applying (M_+, M_-)-theory, they determined all maximal totally geodesic spheres in $SU(n)$.

2.3. Centrosomes $C(o,p)$. The notion of centrosomes was introduced and studied by the author and T. Nagano in [**29**]. Centrosomes play significant roles in topology as well, e.g., they were used by J. M. Burns [**10**] to compute homotopy of compact symmetric spaces (see §4).

Let o be a given point of a compact symmetric space M. If there is a pole p of o in M, then the set consisting of the midpoints of all geodesics joining o and p is called the *centrosome* of $\{o,p\}$; denoted by $C(o,p)$. Every connected component of $C(o,p)$ is again a totally geodesic submanifold of M.

For a compact symmetric space M, the *Cartan quadratic morphism* is defined by $Q = Q_o : M \to G_M$ which carries a point $x \in M$ into $s_x s_o \in G_M$.

From [**29**, pages 279-280] we have the following result for centrosomes.

THEOREM 2.10. *The following five conditions are equivalent to each other for two distinct points o, q of a connected compact symmetric space M.*

(i) $s_o s_q = s_q s_o$;
(ii) $Q(q)^2 = 1_{G_M}$, *where $Q = Q_o$ is Cartan quadratic morphism;*
(iii) *either s_o fixes q or q is a point in the centrosome $C(o,p)$ for some pole p of o;*
(iv) *either $s_o(q) = q$ or $s_o(q) = \gamma(q)$ for the covering transformation γ for some pole $p = \gamma(o)$ of o;*
(v) *either $s_o(q) = q$ or there is a double covering morphism $\pi : M \to M''$ such that $s_{o''}$ fixes q'', where $o'' = \pi(o)$ and $q'' = \pi(q)$.*

REMARK 2.3. P. Quast [**76**] classified centrosomes of simply-connected irreducible symmetric spaces of compact type in terms of the root system. He distinguished four types of centrosomes according to some algebraic properties involving the highest root and the Cartan matrix of the root system.

2.4. Weyl group and Orbit set $\mathcal{P}(M)$. Given a pair of antipodal points (o,p) in a compact symmetric space $M = G/H$, we have the pair $(M_+^o(p), M_-^o(p))$ of polar $M_+^o(p)$ and meridian $M_-^o(p)$. The isometry group G acts on the set of all such pairs in the natural fashion. We denote its orbit set by $\mathcal{P}(M)$. In this way, two pairs $(M_+^o(p), M_-^o(p))$ and $(M_+^{o'}(p'), M_-^{o'}(p'))$ are identified if there is an isometry $g \in G_M$ which carries $o, p, M_+^o(p), M_-^o(p)$ into $o', p', M_+^{o'}(p'), M_-^{o'}(p')$, respectively. The $\mathcal{P}(M)$ is a finite set and its cardinal number $\#\mathcal{P}(M)$ is a global invariant.

PROPOSITION 2.1. [**26**] $\#\mathcal{P}(M) \leq 2^{rk(M)} - 1$ *for each compact symmetric space M. In particular, if $rk(M) = 1$, we have $\#\mathcal{P}(M) = 1$.*

One important property of $\mathcal{P}(M)$ is the following result from [**26**].

THEOREM 2.11. *An irreducible compact symmetric space M is globally determined by $\mathcal{P}(M)$, i.e., the set of the global isomorphism classes of compact irreducible symmetric spaces are in one-to-one correspondence with the set of the corresponding $\mathcal{P}(M)$.*

REMARK 2.4. Note that Satake and Dynkin diagrams for symmetric spaces do not distinguish symmetric spaces globally, e.g., in their diagrams the n-sphere S^n and the real projective n-space RP^n have the same diagrams. However, $\mathcal{P}(S^n)$ and $\mathcal{P}(RP^n)$ are quite different as we can see in Table I given in §2.5.

Let (o, p) be an antipodal pair of a compact symmetric space M. Assume that $\phi : B \to M$ is a totally geodesic embedding, then ϕ gives rise to a mapping

$$P(\phi) : \mathcal{P}(B) \to \mathcal{P}(M)$$

induced from the mapping which carries $(B_+^o(p), B_-^o(p))$ to $(M_+^o(p), M_-^o(p))$. This mapping $P(\phi)$ is well-defined since every isometry g in G_B "extends" to an isometry $\bar{g}$ in G_M so that $\phi \circ g = \bar{g} \circ \phi$. It is easy to see that $\phi(B_+^o(p)) \to M_+^o(p)$ and $\phi(B_-^o(p)) \to M_-^o(p)$ are totally geodesic submanifolds as well. Because this is an important, we express it by saying that $P(\phi)$ is a *pairwise totally geodesic immersion*. Now, we may state this important fact as the following.

THEOREM 2.12. **[26]** *Every totally geodesic embedding of a compact symmetric space B into another compact symmetric space M induces a pairwise totally geodesic immersion $P(\phi) : \mathcal{P}(B) \to \mathcal{P}(M)$.*

We also have the following.

THEOREM 2.13. **[26]** *Let $\phi : B \to M$ be a totally geodesic embedding of a compact symmetric space into another. If $rk(B) = rk(M)$, then*

(a) *$P(\phi) : \mathcal{P}(B) \to \mathcal{P}(M)$ is surjective.*
(b) *The Weyl group $W(B)$ of B is a subgroup of $W(M)$.*
(c) *If $W(B)$ is isomorphic with $W(M)$ by the natural homomorphism, then $P(\phi)$ is bijective.*

2.5. Examples. For polars, meridians and centrosomes of rank one compact symmetric spaces M, we have the following.

TABLE I

M	$M_+^o(p)$	$M_-^o(p)$	$C(o,p)$
S^n	$\{p\}$	S^n	S^{n-1}
RP^n	RP^{n-1}	S^1	none
CP^n	CP^{n-1}	S^2	none
HP^n	HP^{n-1}	S^4	none
$\mathcal{O}P^2$	S^8	S^8	none

REMARK 2.5. For polars, meridians and centrosomes of compact symmetric spaces of higher rank, see **[18, 29, 62, 63]**.

3. Applications to totally geodesic embeddings

One of applications of (M_+, M_-)-theory discussed in **[26]** is the applications to totally geodesic submanifolds. If B is a complete totally geodesic submanifold of a compact symmetric space M, then B is also a compact symmetric space. It follows from Theorem 2.12 that, for a pair (B_+, B_-) in $\mathcal{P}(B)$, there is a pair (M_+, M_-) in $\mathcal{P}(M)$ such that B_+ and B_- are totally geodesic in M_+ and M_-, respectively. By applying the same argument to $B_+ \subset M_+$ and $B_- \subset M_-, \cdots$, etc., we obtain a finite sequence of pairwise totally geodesic immersions as follows:

$$
\begin{array}{ccccccc}
 & & & & & \nearrow & \cdots\cdots \\
 & & & & (B_+)_+ \subset (M_+)_+ & & \\
 & & & \nearrow & & \searrow & \cdots\cdots \\
 & & B_+ \subset M_+ & & & & \\
 & & & \searrow & & \nearrow & \cdots\cdots \\
 & \nearrow & & & (B_+)_- \subset (M_+)_- & \searrow & \cdots\cdots \\
B \subset M & & & & & \nearrow & \cdots\cdots \\
 & \searrow & & & (B_-)_+ \subset (M_-)_+ & \searrow & \cdots\cdots \\
 & & & \nearrow & & & \\
 & & B_- \subset N_- & & & & \\
 & & & \searrow & & \nearrow & \cdots\cdots \\
 & & & & (B_-)_- \subset (M_-)_- & & \\
 & & & & & \searrow & \cdots\cdots
\end{array}
$$

This argument can be applied to investigate totally geodesic submanifolds of compact symmetric spaces. For instance, although totally geodesic submanifolds of rank one compact symmetric spaces were completely classified by J. A. Wolf [**101**] via Lie triple system, totally geodesic submanifolds of rank one compact symmetric spaces can be classified much easily by using (M_+, M_-)-theory via this argument.

THEOREM 3.1. [**101**] *The maximal totally geodesic submanifold of S^n is S^{n-1}; of RP^n is RP^{n-1}; of CP^n are CP^{n-1} and RP^n; of HP^n are HP^{n-1} and CP^n; of $\mathbb{O}P^2$ are HP^2 and S^8.*

PROOF. Existence of the totally geodesic immersions stated in the theorem are well known. The remaining part follows easily from Theorem 2.12 and Table I. □

REMARK 3.1. For some further applications in this respect, see [**26**].

4. Application to homotopy

A famous work of R. Bott is his periodicity theorem which describes periodicity of homotopy groups of classical groups (cf. [**8**]). R. Bott's original results may be succinctly summarized as

THEOREM 4.1. *The homotopy groups of the classical groups are periodic:*

$$\pi_i(U) = \pi_{i+2}(U),\ \pi_i(O) = \pi_{i+4}(Sp),\ \pi_i(Sp) = \pi_{i+4}(O)$$

for $i = 0, 1, \cdots$, where U is the direct limit defined by $U = \cup_{k=1}^{\infty} U(k)$ and similarly for O and Sp.

The second and third of the isomorphisms given in Theorem 4.1 imply the *8-fold periodicity*:

$$\pi_i(O) = \pi_{i+8}(O),\ \ \pi_i(Sp) = \pi_{i+8}(Sp),\ \ i = 0, 1, \cdots .$$

Bott's proof relies on the observation that in a compact Riemannian symmetric space M one can choose two points p, q in "special position" such that the connected components of the space of shortest geodesics in M joining p, q are again compact symmetric spaces. By putting $M = M_0$ and M_1 being one of the resulting connected components, this construction can be repeated inductively (see [**7, 8**] for details).

Recall that the *index of a geodesic* γ in a compact symmetric space M from p to q is the number of conjugate points of p counted with their multiplicities, in the open geodesic segment from p to q. Denote the space of shortest geodesics from p to q by the *symbol* Ω^d, which relates closely with the notion of centrosomes. The proof of Bott's periodicity theorem relied on the following.

BOTT'S THEOREM. *If Ω^d is a topological manifold and if every non-shortest geodesic from p to q has index greater than or equal to λ_0, then the $(i+1)$-th homotopy groups of M satisfies $\pi_{i+1}(M) \cong \pi_i(\Omega^d)$ for $i < \lambda_0 - 1$.*

In [**10**], the (M_+, M_-)–theory was applied to compute homotopy groups of compact symmetric spaces. J. M. Burns explained in [**10**] that how the (M_+, M_-)–theory in conjunction with Bott's theorem can be used to compute the homotopy of compact symmetric spaces. He carried out the computation of the homotopy in compact symmetric spaces of types: AI, AII and CI. Also, he computed the homotopy of compact exceptional symmetric spaces: $EIII - EIX$, F_4, FI and FII. For further applications in this respect, see also [**56, 75**].

5. An algorithm for stability and its applications

Now, we present a general theory for determining the stability of totally geodesic submanifolds in symmetric spaces introduced in 1980 by the author, P.-F. Leung and T. Nagano (see [**24**] and the last chapter of author's book [**19**]).

5.1. Stable submanifolds. Let $\phi : N \to M$ be a minimal immersion from a compact Riemannian manifold N into another Riemannian manifold M. Consider a normal variation $\{\phi_t\}$ of f such that $\phi_0 = \phi$. Let ξ denote the variation vector field associated $\{\phi_t\}$. Then the compact minimal submanifold N in M is called *stable* if the second variation of the volume integral is non-negative for each normal variation of f. The second variational formula of $\{\phi_t\}$ is given by (cf. [**83**])

$$\mathcal{V}''(\xi) = \int_N \{\|D\xi\|^2 - \bar{S}(\xi,\xi) - \|A_\xi\|^2\} * 1, \tag{5.1}$$

where D is the normal connection, $\bar{S}(\xi,\xi)$ is defined by $\sum_i \langle R^M(e_i,\xi)\xi, e_i\rangle$ for a local orthonormal frame $\{e_1,\dots,e_n\}$ of TN, R^M denotes the curvature tensor of M, and A is the shape operator. If N is totally geodesic, then $A = 0$. Hence, the stability obtained trivially when $\bar{S}$ is non-positive. Hence, we are only interested in the case in which M is a symmetric space of compact type.

Now, consider the *Jacobi operator* L of N in M, which is a self-adjoint strongly elliptic linear differential operator L of the second order acting on the space of sections of the normal bundle given by

$$L = -\Delta^D - \hat{A} - \hat{S}, \tag{5.2}$$

where Δ^D is the Laplace operator associated with the normal connection D and

$$\langle \hat{A}\xi, \eta\rangle = \langle A_\xi, A_\eta\rangle, \quad \langle \hat{S}\xi, \eta\rangle = \bar{S}(\xi,\eta).$$

After applying Stokes' theorem to (5.1), we find as did in [**83**] that

$$\mathcal{V}''(\xi) = \int_N \langle Lv, v\rangle * 1. \tag{5.3}$$

The Jacobi operator L has discrete eigenvalues $\lambda_1 < \lambda_2 < \cdots \nearrow \infty$. Put

$$E_\lambda = \{\xi \in \Gamma(T^\perp N) \,:\, L(\xi) = \lambda\xi\}.$$

Then the number of $\sum_{\lambda<0} dim(E_\lambda)$ is called the *index* of N in M. A vector field ξ in E_0 is called a *Jacobi vector field*. Clearly, the minimal submanifold N is stable if and only if the eigenvalues of L are non-negative. Otherwise, N is unstable.

5.2. Algorithm for stability. Assume N is a complete totally geodesic submanifold of a symmetric space M. Since N is totally geodesic, there exists a finitely covering group G_N of the connected isometry group G_N^o of N such that G_N is a subgroup of the connected isometry group G_M of M which leaves N invariant, provided that G_N^o is semi-simple. Let $\mathcal{P}$ denote the orthogonal complement of the Lie algebra $g_{_N}$ in the Lie algebra $g_{_M}$ with respect to the bi-invariant inner product on $g_{_M}$ which is compatible with the metric of M. Here, every member of $g_{_M}$ is though of as a Killing vector field because of the action G_M on M. Let $\hat{P}$ denote the space of the vector fields corresponding to the member of $\mathcal{P}$ restricted to N.

We have the following.

LEMMA 5.1. [**24**] *To every member of $\mathcal{P}$ there corresponds a unique (but not canonical) vector field $v \in \hat{P}, v$ is a normal vector field and hence $\hat{P}$ is a G_N-invariant subspace of the space $\Gamma(T^\perp N)$ of sections of the normal bundle to N. Moreover, $\hat{P}$ is homomorphic with $\mathcal{P}$ as a G_N-module.*

Next, we explain the algorithm for determining the stability of the totally geodesic submanifold N in M as follows. Since group G_N acts on sections in $\Gamma(T^\perp N)$, hence acts on the differential operators: $\Gamma(T^\perp N) \to \Gamma(T^\perp N)$. Because L is defined with N and the metric of M only, G_N leaves L fixed. Hence, each eigenspace of L is left invariant by G_N. If V is one of the G_N-invariant irreducible subspaces, then we have a representation $\rho : G_N \to GL(V)$. Let us denote by $c(V)$ or by $c(\rho)$ the eigenvalue of the corresponding Casimir operator, which we will explain shortly. Then, modulo details, Theorem 5.1 below says that N is stable if and only if we have $c(V) \geq c(\mathcal{P})$ for every such V.

To define the Casimir operator $c(V)$, we fix an orthonormal basis (e_λ) for $g_{_N}$ and consider the linear endomorphism C or C_V of V defined by

$$C = -\sum \rho(e_\lambda)^2. \tag{5.4}$$

Then C is $c(V)I_V$, where I_V is the identity map on V (see Chapter 8 of [**9**]). In our case, the Casimir operator is given by

$$C_V(\xi) = -\sum_\lambda [e_\lambda, [e_\lambda, \xi]]$$

for every member ξ of V (after extending to a neighborhood of N).

The following theorem from [**19, 24**] gives a necessary and sufficient condition for the stability of totally geodesic submanifolds in compact symmetric spaces.

THEOREM 5.1. *A compact connected totally geodesic submanifold $N = G_N/K_N$ of a locally symmetric space $M = G_M/K_M$ is stable as a minimal submanifold if and only if $c(V) \geq c(P')$ for the eigenvalue of the Casimir operator of every simple G_N-module V which shares as a K_N-module some simple K_N-submodule of the K_N-module $T_o^\perp N$ in common with some simple G_N-submodule P' of P.*

REMARK 5.1. For the proof of Theorem 5.1, see [**24**] or [**19**, pages 106–108].

By applying Theorem 5.1, we have the *algorithm for stability* goes like this. One can compute $c(V)$ by the Freudenthal formula (cf. [**9**, Chapter 8, page 120]).

Once one knows the action ρ of G_N on V. Then the rest is to know all the simple G_N-modules V in $\Gamma(T^\perp N)$, which can be done by means of the Frobenius theorem as reformulated by R. Bott. This asserts in our case that a simple G_N-module V appears in $\Gamma(T^\perp N)$ if and only if V as a K_N-module contains a simple K_N-module which is isometric with a K_N-module of $T_o^\perp N$.

5.3. Stability of polars and meridians. By applying this algorithm for stability, we have the following results for polars and meridians.

THEOREM 5.2. [**24**] *We have:*

(a) *Among the compact connected simple Lie groups M^*, the only ones that have unstable M_+^* are $SU(n)^*$, $SO(2n)^*$ with n odd, E_6^* and G_2.*
(b) *The unstable polars M_+ are $G^{\mathbb{C}}(k, n-k)$, $0 < k < n-k$, for $SU(n)^*$; $SO(2n)/U(n)^*$ for $SO(2n)^*$; $EIII^*$ for E_6^*; and M_+^* for G_2.*
(c) *Every meridian M_- is stable for the group M^*.*

THEOREM 5.3. *Let M^* be a compact symmetric space G/K with G simple. Then, among M_+ and M_-, the unstable minimal submanifolds are $G^{\mathbb{R}}(k, n-k)$, $k < n-k$, in $AI(n)^*$; $G^{\mathbb{H}}(k, n-k)$, $k < n-k$, in $AII(n)^*$; $SO(k)$ in $G^{\mathbb{R}}(k,k)$ with k odd; $M_+ = M_- = SO(2) \times AI(n)$ in $CI(n)^*$; $M_+ = M_- = SO(2) \times AII(\frac{n}{2})$ in $DIII^* = SO(2n)/U(n)$ with n even; $G^{\mathbb{H}}(2,2)$ in EI^*; FII in EIV^*; $AII(4)$ in EV^*; and $M_+ = M_- = S^2 \cdot S^2$ in GI.*

5.4. Ohnita's formulas. The algorithm of [**24**] given in §5.2 was reformulated and improved by Y. Ohnita [**72**] in 1987 to include the following formulas for the index $i(\phi)$, the nullity $n(\phi)$ and the Killing nullity $n_k(\phi)$ of a compact totally geodesic embedding $\phi : B \to M$ in a compact symmetric space (see [**72**] for details).

(a) $i(\phi) = \sum_{i=1}^t \sum_{\lambda \in D(G), a_\lambda < a_i} m(\lambda) d_\lambda$,
(b) $n(\phi) = \sum_{i=1}^t \sum_{\lambda \in D(G), a_\lambda = a_i} m(\lambda) d_\lambda$,
(c) $n_k(\phi) = \sum_{i=1, \mathfrak{m}_i^\perp \neq \{0\}}^t \dim \mathfrak{g}_i^\perp$,

where $m(\lambda) = \dim \operatorname{Hom}_K(V_\lambda, (\mathfrak{m}_i^\perp)^{\mathbb{C}})$ and d_λ is the dimension of the representation λ.

In [**72**], Y. Ohnita also proved that Helgason spheres in compact symmetric spaces are stable.

5.5. Further applications to stability. The following stability results for totally geodesic submanifolds were obtained by applying the algorithm in [**24**] and Ohnita's formulas in [**72**]:

(1) The stability of symmetric R-spaces in Hermitian symmetric spaces by M. Takeuchi [**85**] (1984).
(2) A detailed study of the stability of polars and meridians by M. S. Tanaka [**87**] (1995).
(3) K. Mashimo and H. Tasaki [**58**] proved that every closed Lie subgroup of Dykin index 1 in a compact simple connected Lie group is stable (1990).
(4) Mashimo [**57**] determined all of unstable Cartan embeddings of compact symmetric spaces (1992).
(5) T. Kimura and Tanaka [**48**] determined the stability of the maximal totally geodesic submanifolds in compact symmetric spaces of rank two (2009).

(6) There is a larger class of totally geodesic submanifolds, called the *class of basic pairs*. This class contains the class of polars and meridians (see [**45**] for details). By applying Ohnita's formulas, Q. Jin studied the stability of the basic pairs in [**45, 46**].

6. 2-numbers and maximal antipodal sets

By applying the (M_+, M_-)-theory, the author and T. Nagano [**27, 29**] were able to study 2-numbers and maximal antipodal sets initiated in 1982.

6.1. 2-number of Riemannian manifolds. The notion of 2-numbers are defined in [**27, 29**] as follows: A subset S of a Riemannian manifold M is called an *antipodal set* if any two points in S are antipodal on some circle connecting them in M. An antipodal set A_2M in a connected Riemannian manifold M is called a *maximal antipodal set* if it doesn't lie in any antipodal set as a proper subset. The supremum of the cardinality of all maximal antipodal set of M is called the *2-number* of M, denote by $\#_2M$. An antipodal set in M is called a *great antipodal set* if its cardinality attains $\#_2M$. When a Riemannian manifold M contains no closed geodesics, e.g., a Euclidean space, we simply put $\#_2M = 0$.

Because every compact connected Riemannian manifold contains at least one closed geodesic (cf. [**55**]), we have

$$\#_2M \geq 2 \tag{6.1}$$

for compact Riemannian manifolds. Inequality (6.1) is sharp due to $\#_2S^n = 2$.

6.2. 2-number of compact symmetric spaces. For compact symmetric spaces we have the following.

PROPOSITION 6.1. [**29**, page 275] *For a compact symmetric space M, $\#_2M$ equals the maximal possible cardinality $\#(A_2M)$ of a subset A_2M of M such that the point symmetry s_x fixes every point of A_2M for each $x \in A_2M$.*

The following sharp inequality is important for the study of 2-numbers.

THEOREM 6.1. [**29**, page 276] *The 2-number $\#_2M$ of a compact symmetric space M does not exceed one plus the sum of the 2-numbers of all the polars of a point in a compact symmetric space M, i.e., $\#_2M \leq 1 + \sum \#_2M_+$.*

A clear proof of the finiteness of 2-number for a compact symmetric space was given in [**92**] by M. S. Tanaka and H. Tasaki.

REMARK 6.1. The author and Nagano [**29**] have determined 2-numbers of irreducible compact symmetric spaces via maximal antipodal sets (see also Appendices II and III of [**18**, pages 67–71]). But the antipodal sets of *oriented* real Grassmannians were not discussed in their articles.

On the other hand, H. Tasaki [**98, 99**] described in details the maximal antipodal sets of oriented real Grassmannians. Further, H. Kurihara and T. Okuda have studied in [**53**] maximal antipodal sets of complex Grassmannian manifolds and applied them to design theory (see also [**52**]).

REMARK 6.2. For further studies of the great antipodal sets of some compact symmetric spaces and also of compact Lie groups, see [**92, 94–96**] by M. S. Tanaka, H. Tasaki and O. Yasukura.

6.3. 2-number and totally geodesic embeddings. If B is a totally geodesic submanifold of a Riemannian manifold M, then every geodesic of B is a geodesic of M. Hence

$$\#_2 B \leq \#_2 M. \tag{6.2}$$

Consequently, the invariant $\#_2 M$ is a geometric obstruction to the existence of a totally geodesic embedding. For example, although the real projective n-space RP^n can be topologically embedded in S^m for sufficient high m, but inequality (6.2) implies that RP^k with $k \geq 2$ cannot totally geodesic embedded into any m-sphere S^m regardless of codimension due to $\#_2 RP^k = k + 1 > 2 = \#_2 S^m$.

7. Links between 2-number and topology

The 2-number links closely to several topics in topology.

7.1. 2-number and Euler number. The following links between 2-number and Euler number of a compact symmetric space were proved in [**29**].

THEOREM 7.1. *For any compact symmetric space M, we have $\#_2 M \geq \chi(M)$, where $\chi(M)$ denotes the Euler number of M.*

THEOREM 7.2. *If M is a compact Hermitian symmetric space of semi-simple type, we have $\#_2 M = \chi(M) = 1 + \sum \#_2 M_+$.*

An immediate consequence of this theorem is the following.

COROLLARY 7.1. *For every complete totally geodesic Hermitian submanifold B of a semi-simple Hermitian symmetric space M, we have $\chi(B) \leq \chi(M)$.*

7.2. 2-numbers and homology. Symmetric R-spaces were introduced by T. Nagano [**61**] and M. Takeuchi [**84**] in 1965 independently as compact symmetric spaces which are at the same time R-spaces, i.e., they also admit a transitive action of a center-free noncompact semisimple Lie group and the corresponding stabilizer of a point is a certain maximal parabolic subgroup. The terminology "*symmetric R-space*" was first used by Takeuchi in [**84**] (see [**90**] for a recent detailed survey on symmetric R-spaces).

Symmetric R-spaces were completely classified by S. Kobayashi and T. Nagano in [**49**] which includes:

(a) All Hermitian symmetric spaces of compact type.
(b) Grassmann manifolds $O(p+q)/O(p) \times O(q)$, $Sp(p+q)/Sp(p) \times Sp(q)$.
(c) The classical groups $SO(m)$, $U(m)$, $Sp(m)$.
(d) $U(2m)/Sp(m)$, $U(m)/O(m)$.
(e) $(SO(p+1) \times SO(q+1))/S(O(p) \times O(q))$, where $S(O(p) \times O(q))$ is the subgroup of $SO(p+1) \times SO(q+1)$ consisting of matrices of the form:

$$\begin{pmatrix} \epsilon & 0 & & \\ 0 & A & & \\ & & \epsilon & 0 \\ & & 0 & B \end{pmatrix}, \quad \epsilon = \pm 1, \quad A \in O(p), \quad B \in O(q),$$

(This symmetric R-space is covered twice by $S^p \times S^q$).
(f) The Cayley projective plane $FII = \mathbb{O}P^2$.
(g) The three exceptional spaces $E_6/Spin(10) \times T$, $E_7/E_6 \times T$, and E_6/F_4.

For symmetric R-spaces, M. Takeuchi proved the following.

THEOREM 7.3. [**86**] *$\#_2 M = \sum_{i\geq 0} B_i(M, \mathbb{Z}_2)$ for every symmetric R-space M, where $B_i(M, \mathbb{Z}_2)$ is the i-th Betti number of M with coefficients in $\mathbb{Z}_2$.*

7.3. Holomorphic 2-numbers and Euler number. C. U. Sánchez defined in [**80**] the *holomorphic 2-number* $\#_2^H M$ for a compact Hermitian symmetric space M as the maximal possible cardinality of a subset A_2 such that for every pair of points x and y of A_2, there exists a totally geodesic complex curve of genus 0 in M on which x and y are antipodal to each other. Sánchez [**80**] proved the following.

THEOREM 7.4. *$\#_2^H M = \#_2 M$ for every compact Hermitian symmetric space.*

The following corollary follows immediately from Theorem 7.4 and the equality $\#_2 M = \chi(M)$ from Theorem 7.2.

COROLLARY 7.2. *$\#_2^H M = \chi(M)$ for each compact Hermitian symmetric space.*

7.4. Two-numbers and covering maps. The following links between 2-numbers and covering maps were obtained in [**29**].

THEOREM 7.5. *Let M and M'' be compact symmetric spaces. If M is a double covering of M'', then $\#_2 M \leq 2 \cdot \#_2(M'')$.*

REMARK 7.1. The inequality in Theorem 7.5 is sharp, since the equality case holds for $M = SO(2m)$ with $m > 2$.

For k-fold coverings with odd k, we have

THEOREM 7.6. *Let $\phi : M \to N$ is a k-fold covering between compact symmetric spaces. If k is odd, then $\#_2 M = \#_2(N)$.*

8. Applications to compact Lie groups

One important application of (M_+, M_-)-theory is its application to group theory via 2-number.

8.1. 2-ranks of Borel and Serre. The 2-rank, $r_2 G$, of a compact Lie group G was defined and studied by A. Borel and J.-P. Serre in [**6**]. By definition the 2-rank of a compact connect Lie group G is the maximal possible rank of the elementary 2-subgroups of G. Borel and Serre proved in [**6**] the following two results for a compact connected Lie group G.

(1) $rk(G) \leq r_2(G) \leq 2 \cdot rk(G)$ and
(2) G has (topological) 2-torsion if $rk(G) < r_2(G)$.

In [**6**], Borel and Serre were able to determine the 2-rank of the simply-connected simple Lie groups $SO(n), Sp(n), U(n), G_2$ and F_4. They also proved that G_2, F_4 and E_8 have 2-torsion. Also, they mentioned in [**6**, page 139] that they were unable to determine the 2-rank for E_6 and E_7. After Borel and Serre's paper [**6**], 2-ranks of compact Lie groups G have been investigated by several mathematicians; mostly, using the theories in algebras and/or topology (see e.g., [**31**, **50**, **78**]).

8.2. Link between 2-number and 2-rank. Let G be a connected compact Lie group. By assigning $s_x(y) = xy^{-1}x$ to each $x \in G$, we have $s_x^2 = id_G$ for every $x \in G$. Thus, G is a compact symmetric space with respect to a bi-invariant Riemannian metric.

The following link between 2-rank and 2-number was proved in [**27**].

THEOREM 8.1. *Let G be a connected compact Lie group. Then $\#_2 G = 2^{r_2 G}$.*

For the product of two compact Lie groups G_1 and G_2, we have

THEOREM 8.2. [**27**] *Let G_1 and G_2 be two connected compact Lie groups. Then $\#_2(G_1 \times G_2) = 2^{r_2 G_1 + r_2 G_2}$.*

Based on these two theorems together with the (M_+, M_-)-theory, Nagano and I were able to determine the 2-ranks of all compact connected Lie groups in [**29**, pages 289-293]. Consequently, we have settled the problem of determination of 2-ranks of all connected compact simple Lie groups as follows.

8.3. Classical groups. For classical groups we have:

THEOREM 8.3. *Let $U(n)/\mathbb{Z}\mu$ be the quotient group of the unitary group $U(n)$ by the cyclic normal subgroup $\mathbb{Z}\mu$ of order μ. Then we have*

$$r_2(U(n)/\mathbb{Z}\mu) = \begin{cases} n+1 & \text{if } \mu \text{ is even and } n = 2 \text{ or } 4; \\ n & \text{otherwise.} \end{cases}$$

THEOREM 8.4. *For $SU(n)/\mathbb{Z}\mu$, we have*

$$r_2(SU(n)/\mathbb{Z}\mu) = \begin{cases} n+1 & \text{for } (n,\mu) = (4,2); \\ n & \text{for } (n,\mu) = (2,2) \text{ or } (4,4); \\ n-1 & \text{for the other cases.} \end{cases}$$

THEOREM 8.5. *One has $r_2(SO(n)) = n-1$, and for $SO(n)^*$ we have*

$$r_2(SO(n)^*) = \begin{cases} 4 & \text{for } n = 4; \\ n-2 & \text{for } n \text{ even} > 4. \end{cases}$$

THEOREM 8.6. *Let $O(n)^* = O(n)/\{\pm 1\}$. We have*
(a) *$r_2(O(n)) = n$;*
(b) *$r_2(O(n)^*)$ is $n = 2$ or 4, while it is $n-1$ otherwise.*

THEOREM 8.7. *One has $r_2(Sp(n)) = n$, and for $Sp(n)^*$ we have*

$$r_2(Sp(n)^*) = \begin{cases} n+2 & \text{for } n = 2 \text{ or } 4 \\ n+1 & \text{otherwise.} \end{cases}$$

Thus, for every n we also have $r_2(Sp(n)^*) = r_2(U(n)/\mathbb{Z}_2) + 1$.

8.4. Spinors, semi-spinors and *Pin*(n). We consider the spinor $Spin(n)$ and its related groups. Recall that $Spin(n)$ is a subset of the Clifford algebra $Cl(n)$ which is generated over the real field $\mathbb{R}$ by the vectors $\mathbf{e}_i$ in the fixed orthonormal basis of $\mathbb{R}^n$; subject to the conditions $\mathbf{e}_i\mathbf{e}_j = -\mathbf{e}_j\mathbf{e}_i$ and $\mathbf{e}_i\mathbf{e}_i = -1$, $i \neq j$.

For $Spin(n)$ we proved the next two theorems.

THEOREM 8.8. *Let $r = [\frac{n}{2}]$ be the rank of $Spin(n)$. Then*

$$r_2(Spin(n)) = \begin{cases} r+1 & \text{if } n \equiv -1, 0 \text{ or } 1 \pmod 8 \\ r & \text{otherwise.} \end{cases}$$

THEOREM 8.9. (Periodicity) *For $n \geq 0$, we have*

$$r_2(Spin(n+8)) = r_2(Spin(n)) + 4.$$

REMARK 8.1. Theorem 8.8 and Theorem 8.9 have used by J. A. Wood [**102**] in his study in algebraic coding theory.

For semi-spinor group $SO(4m)^{\#} = Spin(4m)/\{1, e_{((4m))}\}$, we have

THEOREM 8.10. *Let r be the rank $2m$ of $SO(4m)^{\#}$. Then*

$$r_2(SO(4m)^{\#}) = \begin{cases} 3 & \text{if } m = 1 \\ 6 & \text{if } m = 2, \\ r+1 & \text{if } m \text{ is even} > 2, \\ r & \text{if } m \text{ is odd} > 1. \end{cases}$$

REMARK 8.2. The 2-rank of $Spin(16)$ and of $SO(16)^{\#}$ have been obtained independently by J. F. Adams in [**1**]. However, the theory of his proof is completely different from ours given in [**29**].

M. F. Atiyah, R. Bott and A. Shapiro introduced the group $Pin(n)$ in [**3**] while they studied Clifford modules. $Pin(n)$ is a group in Clifford algebra $Cl(n)$ and it double covers $O(n)$ and whose connected component $Spin(n)$ double covers $SO(n)$.

THEOREM 8.11. *For $Pin(n)$, we have $r_2(Pin(n)) = r_2(Spin(n+1))$, $n \geq 0$.*

8.5. Exceptional groups. Also in [**29**], we have determined the 2-ranks of exceptional Lie groups as follows.

THEOREM 8.12. *We have $r_2G_2 = 3$, $r_2F_4 = 5$, $r_2E_6 = 6$, $r_2E_7 = 7$, $r_2E_8 = 9$ for the simply-connected exceptional simple Lie groups.*

THEOREM 8.13. *We have $r_2E_6^* = 6$.*

REMARK 8.3. $r_2G_2 = 3$ and $r_2F_4 = 5$ are due to A. Borel and J. P. Serre.

9. Applications to algebraic geometry

9.1. Arithmetic distance for classical Hermitian symmetric spaces. The notion of *arithmetic distance* for classical Hermitian symmetric spaces M was introduced by W.-L. Chow in [**30**]. In case that M is the complex Grassmannian $G_p(\mathbb{C}^n)$, the arithmetic distance $d(V, W)$ between two elements $V, W \in G_p(\mathbb{C}^n)$ is

$$d(V, W) = \dim_{\mathbb{C}} V/V \cap W, \tag{9.1}$$

i.e., the codimension of their intersection $V \cap W$ in V. Equivalently, $d(V, W)$ is the smallest integer s such that there is a finite set $\{U_i\}_{1 \leq i \leq s+1}$ of linear subspaces in $G_p(\mathbb{C}^n)$ such that (i) $U_1 = V$ and $U_{s+1} = W$, and (ii) $d(U_i, U_{i+1}) = 1$ for $1 \leq i \leq s$.

In 1949, Chow [**30**] proved that a *d-preserving transformation of M is either holomorphic or anti-holomorphic, provided $rk(M) > 1$.*

9.2. Arithmetic distance for compact symmetric spaces. By applying Helgason spheres, T. Nagano [**62**] and S. Peterson [**73**] extended Chow's arithmetic distance d to arbitrary irreducible compact symmetric spaces M. In this general case, $d(x, y)$ satisfies $d(x, y) \leq j$ if the two points $x, y \in M$ are joined by a chain of j Helgason spheres in M. More precisely, put $d(x, y) = 0$ if $x = y$; and $d(x, y) = 1$ if $x \neq y$ and x, y lie in a Helgason sphere. Otherwise, $d(x, y)$ is defined to be the smallest j such that there is a chain of j Helgason spheres joining x and y. When a compact symmetric space M is Hermitian and irreducible, then the holomorphic transformations permute the Helgason spheres and thus preserve the new arithmetic distance. Further, it equals Chow's arithmetic distance if $M = G_p(\mathbb{C}^n)$ (see [**62**]).

By applying the (M_+, M_-)-theory and a formulation of Radon's duality given in [**62**], Peterson proved the following rigidity theorem .

THEOREM 9.1. [**73**] *Let M be a $G_d(\mathbb{F}^n)$, $\mathbb{F} = \mathbb{R}, \mathbb{C}, \mathbb{H}$, or $M = AI(n)$ with $\dim M \geq 3$ and let $L = \{\varphi : M \to M, \varphi$ a diffeomorphism preserving the arithmetic distance $d\}$. Then $L = L'$, where L' be the geometric transformation group of M.*

For further results in this respect, see [**100**] by G. Thorbergsson.

9.3. 2-number and projective rank. By [**35**], the *projective rank*, $Pr(M)$, of a compact Hermitian symmetric space M is the maximal complex dimension of totally geodesic complex projective spaces N in M. A subset $\mathcal{E}$ of a maximal antipodal set of a compact symmetric space M is called *equidistant* if there exists a real number $a > 0$ such that $d_M(x, y) = a$ for any $x, y \in \mathcal{E}$. Put

$$\gamma = \gamma_M = \min\{d_M(x, y) : x, y \in \mathcal{E}, x \neq y\} \tag{9.2}$$

and let $A_\gamma \subset \mathcal{E}$ be an equidistance set (for the distance γ) of maximal cardinality. Let $\mu = \#(A_\gamma)$ denote the cardinality of A_γ.

C. U. Sánchez and A. Guinta [**81**] proved the following.

THEOREM 9.2. *Let M be a compact irreducible Hermitian symmetric space.*

(a) *If $M \neq CI(n)$, then $Pr(M) = \mu - 1$.*
(b) *If $M = CI(n)$ and $A_{\sqrt{2}\gamma}$ is an equidistant set of maximal cardinality (for the distance $\sqrt{2}\gamma(M)$) and μ is its cardinality, then $Pr(M) = \mu - 1$.*

By applying Theorem 9.2, Sánchez and Guinta proved the following.

THEOREM 9.3. [**81**] *For every compact irreducible Hermitian symmetric space M, we have $\#_2 M \geq Pr(M) \cdot rk(M)$.*

10. Applications of (M_+, M_-)-theory via real forms

In this section we present some further applications of (M_+, M_-)-theory via real forms, maximal antipodal sets, and 2-numbers.

10.1. Real forms and great antipodal sets. If M is a Hermitian symmetric space of compact type and τ is an involutive anti-holomorphic isometry of M, then the fixed point set $F(\tau, M) = \{x \in M : \tau(x) = x\}$ is called a *real form* of M, which is totally geodesic and Lagrangian in M.

M. Takeuchi [**85**] proved the link between real forms and symmetric R-spaces.

THEOREM 10.1. *Every real form of a Hermitian symmetric space of compact type is a symmetric R-space. Conversely, every symmetric R-space can be realized as a real form of a Hermitian symmetric space of compact type.*

Furthermore, the correspondence is one-to-one.

M. S. Tanaka and H. Tasaki studied in [**91**] the intersection of two real forms in a Hermitian symmetric space of compact type. They proved that the intersection of two real forms is an antipodal set whenever the intersection is discrete.

Tanaka and Tasaki proved the following.

THEOREM 10.2. [**91**] *Let L_1, L_2 be two real forms of a Hermitian symmetric space of compact type whose intersection is discrete. Then $L_1 \cap L_2$ is an antipodal set in L_1 and L_2. Furthermore, if L_1 and L_2 are congruent, then $L_1 \cap L_2$ is a great antipodal set. Thus, $\#(L_1 \cap L_2) = \#_2 L_1 = \#_2 L_2$.*

THEOREM 10.3. [**91**] *Any real form of a Hermitian symmetric space of compact type is a globally tight Lagrangian submanifold in the sense of Y.-G. Oh* [**70**].

Also, by studying the real forms in the compact Hermitian symmetric space $\tilde{G}_2(\mathbb{R}^{n+2})$, H. Tasaki [**97**] proved the following two theorems.

THEOREM 10.4. *Let k, ℓ be integers with $0 \le k \le \ell \le [\frac{n}{2}]$, and let L_1 and L_2 be real forms which are congruent to $S^{k,n-k}$ and $S^{\ell,n-\ell}$, respectively. If L_1 and L_2 intersect transversally, then $L_1 \cap L_2$ is a 2-set of L_1 and an antipodal set of L_2. Moreover, if $k = \ell = [\frac{n}{2}]$, then $\#_2(\tilde{G}_2(\mathbb{R}^{n+2})) = \#(L_1 \cap L_2)$.*

THEOREM 10.5. *Any real form of the oriented real Grassmannian $\tilde{G}_2(\mathbb{R}^{n+2})$ is a globally tight Lagrangian submanifold.*

For further results in this respect, see [**42, 43, 77, 91–93**].

10.2. Application to convexity. Let M be a Riemannian manifold and let τ be an involutive isometry of M. A connected component of the fixed point set of τ with positive dimension is called a *reflective submanifold*, which is a totally geodesic submanifold of M. A connected submanifold S of M is called (geodetically) *convex* if any shortest geodesic segment in S is still shortest in M.

In [**77**] P. Quast and M. S. Tanaka proved the following.

THEOREM 10.6. *Every reflective submanifold of a symmetric R-space is convex.*

10.3. Applications to Floer homology. Let (M, ω) be a symplectic manifold, i.e., M is a manifold endowed with a closed non-degenerate 2-form ω. Let L be a Lagrangian submanifold of M. Hence, ω vanishes on L. The symplectic *Floer homology* is a homology theory associated to a symplectic manifold and a non-degenerate symplectomorphism of it. If the symplectomorphism is Hamiltonian, then the homology arises from studying the symplectic action functional on the universal cover of the free loop space of a symplectic manifold. Symplectic Floer homology is invariant under Hamiltonian isotropy of the symplectomorphism.

Let $Ham(M, \omega)$ denote the set of all Hamiltonian diffeomorphisms of M. A. Floer [**36**] defined the homology when $\pi_2(M, L_i) = 0\,(i = 0, 1)$ and he proved that it is isomorphic to the singular homology group $H_*(L_0, \mathbb{Z}_2)$ of L_0 in the case where L_0 is Hamiltonian isotopic to L_1. As a result, Floer solved affirmatively the so called *Arnold conjecture* for Lagrangian intersections in that case.

In [**37**], A. Givental posed the following conjecture which generalized the results of Floer and himself.

ARNOLD-GIVENTAL'S CONJECTURE. *Let (M, ω) be a symplectic manifold and $\tau : M \to M$ be an anti-symplectic involution. If the fixed point set $L = F(M, \tau)$ is not empty and compact, then for any $\phi \in Ham(M, \omega)$ such that the Lagrangian submanifold L and its image $\phi(L)$ intersect transversally, the inequality*

$$\#(L \cap \phi(L)) \ge B(L, \mathbb{Z}_2) \tag{10.1}$$

holds, where $B(L, \mathbb{Z}_2) = \sum_{i \ge 0} B_i(L, \mathbb{Z}_2)$ is the total Betti number over $\mathbb{Z}_2$.

H. Iriyeh, T. Sakai and H. Tasaki [**44**] computed Lagrangian Floer homology $HF(L_0, L_1; \mathbb{Z}_2)$ of a pair of real forms (L_0, L_1) in a Hermitian symmetric space M of compact type, where L_0 is not necessarily congruent to L_1. Thus, they obtained a generalization of Arnold-Givental's inequality (10.1) in the case that M is irreducible. As an application, Iriyeh, Sakai and Tasaki obtained the following.

THEOREM 10.7. [**44**] *Every totally geodesic Lagrangian sphere in the complex hyperquadric is globally volume minimizing under Hamiltonian deformations.*

11. Index numbers and flag manifolds

11.1. k-numbers and flag manifolds. A *real flag manifold*, also known as a *R-space*, is a homogeneous space of the form G/P, where G is a real semisimple Lie group without compact factors and P is a parabolic subgroup. Here $G/P = K/K \cap P$ is a K-orbit on P. Every complex flag manifold may be considered as an R-space. In fact, let U be a compact connected semisimple centerless Lie group and with $\mathfrak{u}$ being its Lie algebra. Then the *complex flag manifolds* of U are the orbits of the adjoint action of U on $\mathfrak{u}$. Take $M = Ad(U)Y$ for $Y \neq 0$ in U and let $\mathfrak{g} = \mathfrak{u}_c = \mathfrak{u} + i\mathfrak{u}$. Then there exists a Cartan decomposition of the realization $\mathfrak{g}_{\mathbb{R}}$ of $\mathfrak{g}$ and one may consider M as the orbit of iY in $i\mathfrak{u}$ by the adjoint action of U.

C. U. Sánchez proved the following.

PROPOSITION 11.1. [**80**] *Let M be a real flag manifold. Then there is a complex flag manifold M_c such that M is isometrically imbedded in M_c. If M is a symmetric R-space, then M_c is a Hermitian symmetric space and the isometric embedding is totally geodesic. If M is already a complex flag manifold, then $M_c = M$.*

For a complex flag manifold M_c, there is a positive integer $k_0 = k_o(M_c) \geq 2$ such that for each integer $k \geq k_0$ there exists a *k-symmetric structure* on M_c, i.e., for each point $x \in M_c$ there is an isometry θ_x satisfying $\theta_x^k = id$ with x as an isolated fixed point. A k-symmetric structure is called *regular* it satisfies $\theta_x \circ \theta_y = \theta_z \circ \theta_z$ with $z = \theta_x(y)$ (cf. [**51**]). If M has a k-structure of order 2, then this structure is automatically regular and M is a symmetric space in the usual sense.

As an extension of 2-numbers for compact symmetric spaces, Sánchez defined the *k-number* $\#_k(M_c)$ of a complex flag manifold M_c as the maximal possible cardinality of a k-sets $A_k \subset M_c$ such that, for each $x \in A_k$, the corresponding k-symmetry fixes every point of A_k (cf. Proposition 6.1).

Sánchez proved the following.

THEOREM 11.1. [**79**] $\#_k(M_c) = \dim H^*(M_c, \mathbb{Z}_2)$ *for every complex flag manifold M_c.*

11.2. Index number and total Betti number. Using the fact that every real flag manifold M can be isometrically embedded into a complex flag manifold M_c, Sánchez [**80**] defined the *index number*, $\#_I M$, of a real flag manifold M as the maximal possible cardinality of p-sets A_pM (for a prime p), which was defined in terms of fixed points of symmetries of the complex flag manifolds restricted to the real one.

As an extension of Theorem 7.3 of Takeuchi, Sánchez proved the following link between the index number and the total Betti number.

THEOREM 11.2. [**80**] *Let M be a real flag manifold. Then $\#_I M = B(M, \mathbb{Z}_2)$, where $B(M, \mathbb{Z}_2)$ is the total Betti number over $\mathbb{Z}_2$.*

12. Index numbers and CW complex structures

The following conjecture was posed the first time in author's 1987 Report [**18**, page 53].

Chen-Nagano's Conjecture. *For every connected compact symmetric space M, the 2-number $\#_2M$ equals the smallest number of cells that are needed for a CW complex structure on M.*

Remark 12.1. Direct computations show that this conjecture is true for sphere, real projective space and irreducible Hermitian symmetric spaces of compact type. However, this conjecture remains open in general.

Using the convexity Theorems of Atiyah [**2**] and of Guillemin-Sternberg [**38**] for symplectic manifolds with a Hamiltonian torus action and the generalization of Duistermaat [**33**] for fixed point set of antisymplectic involutions, J. Berndt, S. Console and A. Fino presented in [**5**] an alternative proof of Theorem 11.2.

Related to the Chen-Nagano conjecture, Berndt, Console and Fino also proved in [**5**] the following.

Theorem 12.1. *The index number $\#_IM$ of a real flag manifold M equals the smallest number of cells that are needed for a CW complex structure on M.*

Acknowledgment

The author would like to express his thanks to Professor Makiko Sumi Tanaka for her suggestions to improve the presentation of this article.

References

[1] J. F. Adams, *2-tori in E_8*, Math. Ann. **278** (1987), no. 1-4, 29–39, DOI 10.1007/BF01458059. MR909216

[2] M. F. Atiyah, *Convexity and commuting Hamiltonians*, Bull. London Math. Soc. **14** (1982), no. 1, 1–15, DOI 10.1112/blms/14.1.1. MR642416

[3] M. F. Atiyah, R. Bott, and A. Shapiro, *Clifford modules*, Topology **3** (1964), suppl. 1, 3–38, DOI 10.1016/0040-9383(64)90003-5. MR167985

[4] M. F. Atiyah and I. M. Singer, *The index of elliptic operators. III*, Ann. of Math. (2) **87** (1968), 546–604, DOI 10.2307/1970717. MR236952

[5] Jürgen Berndt, Sergio Console, and Anna Fino, *On index number and topology of flag manifolds*, Differential Geom. Appl. **15** (2001), no. 1, 81–90, DOI 10.1016/S0926-2245(01)00050-X. MR1845178

[6] A. Borel and J.-P. Serre, *Sur certains sous-groupes des groupes de Lie compacts* (French), Comment. Math. Helv. **27** (1953), 128–139, DOI 10.1007/BF02564557. MR54612

[7] Raoul Bott, *The stable homotopy of the classical groups*, Ann. of Math. (2) **70** (1959), 313–337, DOI 10.2307/1970106. MR110104

[8] Raoul Bott, *The periodicity theorem for the classical groups and some of its applications*, Advances in Math. **4** (1970), 353–411, DOI 10.1016/0001-8708(70)90030-7. MR259904

[9] N. Bourbaki, *Éléments de mathématique. Fasc. XXXVIII: Groupes et algèbres de Lie. Chapitre VII: Sous-algèbres de Cartan, éléments réguliers. Chapitre VIII: Algèbres de Lie semi-simples déployées* (French), Actualités Scientifiques et Industrielles, No. 1364. Hermann, Paris, 1975. MR0453824

[10] John M. Burns, *Homotopy of compact symmetric spaces*, Glasgow Math. J. **34** (1992), no. 2, 221–228, DOI 10.1017/S0017089500008764. MR1167338

[11] J. M. Burns, *Conjugate loci of totally geodesic submanifolds of symmetric spaces*, Trans. Amer. Math. Soc. **337** (1993), no. 1, 411–425, DOI 10.2307/2154329. MR1091705

[12] John M. Burns and Michael J. Clancy, *Polar sets as nondegenerate critical submanifolds in symmetric spaces*, Osaka J. Math. **31** (1994), no. 3, 533–559. MR1309402

[13] É. Cartan, *Sur une classe remarquable d'espaces de Riemann* (French), Bull. Soc. Math. France **54** (1926), 214–264. MR1504900

[14] É. Cartan, *Sur une classe remarquable d'espaces de Riemann. II* (French), Bull. Soc. Math. France **55** (1927), 114–134. MR1504909

[15] Bang-Yen Chen, *On the G–total curvature and topology of immersed manifolds*, ProQuest LLC, Ann Arbor, MI, 1971. Thesis (Ph.D.)–University of Notre Dame. MR2940079

[16] Bang-yen Chen, *Geometry of submanifolds and its applications*, Science University of Tokyo, Tokyo, 1981. MR627323

[17] Bang-Yen Chen, *Symmetries of compact symmetric spaces*, Geometry and topology of submanifolds (Marseille, 1987), World Sci. Publ., Teaneck, NJ, 1989, pp. 38–48. MR1116130

[18] Bang-Yen Chen, *A new approach to compact symmetric spaces and applications: A report on joint work with Professor T. Nagano*, Katholieke Universiteit Leuven, Louvain, 1987. MR952556

[19] Bang-Yen Chen, *Geometry of slant submanifolds*, Katholieke Universiteit Leuven, Belgium, 1990. MR1099374

[20] Bang-Yen Chen, *Pseudo-Riemannian geometry, δ-invariants and applications*, World Scientific Publishing Co. Pte. Ltd., Hackensack, NJ, 2011. With a foreword by Leopold Verstraelen, DOI 10.1142/9789814329644. MR2799371

[21] Bang-Yen Chen, *The 2-ranks of connected compact Lie groups*, Taiwanese J. Math. **17** (2013), no. 3, 815–831, DOI 10.11650/tjm.17.2013.2606. MR3072263

[22] Bang-Yen Chen, *Borsuk-Ulam theorem and maximal antipodal sets of compact symmetric spaces*, Int. Electron. J. Geom. **10** (2017), no. 2, 11–19. MR3731645

[23] Bang-Yen Chen, *Two-numbers and their applications—a survey*, Bull. Belg. Math. Soc. Simon Stevin **25** (2018), no. 4, 565–596. MR3896273

[24] B.-Y. Chen, P.-F. Leung and T. Nagano, *Totally geodesic submanifolds of symmetric spaces III*, preprint (1980), `arXiv:1307.7325v2` [math.DG].

[25] Bang-yen Chen and Tadashi Nagano, *Totally geodesic submanifolds of symmetric spaces. I*, Duke Math. J. **44** (1977), no. 4, 745–755. MR458340

[26] Bang-yen Chen and Tadashi Nagano, *Totally geodesic submanifolds of symmetric spaces. II*, Duke Math. J. **45** (1978), no. 2, 405–425. MR487902

[27] Bang-yen Chen and Tadashi Nagano, *Un invariant géométrique riemannien* (French, with English summary), C. R. Acad. Sci. Paris Sér. I Math. **295** (1982), no. 5, 389–391. MR684733

[28] Bang-Yen Chen and Tadashi Nagano, *Harmonic metrics, harmonic tensors, and Gauss maps*, J. Math. Soc. Japan **36** (1984), no. 2, 295–313, DOI 10.2969/jmsj/03620295. MR740319

[29] Bang-Yen Chen and Tadashi Nagano, *A Riemannian geometric invariant and its applications to a problem of Borel and Serre*, Trans. Amer. Math. Soc. **308** (1988), no. 1, 273–297, DOI 10.2307/2000963. MR946443

[30] Wei-Liang Chow, *On the geometry of algebraic homogeneous spaces*, Ann. of Math. (2) **50** (1949), 32–67, DOI 10.2307/1969351. MR28057

[31] Arjeh M. Cohen and Gary M. Seitz, *The r-rank of the groups of exceptional Lie type*, Nederl. Akad. Wetensch. Indag. Math. **49** (1987), no. 3, 251–259. MR914084

[32] S. Console, *Geodesics and moments maps of symmetric R-spaces*, Dipart. di Mat. - Università di Torino, Quaderno N. **25**, 2009.

[33] J. J. Duistermaat, *Convexity and tightness for restrictions of Hamiltonian functions to fixed point sets of an antisymplectic involution*, Trans. Amer. Math. Soc. **275** (1983), no. 1, 417–429, DOI 10.2307/1999030. MR678361

[34] Jost-Hinrich Eschenburg, Peter Quast, and Makiko Sumi Tanaka, *Maximal tori of extrinsic symmetric spaces and meridians*, Osaka J. Math. **52** (2015), no. 2, 299–305. MR3326613

[35] Amassa Fauntleroy, *Projective ranks of Hermitian symmetric spaces*, Math. Intelligencer **15** (1993), no. 2, 27–32, DOI 10.1007/BF03024189. MR1212518

[36] Andreas Floer, *Morse theory for Lagrangian intersections*, J. Differential Geom. **28** (1988), no. 3, 513–547. MR965228

[37] A. B. Givental, *Periodic mappings in symplectic topology* (Russian), Funktsional. Anal. i Prilozhen. **23** (1989), no. 4, 37–52, 96, DOI 10.1007/BF01078943; English transl., Funct. Anal. Appl. **23** (1989), no. 4, 287–300 (1990). MR1035373

[38] V. Guillemin and S. Sternberg, *Convexity properties of the moment mapping*, Invent. Math. **67** (1982), no. 3, 491–513, DOI 10.1007/BF01398933. MR664117

[39] Sigurđur Helgason, *Totally geodesic spheres in compact symmetric spaces*, Math. Ann. **165** (1966), 309–317, DOI 10.1007/BF01344015. MR210043

[40] Sigurdur Helgason, *Differential geometry, Lie groups, and symmetric spaces*, Pure and Applied Mathematics, vol. 80, Academic Press, Inc. [Harcourt Brace Jovanovich, Publishers], New York-London, 1978. MR514561

[41] Osamu Ikawa, *The geometry of symmetric triad and orbit spaces of Hermann actions*, J. Math. Soc. Japan **63** (2011), no. 1, 79–136. MR2752433

[42] Osamu Ikawa, Hiroshi Iriyeh, Takayuki Okuda, Takashi Sakai, and Hiroyuki Tasaki, *Antipodal structure of the intersection of real flag manifolds in a complex flag manifold II*, Proceedings of the 19th International Workshop on Hermitian-Grassmannian Submanifolds and 10th RIRCM-OCAMI Joint Differential Geometry Workshop, Natl. Inst. Math. Sci. (NIMS), Taejŏn, 2015, pp. 147–157. MR3699643

[43] Osamu Ikawa, Makiko Sumi Tanaka, and Hiroyuki Tasaki, *The fixed point set of a holomorphic isometry, the intersection of two real forms in a Hermitian symmetric space of compact type and symmetric triads*, Internat. J. Math. **26** (2015), no. 6, 1541005, 32, DOI 10.1142/S0129167X15410050. MR3356876

[44] Hiroshi Iriyeh, Takashi Sakai, and Hiroyuki Tasaki, *Lagrangian Floer homology of a pair of real forms in Hermitian symmetric spaces of compact type*, J. Math. Soc. Japan **65** (2013), no. 4, 1135–1151, DOI 10.2969/jmsj/06541135. MR3127820

[45] Quanqin Jin, *Totally geodesic submanifolds in compact Riemannian symmetric spaces and their stability*, Sci. China Ser. A **43** (2000), no. 12, 1279–1293, DOI 10.1007/BF02880065. MR1813132

[46] Quan Qin Jin, *Totally geodesic submanifolds of symmetric spaces and their stability. II* (Chinese, with English and Chinese summaries), Chinese Ann. Math. Ser. A **25** (2004), no. 6, 725–734. MR2119087

[47] Taro Kimura and Makiko Sumi Tanaka, *Totally geodesic submanifolds in compact symmetric spaces of rank two*, Tokyo J. Math. **31** (2008), no. 2, 421–447, DOI 10.3836/tjm/1233844062. MR2477882

[48] Taro Kimura and Makiko Sumi Tanaka, *Stability of certain minimal submanifolds in compact symmetric spaces of rank two*, Differential Geom. Appl. **27** (2009), no. 1, 23–33, DOI 10.1016/j.difgeo.2008.06.004. MR2488985

[49] Shoshichi Kobayashi and Tadashi Nagano, *On filtered Lie algebras and geometric structures. I*, J. Math. Mech. **13** (1964), 875–907. MR0168704

[50] Akira Kono, *On the 2-rank of compact connected Lie groups*, J. Math. Kyoto Univ. **17** (1977), no. 1, 1–18, DOI 10.1215/kjm/1250522806. MR482791

[51] Oldřich Kowalski, *Generalized symmetric spaces*, Lecture Notes in Mathematics, vol. 805, Springer-Verlag, Berlin-New York, 1980. MR579184

[52] H. Kurihara, *Antipodal sets and designs on unitary groups*, Graphs Combin. **37** (2021), no. 5, 1559–1583, DOI 10.1007/s00373-021-02287-9. MR4331068

[53] Hirotake Kurihara and Takayuki Okuda, *Great antipodal sets on complex Grassmannian manifolds as designs with the smallest cardinalities*, J. Algebra **559** (2020), 432–466, DOI 10.1016/j.jalgebra.2020.05.004. MR4097913

[54] H. Blaine Lawson Jr. and James Simons, *On stable currents and their application to global problems in real and complex geometry*, Ann. of Math. (2) **98** (1973), 427–450, DOI 10.2307/1970913. MR324529

[55] L. A. Lyusternik and A. I. Fet, *Variational problems on closed manifolds* (Russian), Doklady Akad. Nauk SSSR (N.S.) **81** (1951), 17–18. MR0044760

[56] Augustin-Liviu Mare and Peter Quast, *Bott periodicity for inclusions of symmetric spaces*, Doc. Math. **17** (2012), 911–952. MR3007680

[57] Katsuya Mashimo, *On the stability of Cartan embeddings of compact symmetric spaces*, Arch. Math. (Basel) **58** (1992), no. 5, 500–508, DOI 10.1007/BF01190121. MR1156583

[58] Katsuya Mashimo and Hiroyuki Tasaki, *Stability of closed Lie subgroups in compact Lie groups*, Kodai Math. J. **13** (1990), no. 2, 181–203, DOI 10.2996/kmj/1138039218. MR1061920

[59] Katsuya Mashimo and Hiroyuki Tasaki, *Stability of maximal tori in compact Lie groups*, Algebras Groups Geom. **7** (1990), no. 2, 114–126. MR1109565

[60] Mamoru Mimura and Hirosi Toda, *Topology of Lie groups. I, II*, Translations of Mathematical Monographs, vol. 91, American Mathematical Society, Providence, RI, 1991. Translated from the 1978 Japanese edition by the authors, DOI 10.1090/mmono/091. MR1122592

[61] Tadashi Nagano, *Transformation groups on compact symmetric spaces*, Trans. Amer. Math. Soc. **118** (1965), 428–453, DOI 10.2307/1993971. MR182937

[62] Tadashi Nagano, *The involutions of compact symmetric spaces*, Tokyo J. Math. **11** (1988), no. 1, 57–79, DOI 10.3836/tjm/1270134261. MR947946

[63] Tadashi Nagano, *The involutions of compact symmetric spaces. II*, Tokyo J. Math. **15** (1992), no. 1, 39–82, DOI 10.3836/tjm/1270130250. MR1164185

[64] Tadashi Nagano and Sumi Makiko, *The structure of the symmetric space with applications*, Geometry of manifolds (Matsumoto, 1988), Perspect. Math., vol. 8, Academic Press, Boston, MA, 1989, pp. 111–128. MR1040520

[65] Tadashi Nagano and Sumi Makiko, *The spheres in symmetric spaces*, Hokkaido Math. J. **20** (1991), no. 2, 331–352, DOI 10.14492/hokmj/1381413846. MR1114410

[66] Tadashi Nagano and Sumi Makiko, *Stability of p-harmonic maps*, Tokyo J. Math. **15** (1992), no. 2, 475–482, DOI 10.3836/tjm/1270129472. MR1197114

[67] Tadashi Nagano and Makiko Sumi Tanaka, *The involutions of compact symmetric spaces. III*, Tokyo J. Math. **18** (1995), no. 1, 193–212, DOI 10.3836/tjm/1270043621. MR1334718

[68] Tadashi Nagano and Makiko Sumi Tanaka, *The involutions of compact symmetric spaces. IV*, Tokyo J. Math. **22** (1999), no. 1, 193–211, DOI 10.3836/tjm/1270041622. MR1692030

[69] Tadashi Nagano and Makiko S. Tanaka, *The involutions of compact symmetric spaces. V*, Tokyo J. Math. **23** (2000), no. 2, 403–416, DOI 10.3836/tjm/1255958679. MR1806473

[70] Yong-Geun Oh, *Tight Lagrangian submanifolds in* $\mathbf{C}\mathrm{P}^n$, Math. Z. **207** (1991), no. 3, 409–416, DOI 10.1007/BF02571398. MR1115173

[71] Yoshihiro Ohnita, *Stable minimal submanifolds in compact rank one symmetric spaces*, Tohoku Math. J. (2) **38** (1986), no. 2, 199–217, DOI 10.2748/tmj/1178228488. MR843807

[72] Yoshihiro Ohnita, *On stability of minimal submanifolds in compact symmetric spaces*, Compositio Math. **64** (1987), no. 2, 157–189. MR916480

[73] Stephen P. Peterson, *Arithmetic distance on compact symmetric spaces*, Geom. Dedicata **23** (1987), no. 1, 1–14, DOI 10.1007/BF00147387. MR886770

[74] Peter Quast, *'Spindles' in symmetric spaces*, J. Math. Soc. Japan **58** (2006), no. 4, 985–994. MR2276177

[75] Peter Quast, *Homotopy of EVII* (English, with English and French summaries), C. R. Math. Acad. Sci. Paris **350** (2012), no. 7-8, 425–426, DOI 10.1016/j.crma.2012.04.008. MR2922098

[76] Peter Quast, *Centrioles in symmetric spaces*, Nagoya Math. J. **211** (2013), 51–77, DOI 10.1215/00277630-2141499. MR3079279

[77] Peter Quast and Makiko Sumi Tanaka, *Convexity of reflective submanifolds in symmetric R-spaces*, Tohoku Math. J. (2) **64** (2012), no. 4, 607–616, DOI 10.2748/tmj/1356038981. MR3008241

[78] Daniel Quillen, *The spectrum of an equivariant cohomology ring. I, II*, Ann. of Math. (2) **94** (1971), 549–572; ibid. (2) 94 (1971), 573–602, DOI 10.2307/1970770. MR298694

[79] Cristián U. Sánchez, *The invariant of Chen-Nagano on flag manifolds*, Proc. Amer. Math. Soc. **118** (1993), no. 4, 1237–1242, DOI 10.2307/2160084. MR1163336

[80] Cristián U. Sánchez, *The index number of an R-space: an extension of a result of M. Takeuchi's*, Proc. Amer. Math. Soc. **125** (1997), no. 3, 893–900, DOI 10.1090/S0002-9939-97-03517-X. MR1343722

[81] Cristián U. Sánchez and Ana Giunta, *The projective rank of a Hermitian symmetric space: a geometric approach and consequences*, Math. Ann. **323** (2002), no. 1, 55–79, DOI 10.1007/s002080100296. MR1906908

[82] Cristián U. Sánchez, Ana L. Calí, and José L. Moreschi, *Spheres in Hermitian symmetric spaces and flag manifolds*, Geom. Dedicata **64** (1997), no. 3, 261–276, DOI 10.1023/A:1004913505961. MR1440560

[83] James Simons, *Minimal varieties in riemannian manifolds*, Ann. of Math. (2) **88** (1968), 62–105, DOI 10.2307/1970556. MR233295

[84] Masaru Takeuchi, *Cell decompositions and Morse equalities on certain symmetric spaces*, J. Fac. Sci. Univ. Tokyo Sect. I **12** (1965), 81–192 (1965). MR216517

[85] Masaru Takeuchi, *Stability of certain minimal submanifolds of compact Hermitian symmetric spaces*, Tohoku Math. J. (2) **36** (1984), no. 2, 293–314, DOI 10.2748/tmj/1178228853. MR742600

[86] Masaru Takeuchi, *Two-number of symmetric R-spaces*, Nagoya Math. J. **115** (1989), 43–46, DOI 10.1017/S0027763000001513. MR1018081

[87] Makiko Sumi Tanaka, *Stability of minimal submanifolds in symmetric spaces*, Tsukuba J. Math. **19** (1995), no. 1, 27–56, DOI 10.21099/tkbjm/1496162798. MR1346752

[88] Makiko Sumi Tanaka, *Fixed point sets of isometries and the intersection of real forms in a Hermitian symmetric space of compact type*, Proceedings of the 17th International Workshop on Differential Geometry and the 7th KNUGRG-OCAMI Differential Geometry Workshop [Vol. 17], Natl. Inst. Math. Sci. (NIMS), Taejŏn, 2013, pp. 31–39. MR3203379

[89] Makiko Sumi Tanaka, *Antipodal sets of compact symmetric spaces and the intersection of totally geodesic submanifolds*, Differential geometry of submanifolds and its related topics, World Sci. Publ., Hackensack, NJ, 2014, pp. 205–219. MR3203484

[90] Makiko Sumi Tanaka, *Geometry of symmetric R-spaces*, Geometry and analysis on manifolds, Progr. Math., vol. 308, Birkhäuser/Springer, Cham, 2015, pp. 471–481, DOI 10.1007/978-3-319-11523-8_21. MR3331413

[91] Makiko Sumi Tanaka and Hiroyuki Tasaki, *The intersection of two real forms in Hermitian symmetric spaces of compact type*, J. Math. Soc. Japan **64** (2012), no. 4, 1297–1332, DOI 10.2969/jmsj/06441297. MR2998924

[92] Makiko Sumi Tanaka and Hiroyuki Tasaki, *Antipodal sets of symmetric R-spaces*, Osaka J. Math. **50** (2013), no. 1, 161–169. MR3080635

[93] Makiko Sumi Tanaka and Hiroyuki Tasaki, *The intersection of two real forms in Hermitian symmetric spaces of compact type II*, J. Math. Soc. Japan **67** (2015), no. 1, 275–291, DOI 10.2969/jmsj/06710275. MR3304022

[94] Makiko Sumi Tanaka and Hiroyuki Tasaki, *Maximal antipodal subgroups of some compact classical Lie groups*, J. Lie Theory **27** (2017), no. 3, 801–829. MR3611097

[95] Makiko Sumi Tanaka and Hiroyuki Tasaki, *Maximal antipodal subgroups of the automorphism groups of compact Lie algebras*, Hermitian-Grassmannian submanifolds, Springer Proc. Math. Stat., vol. 203, Springer, Singapore, 2017, pp. 39–47, DOI 10.1007/978-981-10-5556-0_4. MR3710831

[96] Makiko Sumi Tanaka, Hiroyuki Tasaki, and Osami Yasukura, *Maximal antipodal sets of* G_2 *and* $G_2/SO(4)$ *and related geometry*, Proceedings of the 22nd International Workshop on Differential Geometry of Submanifolds in Symmetric Spaces & the 17th RIRCM-OCAMI Joint Differential Geometry Workshop, Publisher not identified, place of publication not identified, 2019, pp. 153–159. MR3971856

[97] Hiroyuki Tasaki, *The intersection of two real forms in the complex hyperquadric*, Tohoku Math. J. (2) **62** (2010), no. 3, 375–382, DOI 10.2748/tmj/1287148617. MR2742014

[98] Hiroyuki Tasaki, *Antipodal sets in oriented real Grassmann manifolds*, Internat. J. Math. **24** (2013), no. 8, 1350061, 28, DOI 10.1142/S0129167X13500614. MR3103877

[99] Hiroyuki Tasaki, *Estimates of antipodal sets in oriented real Grassmann manifolds*, Internat. J. Math. **26** (2015), no. 6, 1541008, 12, DOI 10.1142/S0129167X15410086. MR3356879

[100] Gudlaugur Thorbergsson, *Classical symmetric R-spaces*, Rend. Semin. Mat. Univ. Politec. Torino **74** (2016), no. 1, 329–354. MR3772594

[101] Joseph A. Wolf, *Elliptic spaces in Grassmann manifolds*, Illinois J. Math. **7** (1963), 447–462. MR156295

[102] Jay A. Wood, *Spinor groups and algebraic coding theory*, J. Combin. Theory Ser. A **51** (1989), no. 2, 277–313, DOI 10.1016/0097-3165(89)90053-8. MR1001269

Department of Mathematics, Michigan State University, East Lansing, MI 48824, USA

Email address: chenb@msu.edu

Contemporary Mathematics
Volume **777**, 2022
https://doi.org/10.1090/conm/777/15628

Biharmonic and biconservative hypersurfaces in space forms

Dorel Fetcu and Cezar Oniciuc

ABSTRACT. We present some general properties of biharmonic and biconservative submanifolds and then survey recent results on such hypersurfaces in space forms. We also propose an alternative version of a well-known result of Nomizu and Smyth for hypersurfaces, by replacing the CMC hypothesis with the more general condition of biconservativity.

1. Introduction

Suggested in 1964 by J. Eells and J. H. Sampson (see [**25**]) as a generalization of harmonic maps, the biharmonic maps represent nowadays a well-established and dynamic topic in Differential Geometry.

A biharmonic map $\phi : M \to N$ between two Riemannian manifolds is a critical point of the bienergy functional

$$E_2 : C^\infty(M, N) \to \mathbb{R}, \quad E_2(\phi) = \frac{1}{2} \int_M |\tau(\phi)|^2 \, dv,$$

where M is compact and $\tau(\phi) = \operatorname{trace} \nabla d\phi$ is the tension field of ϕ. These maps are characterized by the Euler-Lagrange equation, also known as the biharmonic equation, obtained by G.-Y. Jiang in 1986 (see [**46**]):

$$\tau_2(\phi) = -\Delta\tau(\phi) - \operatorname{trace} R^N(d\phi(\cdot), \tau(\phi))d\phi(\cdot) = 0, \tag{1.1}$$

where $\tau_2(\phi)$ is the bitension field of ϕ.

Since any harmonic map is biharmonic, our interest is to study non-harmonic biharmonic maps, which are called proper biharmonic. A biharmonic submanifold of N is a biharmonic isometric immersion $\phi : M \to N$. Sometimes, throughout this paper, we will identify a submanifold as M rather than mentioning the immersion ϕ.

In Euclidean spaces, B.-Y. Chen (see [**16**]) proposed an alternative definition of biharmonic submanifolds that coincides with the previous one when the ambient space is $\mathbb{R}^n$. B.-Y. Chen also conjectured that *there are no proper biharmonic submanifolds in* $\mathbb{R}^n$ (see [**16**]). On the other hand, there are many examples and classification results in spaces of non-negative curvature, especially in the Euclidian sphere $\mathbb{S}^n$.

2020 *Mathematics Subject Classification.* Primary 53C42, 53C24, 53C21.

Key words and phrases. Biharmonic hypersurfaces, biconservative hypersurfaces, parabolic surfaces, real space forms.

The second author was partially supported from the grant PN-III-P4-ID-PCE-2020-0794.

From the theory of biharmonic submanifolds a new interesting subject, the study of biconservative submanifolds, arose and keeps gaining ground in today's mathematical research as there are many interesting examples of biconservative submanifolds, even when the biharmonic ones fail to exist. We will present details on this rather new notion in the next sections of this survey.

In this survey, we only discuss the hypersurfaces in space forms. We omit, for space reasons, other very interesting results on submanifolds in Minkowski spaces, product spaces, complex space forms, PNMC submanifolds, etc. (see, for example, [**18**], [**30**]-[**32**], [**37**], [**38**], [**44**], [**45**], [**55**], [**60**], [**68**], [**74**]-[**78**], [**83**]). For the same reason, even in the case of hypersurfaces in space forms, we will not refer to their stability properties and index.

The paper is organized as follows. In Section 2 we present definitions and general properties of biharmonic and biconservative submanifolds in Riemannian manifolds. The third section is devoted to biharmonic and biconservative surfaces in three dimensional space forms, and it is divided accordingly into two subsections. Finally, Section 4, also consisting of two subsections, is concerned with biharmonic and biconservative hypersurfaces (with dimension greater than two) in space forms. We give a much simpler proof for Theorem 4.22, of J. H. Chen in [**21**], on compact proper biharmonic hypersurfaces and we prove a local version of this result, Proposition 4.23. This result can be also seen as a local version of the fact that $|A|^2 =$ constant implies CMC for compact proper biharmonic hypersurfaces in spheres, with additional hypotheses on the curvature. We also prove Theorem 4.31 which, together with Theorem 4.28, provides an alternative version of a result of K. Nomizu and B. Smyth on compact CMC hypersurfaces, by replacing the CMC hypothesis with the biconservative one, which is more general. Throughout this paper a special focus is on the case when the ambient space is a Euclidean sphere. We end by posing two open problems.

Conventions. We use the following definitions and sign conventions. Consider a smooth map $\phi : M \to N$, between two Riemannian manifolds, where M is assumed to be oriented and connected. In general, we will not indicate explicitly the metrics on M or N. Then,

$$\Delta = -\operatorname{trace}(\nabla^{\phi})^2 = -\operatorname{trace}(\nabla^{\phi}\nabla^{\phi} - \nabla^{\phi}_{\nabla})$$

is the rough Laplacian defined on the set of all sections in $\phi^{-1}(TN)$, that is on $C(\phi^{-1}(TN))$, and R^N is the curvature tensor field of N, given by

$$R^N(U,V)W = [\bar{\nabla}_U, \bar{\nabla}_V]W - \bar{\nabla}_{[U,V]}W.$$

Here, ∇^{ϕ} denotes the pull-back connection on $\phi^{-1}(TN)$, while ∇ and $\bar{\nabla}$ are the Levi-Civita connections on TM and TN, respectively. Henceforth, for the sake of simplicity, we will denote all connections on various fiber bundles by ∇, the difference being clear from the context. The n-dimensional Euclidian sphere of radius r will be denoted by $\mathbb{S}^n(r)$ and, when $r = 1$, by $\mathbb{S}^n$.

2. Definitions and general properties

In this section we briefly recall basic and general results on biharmonic and biconservative submanifolds with a focus on surfaces and hypersurfaces.

2.1. Biconservative submanifolds. Consider a given map

$$\phi : M^m \to (N^n, h),$$

where the metric h is also fixed, and define a new functional on the set $\mathcal{G}$ of all Riemannian metrics on M by

$$\mathcal{F}_2 : \mathcal{G} \to \mathbb{R}, \quad \mathcal{F}_2(g) = E_2(\phi).$$

Critical points of this functional are characterized by the vanishing of the stress-energy tensor of the bienergy (see [**49**]). This tensor, denoted by S_2, was introduced in [**47**] as

$$\begin{aligned} S_2(X,Y) =& \frac{1}{2}|\tau(\phi)|^2\langle X,Y\rangle + \langle d\phi, \nabla\tau(\phi)\rangle\langle X,Y\rangle \\ & - \langle d\phi(X), \nabla_Y\tau(\phi)\rangle - \langle d\phi(Y), \nabla_X\tau(\phi)\rangle \end{aligned}$$

and it satisfies

$$\operatorname{div} S_2 = -\langle \tau_2(\phi), d\phi\rangle.$$

We note that, for isometric immersions, $(\operatorname{div} S_2)^\sharp = -\tau_2(\phi)^\top$, where $\tau_2(\phi)^\top$ is the tangent part of the bitension field.

DEFINITION 2.1. A map $\phi : M \to N$ is called biconservative if $\operatorname{div} S_2 = 0$.

DEFINITION 2.2. A submanifold $\phi : M \to N$ is called biconservative if the isometric immersion ϕ is biconservative.

REMARK 2.3. We have the following direct consequences.

(1) Any biharmonic map is also biconservative.
(2) If $\phi : M \to N$ is a submersion, not necessarily Riemannian, then ϕ is a biconservative map if and only if it is biharmonic.
(3) A submanifold M of N is biconservative if and only if $\tau_2(\phi)^\top = 0$.

We have the following general properties of the stress-bienergy tensor of a submanifold.

PROPOSITION 2.4. *For a submanifold $\phi : M^m \to N^n$ we have:*

(i) *the stress-bienergy tensor of ϕ is given by*

$$S_2 = -\frac{m^2}{2}|H|^2 I + 2mA_H,$$

where A_H is the shape operator in the direction of H, the mean curvature vector field;

(ii) $\operatorname{trace} S_2 = m^2|H|^2\left(2 - \frac{m}{2}\right)$;
(iii) *the relation between the divergence of S_2 and the divergence of A_H is given by*

$$\operatorname{div} S_2 = -\frac{m^2}{2}\operatorname{grad}\left(|H|^2\right) + 2m \operatorname{div} A_H;$$

(iv) $|S_2|^2 = m^4|H|^4\left(\frac{m}{4} - 2\right) + 4m^2\left|A_H\right|^2$.

2.2. Properties of biharmonic submanifolds. Next, we present some non-existence and unique-continuation properties of biharmonic maps, derived from a Weitzenböck formula and the fact that these maps are characterized by an elliptic equation.

THEOREM 2.5 ([**46**]). *Let $\phi : M \to N$ be a smooth map, where M is compact and* $\mathrm{Riem}^N \leq 0$. *Then ϕ is biharmonic if and only if it is harmonic.*

Given up to compactness, we can state the following proposition.

PROPOSITION 2.6 ([**69**]). *Let $\phi : M \to N$ be an isometric immersion such that $|\tau(\phi)| = $ constant and assume that* $\mathrm{Riem}^N \leq 0$. *Then ϕ is biharmonic if and only if it is minimal.*

In the case of hypersurfaces the next results hold.

THEOREM 2.7 ([**69**]). *Let $\phi : M \to N$ be a compact hypersurface and assume that* $\mathrm{Ricci}^N \leq 0$. *Then M is biharmonic if and only if it is minimal.*

PROPOSITION 2.8 ([**69**]). *Let $\phi : M \to N$ be a hypersurface such that $|\tau(\phi)| =$ constant, where* $\mathrm{Ricci}^N \leq 0$. *Then ϕ is biharmonic if and only if it is minimal.*

The proofs of these non-existence results are based on a classical Weitzenböck formula

$$\frac{1}{2}\Delta|\sigma|^2 = \langle \Delta\sigma, \sigma\rangle - |\nabla\sigma|^2, \quad \sigma \in C(\phi^{-1}TN),$$

where one considers $\sigma = \tau(\phi)$ and gets $\nabla\tau(\phi) = 0$. Then, using the equation

$$|\tau(\phi)|^2 = -\operatorname{trace}\langle\cdot, \nabla_{\cdot}\tau(\phi)\rangle,$$

which holds for isometric immersions, one obtains that $\tau(\phi)$ vanishes.

PROPOSITION 2.9 ([**69**]). *Let $\phi : M \to N$ be a smooth map such that $|\tau(\phi)| =$ constant and assume that there exists a point $p \in M$ where* $\operatorname{rank}\phi(p) \geq 2$. *If* $\mathrm{Riem}^N < 0$, *then ϕ is biharmonic if and only if it is harmonic.*

Concerning the unique continuation property, we have the following results.

THEOREM 2.10 ([**8**],[**11**]). *Let $\phi : M \to N$ be a biharmonic map. If ϕ is harmonic on an open subset, then it is harmonic everywhere.*

THEOREM 2.11 ([**8**]). *Let $\phi_1, \phi_2 : M \to N$ be two biharmonic maps. If they agree on an open subset, then they are identical.*

THEOREM 2.12 ([**8**]). *Let $\phi : M \to N$ be a biharmonic map and let P be a regular, closed, totally geodesic submanifold of N. If an open subset of M is mapped into P, then all of M is mapped into P.*

REMARK 2.13. Here, the word 'closed' is used in its topological sense, as P is a closed subset of N.

The main idea in the proofs of the unique continuation results is to define new variables such that the biharmonic equation, initially a semi-linear elliptic equation of order four, becomes a second order semi-linear elliptic equation. Then, by making appropriate estimations, one applies the standard result of Aronszajn in [**1**].

An important property of constant mean curvature (CMC) proper biharmonic submanifolds in $\mathbb{S}^n$ is that $|H|$ is bounded from above by 1. More precisely, we have the following proposition.

PROPOSITION 2.14 ([**70**]). *Let $\phi : M^m \to \mathbb{S}^n$ be a CMC proper biharmonic submanifold. Then $|H| \in (0,1]$ and, moreover, $|H| = 1$ if and only if M lies in the small hypersphere $\mathbb{S}^{n-1}(1/\sqrt{2})$ as a minimal submanifold.*

A link between proper biharmonic immersions in spheres and maps of finite type, in the sense of B.-Y. Chen, was established by the next result.

THEOREM 2.15 ([**3**],[**6**],[**51**]). *Let $\phi : M^m \to \mathbb{S}^n$ be a proper biharmonic immersion. Denote by $\psi = i \circ \phi : M \to \mathbb{R}^{n+1}$ the isometric immersion of M in $\mathbb{R}^{n+1}$, where $i : \mathbb{S}^n \to \mathbb{R}^{n+1}$ is the canonical inclusion map. Then*

(i) *The map ψ is of 1-type if and only if $|H| = 1$. In this case, $\psi = \psi_0 + \psi_{t_1}$, with $\Delta\psi_{t_1} = 2m\psi_{t_1}$, ψ_0 is a constant vector. Moreover, $\langle\psi_0, \psi_{t_1}\rangle = 0$ at any point, $|\psi_0| = |\psi_{t_1}| = 1/\sqrt{2}$ and $\phi_{t_1} : M \to \mathbb{S}^{n-1}\left(1/\sqrt{2}\right)$ is a minimal immersion.*

(ii) *The map ψ is of 2-type if and only if $|H|$ is constant, $|H| \in (0,1)$. In this case $\psi = \psi_{t_1} + \psi_{t_2}$, with $\Delta\psi_{t_1} = m(1 - |H|)\psi_{t_1}$, $\Delta\psi_{t_2} = m(1 + |H|)\psi_{t_2}$ and*

$$\psi_{t_1} = \tfrac{1}{2}\psi + \tfrac{1}{2|H|}H\,,\ \psi_{t_2} = \tfrac{1}{2}\psi - \tfrac{1}{2|H|}H.$$

Moreover, $\langle\psi_{t_1}, \psi_{t_2}\rangle = 0$, $|\psi_{t_1}| = |\psi_{t_2}| = 1/\sqrt{2}$ and

$$\phi_{t_i} : M \to \mathbb{S}^n\left(\frac{1}{\sqrt{2}}\right), \quad i = 1, 2,$$

are harmonic maps with constant density energy.

This theorem was the starting point for the studies on biharmonic surfaces with constant Gaussian curvature in space forms $N^n(c)$ (see [**29**], [**51**]), where one used the work of Y. Myiata (see [**56**]) to obtain classification results.

2.3. Properties of biconservative submanifolds. The biconservative submanifolds do not have properties similar to those of biharmonic ones in the previous subsection as they are not characterized by an elliptic equation. However, they do have some very interesting properties of their own, especially when talking about biconservative surfaces. Moreover, the class of biconservative maps is richer than that of biharmonic maps and, in many situations when proper biharmonic submanifolds do not exist, we have examples of biconservative submanifolds.

One of these properties can be viewed as a generalization of a classical Hopf's result in higher codimension, and it shows that the notion of biconservativity is a quite natural one.

THEOREM 2.16 ([**59**],[**63**]). *Let $\phi : M^2 \to N^n$ be a CMC surface. Then the following properties are equivalent:*

(i) *M is biconservative;*
(ii) *$\langle A_H(\partial_z), \partial_z\rangle$ is holomorphic;*
(iii) *A_H is a Codazzi tensor field.*

We note that the above theorem follows by using the properties of any divergence-free symmetric tensor field of type $(1,1)$ defined on a Riemannian surface.

REMARK 2.17. In the case of biharmonic surfaces, the fact that $\langle A_H(\partial_z), \partial_z\rangle$ is holomorphic if and only if M is CMC was proved in [**50**].

REMARK 2.18. We can say that $\langle A_H(\partial_z), \partial_z\rangle$ is a generalization of the Hopf's function as in 3-dimensional space forms we have the following. If M^2 is a surface in $N^3(c)$ which is a topological sphere, then it is CMC, i.e., $|H| =$ constant, if and only if $\langle A_H(\partial_z), \partial_z\rangle$ is holomorphic (see [**59**]). We note that for any surface in $N^3(c)$, $\langle A_H(\partial_z), \partial_z\rangle$ is holomorphic when M is CMC, but, in general, the converse does not hold. All non-CMC surfaces with $\langle A_H(\partial_z), \partial_z\rangle$ holomorphic must have the curvature equal to c (but they are not umbilical). In [**59**], there were found all surfaces in $\mathbb{R}^3$ with $\langle A_H(\partial_z), \partial_z\rangle$ holomorphic that are not CMC, and they are given by

$$X_\alpha(u, v) = (u\cos v, u\sin v, \alpha u),$$

where α is a positive real constant. In $\mathbb{S}^3$ such surfaces are given by

$$X_\alpha(u, v) = \left(\frac{\cos u}{\alpha}\cos v, \frac{\sin u}{\alpha}\cos v, \frac{\sqrt{\alpha^2-1}}{\alpha}\cos v, \sin v\right),$$

where α is a real constant, $\alpha > 1$ (see [**63**]).

REMARK 2.19. If $\phi : M^2 \to N^n$ is a non-pseudo-umbilical CMC biconservative surface, then the set of pseudo-umbilical points has no accumulation points. Also, if M^2 is a CMC biconservative surface and a topological sphere, then it is pseudo-umbilical (see [**59**]); this should be compared with a classical result: a PMC surface M^2, i.e., a surface satisfying $\nabla^\perp H = 0$, of genus zero in a space form is pseudo-umbilical (see [**41**]).

Using Theorem 2.16 and the CMC hypothesis, around non-pseudo-umbilical points, one can obtain an explicit form of the metric on the surface and of the shape operator A_H. It follows that the CMC biconservative surfaces with no pseudo-umbilical points are globally conformally flat. Then, one can prove the following proposition.

PROPOSITION 2.20 ([**63**]). *Let $\phi : M^2 \to N^n$ be a CMC biconservative surface. Assume that M is compact and does not have pseudo-umbilical points. Then M is a topological torus and, moreover, if $K \geq 0$ or $K \leq 0$, we have $\nabla A_H = 0$ and $K = 0$.*

REMARK 2.21. One can compare this result with Theorem 5.4 in [**20**].

We can also deduce that if M^2 is a CMC biconservative surface satisfying certain additional hypotheses, then, up to a global conformal diffeomorphism, M^2 either is $\mathbb{R}^2$, or a cylinder, or a torus.

PROPOSITION 2.22. *Let $\phi : M^2 \to N^n$ be an oriented complete CMC biconservative surface. Denote by λ_1 and λ_2 the eigenvalue functions of A_H and assume that $\mu = \lambda_1 - \lambda_2$ is positive with* $\inf \mu = \mu_0 > 0$. *Then M is parabolic.*

The proof is a direct consequence of Theorem 4.5 in [**63**] (see also Theorem 12 in [**50**]) as μg is a flat complete metric globally conformally equivalent to g.

Another important property of biconservative surfaces is given by the following result.

THEOREM 2.23 ([**63**]). *Let $\phi : M^2 \to N^n$ be a compact CMC biconservative surface. If the Gaussian curvature K of the surface is nonnegative, then $\nabla A_H = 0$ and M is flat or pseudo-umbilical.*

The proof of this result relies on a formula for the Laplacian of a divergence-free symmetric tensor of type $(1,1)$ defined on a Riemannian surface, not necessarily satisfying the Codazzi condition.

REMARK 2.24. We note that Proposition 2.20, only in the case $K \geq 0$, can be viewed as a consequence of Theorem 2.23.

In codimension two we have the following rigidity result, since any PMC submanifold in a space form is biconservative.

THEOREM 2.25 ([**59**]). *Let $\phi : M^2 \to N^4(c)$ be a CMC biconservative surface in a space form of constant sectional curvature $c \neq 0$. Then M^2 is PMC.*

In the special case of biconservative surfaces in $\mathbb{R}^4$, the situation is a bit less rigid.

PROPOSITION 2.26 ([**59**]). *Let $\phi : M^2 \to \mathbb{R}^4$ be a biconservative surface with constant mean curvature different from zero, which is not PMC. Then, locally, the surface is given by*

$$X(u,v) = (\gamma(u), v + a) = (\gamma^1(u), \gamma^2(u), \gamma^3(u), v + a), \quad a \in \mathbb{R} ,$$

where $\gamma : I \to \mathbb{R}^3$ is a curve in $\mathbb{R}^3$ parametrized by arc-length, with constant non-zero curvature, and non-zero torsion.

2.4. Characterization formulas for biharmonic submanifolds. As we have seen, the tension field $\tau(\phi)$ of a map $\phi : M \to N$ plays a central role in the biharmonic equation (1.1). In the case of a submanifold M^m, we have $\tau(\phi) = mH$ and the tension field is a normal vector field. Therefore it is just natural to identify the tangent and normal parts of (1.1) in order to better characterize bihamornic submanifolds.

The simplest case occurs when the ambient manifold is a space form $N^n(c)$. In this context, the decomposition of $\tau_2(\phi)$ was written for the first time for any constant sectional curvature c in [**3**], although for $c = 0$, it had been given explicitly in [**15**], [**17**], for $c = 1$ in [**69**], and for $c = -1$ in [**11**].

THEOREM 2.27. *A submanifold $\phi : M^m \to N^n(c)$ is biharmonic if and only if*

$$\begin{cases} \Delta^\perp H + \operatorname{trace} B(\cdot, A_H \cdot) - mc\, H = 0 \\ 4 \operatorname{trace} A_{\nabla^\perp_{(\cdot)} H}(\cdot) + m \operatorname{grad}(|H|^2) = 0, \end{cases}$$

where B is the second fundamental form, $\nabla^\perp$ and $\Delta^\perp$ the connection and the Laplacian in the normal bundle, respectively.

When M is a hypersurface, we can consider the mean curvature function $f = (1/m) \operatorname{trace} A$, where $A = A_\eta$, η being a unit section in the normal bundle, and we have the following corollary.

COROLLARY 2.28. *Let M^m be a hypersurface in a space form $N^{m+1}(c)$. Then M is biharmonic if and only if*

$$\begin{cases} \Delta f + (|A|^2 - mc) f = 0 \\ 2A(\operatorname{grad} f) + mf \operatorname{grad} f = 0. \end{cases}$$

In this last form, for $c = 1$, the decomposition of $\tau_2(\phi)$ was used in [**21**].

The biharmonic equation was also decomposed in its tangent and normal parts when the ambient space is a complex space form (see [**26**]) or a Sasakian space form (see [**29**]).

Finally, in the case when the ambient space is an arbitrary Riemannian manifold, the splitting of the biharmonic equation was obtained in [**49**].

THEOREM 2.29. *A submanifold M^m in a Riemannian manifold N is biharmonic if and only if*

$$\begin{cases} \Delta^{\perp} H + \operatorname{trace} B(\cdot, A_H \cdot) + \operatorname{trace}(R^N(\cdot, H)\cdot)^{\perp} = 0 \\ 4 \operatorname{trace} A_{\nabla^{\perp}_{(\cdot)} H}(\cdot) + m \operatorname{grad}(|H|^2) + 4 \operatorname{trace}(R^N(\cdot, H)\cdot)^{\top} = 0, \end{cases}$$

where R^N is the curvature tensor of N. Moreover, the tangent part can be written as

$$4 \operatorname{trace}(\nabla A_H)(\cdot, \cdot) - m \operatorname{grad}(|H|^2) = 0.$$

While these formulas were used to study biharmonic hypersurfaces in a Riemannian manifold, for the first time, in [**69**] (see Theorem 2.7 and Proposition 2.8), they were explicitly written for this case in [**72**].

THEOREM 2.30. *Let M^m be a hypersurface in a Riemannian manifold N. Then M is biharmonic if and only if*

$$\begin{cases} \Delta f + (|A|^2 - \operatorname{Ricci}^N(\eta, \eta)) f = 0 \\ 2A(\operatorname{grad} f) + m f \operatorname{grad} f - 2f (\operatorname{Ricci}^N(\eta))^{\top} = 0, \end{cases}$$

where η is a unit normal vector field.

Many examples of proper-biharmonic submanifolds in spheres are provided by the following results.

THEOREM 2.31 ([**7**]). *Let n_1, n_2 be two positive integers such that $n_1 + n_2 = n - 1$, and let M_1 be a submanifold in $\mathbb{S}^{n_1}(1/\sqrt{2})$ of dimension m_1, with $0 \leq m_1 \leq n_1$, and let M_2 be a submanifold in $\mathbb{S}^{n_2}(1/\sqrt{2})$ of dimension m_2, with $0 \leq m_2 \leq n_2$. Then $M_1 \times M_2$ is proper biharmonic in $\mathbb{S}^n$ if and only if*

$$\begin{cases} m_1 \neq m_2, \quad \text{or} \quad |\tau(\phi_1)| > 0 \\ \tau_2(\phi_1) + 2(m_2 - m_1)\tau(\phi_1) = 0 \\ \tau_2(\phi_2) - 2(m_2 - m_1)\tau(\phi_2) = 0 \\ |\tau(\phi_1)| = |\tau(\phi_2)| = \text{constant}, \end{cases}$$

where $\phi_1 : M_1 \to \mathbb{S}^{n_1}(1/\sqrt{2})$ and $\phi_2 : M_2 \to \mathbb{S}^{n_2}(1/\sqrt{2})$ are the associated isometric immersions.

REMARK 2.32. We note that here we also correct a small inaccuracy in the original version of the theorem (see [**11**]).

COROLLARY 2.33 ([**11**]). *Let $M_1^{m_1}$ and $M_2^{m_2}$ be two minimal submanifolds in $\mathbb{S}^{n_1}(1/\sqrt{2})$ and $\mathbb{S}^{n_2}(1/\sqrt{2})$, respectively, with $n_1 + n_2 = n$. Then the product $M_1 \times M_2$ is a proper biharmonic submanifold of $\mathbb{S}^{n+1}$ if and only if $m_1 \neq m_2$.*

2.5. Characterization formulas for biconservative submanifolds. Since a submanifold is biconservative if and only if the tangent part of its bitension field vanishes, from the splitting of the biharmonic equation and looking only to the tangent part, one obtains formulas characterizing the biconservativity.

PROPOSITION 2.34. *Let M^m be a submanifold in a Riemannian manifold N^n. Then the following properties are equivalent:*

(i) *M is biconservative;*

(ii) $\operatorname{trace} A_{\nabla^{\perp}_{(\cdot)} H}(\cdot) + \operatorname{trace}(\nabla A_H)(\cdot,\cdot) + \operatorname{trace}\left(R^N(\cdot, H)\cdot\right)^{\top} = 0$;

(iii) $4 \operatorname{trace} A_{\nabla^{\perp}_{(\cdot)} H}(\cdot) + m \operatorname{grad}\left(|H|^2\right) + 4 \operatorname{trace}\left(R^N(\cdot, H)\cdot\right)^{\top} = 0$;

(iv) $4 \operatorname{trace}(\nabla A_H)(\cdot,\cdot) - m \operatorname{grad}\left(|H|^2\right) = 0$.

From this proposition, we have the following consequences, some of which are quite straightforward.

PROPOSITION 2.35. *Let M^m be a submanifold of a Riemannian manifold N^n. If $\nabla A_H = 0$, then M is biconservative.*

REMARK 2.36. A converse of Proposition 2.35 is given by Theorem 2.23.

As we have already mentioned, we have the following consequence.

PROPOSITION 2.37. *Let M^m be a submanifold of a Riemannian manifold N^n. Assume that N is a space form and M is PMC. Then M is biconservative.*

PROPOSITION 2.38 ([**6**]). *Let M^m be a pseudo-umbilical submanifold of a Riemannian manifold N^n with $m \neq 4$. Then M is biconservative if and only if it is a CMC submanifold.*

In the particular case of hypersurfaces in space forms, we have the following result.

PROPOSITION 2.39. *A hypersurface M^m in a space form $N^{m+1}(c)$ is biconservative if and only if*

$$A(\operatorname{grad} f) = -\frac{m}{2} f \operatorname{grad} f. \tag{2.1}$$

COROLLARY 2.40. *Any CMC hypersurface in $N^{m+1}(c)$ is biconservative.*

COROLLARY 2.41 ([**64**]). *Let M^m be a biconservative hypersurface in $N^{m+1}(c)$. Then*

$$m f \Delta f - 3m |\operatorname{grad} f|^2 - 2\langle A, \operatorname{Hess} f\rangle = 0.$$

REMARK 2.42. In [**64**], there was used a different definition for the mean curvature function f, i.e., there $f = \operatorname{trace} A$ instead of $f = (1/m) \operatorname{trace} A$.

Moreover, when dealing with biconservative hypersurfaces in space forms, the two distributions determined by $\operatorname{grad} f$ are completely integrable, as showed by the next theorem.

THEOREM 2.43 ([**64**]). *Let M^m be a biconservative hypersurface in $N^{m+1}(c)$ with $\operatorname{grad} f \neq 0$ at any point of M. Then the distribution $\mathcal{D}$, orthogonal to that determined by $\operatorname{grad} f$, is completely integrable. Moreover, any integral manifold of maximal dimension of $\mathcal{D}$ has flat normal connection as a submanifold in $N^{m+1}(c)$.*

REMARK 2.44. The last result actually extends the one obtained in [**40**], in the case of Euclidean space.

3. Surfaces in three dimensional space forms

3.1. Biharmonic surfaces. The first result in this section is a non-existence result, that was proved in [**19**] and [**48**] when working in the Euclidean space, and then in [**11**] in the case of space forms with negative constant sectional curvature.

THEOREM 3.1. *Any biharmonic surface in a space form $N^3(c)$, with $c \leq 0$, is minimal.*

The situation is different when one considers biharmonic surfaces in the Euclidian sphere, as shown by the following theorem in [**10**].

THEOREM 3.2. *Let $\phi : M \to \mathbb{S}^3$ be a proper-biharmonic surface. Then $\phi(M)$ is an open part of the small hypersphere $\mathbb{S}^2(1/\sqrt{2})$. If, moreover, M is complete, then $\phi(M) = \mathbb{S}^2(1/\sqrt{2})$ and ϕ is an embedding.*

3.2. Biconservative surfaces. As we have seen, CMC surfaces in a 3-dimensional space form are biconservative and, therefore, from this point of view, the study of non-CMC biconservative surfaces is more interesting. The maximal biconservative surfaces with $\operatorname{grad} f \neq 0$ at any point are called standard biconservative surfaces, and the domains of their defining immersions endowed with the induced metrics are called abstract standard biconservative surfaces.

The explicit parametric equations of standard biconservative surfaces in $N^3(c)$ were obtained in [**12**], while they had already been found, in a slightly different manner, when working in $\mathbb{R}^3$ (see [**40**]). In the hyperbolic three dimensional space, one of the possible situations was omitted in [**12**], the one corresponding to $C_{-1} = 0$, as we will see later, but it was then treated in [**34**] (see also [**65**]).

We present here only the case when the ambient space is the Euclidean sphere (that is the sectional curvature c is equal to 1).

THEOREM 3.3 ([**12**]). *Let M^2 be a biconservative surface in $\mathbb{S}^3$ such that $(\operatorname{grad} f)(p) \neq 0$ at any point $p \in M$. Then, the surface, viewed in $\mathbb{R}^4$, can be parametrized locally by*

$$Y_{\tilde{C}_1}(u,v) = \sigma(u) + \frac{4\kappa(u)^{-3/4}}{3\sqrt{\tilde{C}_1}} \left(\overline{f}_1(\cos v - 1) + \overline{f}_2 \sin v\right), \tag{3.1}$$

where $\tilde{C}_1 \in \left(64/\left(3^{5/4}\right), \infty\right)$ is a positive constant; $\overline{f}_1, \overline{f}_2 \in \mathbb{R}^4$ are two constant orthonormal vectors; $\sigma(u)$ is a curve parametrized by arc-length that satisfies

$$\langle \sigma(u), \overline{f}_1 \rangle = \frac{4\kappa(u)^{-3/4}}{3\sqrt{\tilde{C}_1}}, \qquad \langle \sigma(u), \overline{f}_2 \rangle = 0, \tag{3.2}$$

and, as a curve in $\mathbb{S}^2$, its curvature $\kappa = \kappa(u)$ is a positive non-constant solution of the following ODE

$$\kappa''\kappa = \frac{7}{4}\left(\kappa'\right)^2 + \frac{4}{3}\kappa^2 - 4\kappa^4 \tag{3.3}$$

such that

$$\left(\kappa'\right)^2 = -\frac{16}{9}\kappa^2 - 16\kappa^4 + \tilde{C}_1\kappa^{7/2}. \tag{3.4}$$

REMARK 3.4. The curve σ lies in the totally geodesic $\mathbb{S}^2 = \mathbb{S}^3 \cap \Pi$, where Π is the linear hyperspace of $\mathbb{R}^4$ orthogonal to $\overline{f}_2$. The constant $\tilde{C}_1$ determines uniquely the curvature κ, up to a translation and change of sign of u, and then κ,

$\overline{f}_1$ and $\overline{f}_2$ determine uniquely the curve σ. Thus, up to isometries of $\mathbb{S}^3$, we have a one parametric family of standard biconservative surfaces in $\mathbb{S}^3$ indexed by $\tilde{C}_1$. Moreover, σ is a geodesic of M.

REMARK 3.5. From (3.3) it follows that σ is a critical point of the curvature energy functional $\Theta(\gamma) = \int_\gamma \kappa^{1/4}$, where γ is a curve in $\mathbb{S}^2$ parametrized by arc-length, in the sense of classical elastic curves (see [**61**]).

None of these standard biconservative surfaces in space forms $N^3(c)$ is complete. Therefore, the next step was finding complete non-CMC biconservative surfaces with $\operatorname{grad} f \neq 0$ on an open subset.

The $c = 0$ case proved to be the simplest one and one was able to construct such surfaces by using both an extrinsic and an intrinsic approach. Thus, working extrinsically, i.e., with the explicit parametric equations, as a standard biconservative surface in $\mathbb{R}^3$ is a rotational surface with the boundary (in its topogical closure) connected and given by one circle, one glued two profile curves thus obtaining a regular closed surface. In this case, the gluing was made along the shared boundary that becomes a geodesic on the resulting surface, where $\operatorname{grad} f = 0$. In the intrinsic way, one first constructed an abstract simply connected complete surface and then the respective non-CMC biconservative immersion. We note that to obtain the abstract surface, one used isothermal coordinates and, again, a gluing process of two abstract standard biconservative surfaces. We note that this surface cannot be factorized to a (non-flat) torus.

THEOREM 3.6 ([**62**],[**64**]). *Let* $\left(\mathbb{R}^2, g_{C_0} = C_0 (\cosh u)^6 \left(du^2 + dv^2\right)\right)$ *be a surface, where* $C_0 \in \mathbb{R}$ *is a positive constant. Then we have:*

(i) *the metric on* $\mathbb{R}^2$ *is complete;*
(ii) *the Gaussian curvature is given by*

$$K_{C_0}(u,v) = K_{C_0}(u) = -\frac{3}{C_0 (\cosh u)^8} < 0, \quad K'_{C_0}(u) = \frac{24 \sinh u}{C_0 (\cosh u)^9},$$

and therefore $\operatorname{grad} K_{C_0} \neq 0$ *at any point of* $\mathbb{R}^2 \setminus Ov$*;*
(iii) *the immersion* $\phi_{C_0} : \left(\mathbb{R}^2, g_{C_0}\right) \to \mathbb{R}^3$ *given by*

$$\phi_{C_0}(u,v) = \left(\sigma^1_{C_0}(u)\cos(3v), \sigma^1_{C_0}(u)\sin(3v), \sigma^2_{C_0}(u)\right)$$

is biconservative in $\mathbb{R}^3$*, where*

$$\sigma^1_{C_0}(u) = \frac{\sqrt{C_0}}{3}(\cosh u)^3, \quad \sigma^2_{C_0}(u) = \frac{\sqrt{C_0}}{2}\left(\frac{1}{2}\sinh(2u) + u\right), \qquad u \in \mathbb{R}.$$

To intrinsically approach the $c \neq 0$ cases, as well as in the $c = 0$ case actually, one needs to use the following intrinsic properties of the standard biconservative surfaces.

THEOREM 3.7 ([**12**]). *Let* $\phi : M^2 \to N^3(c)$ *be a biconservative surface with* $\operatorname{grad} f \neq 0$ *at any point of* M*. Then the Gaussian curvature* K *satisfies*

(i)

$$K = \det A + c = -3f^2 + c;$$

(ii) $c - K > 0$, $\operatorname{grad} K \neq 0$ *at any point of* M*, and the level curves of* K *are circles in* M *with constant curvature*

$$\kappa = \frac{3|\operatorname{grad} K|}{8(c-K)};$$

(iii)
$$(c-K)\Delta K - |\operatorname{grad} K|^2 - \frac{8}{3}K(c-K)^2 = 0.$$

REMARK 3.8. If M^2 is a biconservative surface in $N^3(c)$ with $\operatorname{grad} f \neq 0$ at any point, then it is a linear Weingarten surface. Indeed, from $A(\operatorname{grad} f) = -f \operatorname{grad} f$, we have $3\lambda_1 + \lambda_2 = 0$, where λ_1 and λ_2 are the principal curvatures of M^2.

Now, we present a uniqueness result for biconservative surfaces with nowhere vanishing $\operatorname{grad} f$.

THEOREM 3.9 ([**28**]). *Let (M^2, g) be an abstract surface and $c \in \mathbb{R}$ an arbitrarily fixed constant. If M admits two biconservative immersions in $N^3(c)$ such that the gradients of their mean curvature functions are different from zero at any point of M, then the two immersions differ by an isometry of $N^3(c)$.*

Even if the notion of a biconservative submanifold belongs, obviously, to extrinsic geometry, in the particular case of biconservative surfaces in $N^3(c)$ one can give an intrinsic characterization of such surfaces.

THEOREM 3.10 ([**28**]). *Let (M^2, g) be an abstract surface. Then M can be locally isometrically embedded in a space form $N^3(c)$ as a biconservative surface with the gradient of the mean curvature different from zero everywhere if and only if the Gaussian curvature K satisfies $c - K(p) > 0$, $(\operatorname{grad} K)(p) \neq 0$, for any $p \in M$, and its level curves are circles in M with constant curvature*
$$\kappa = \frac{3|\operatorname{grad} K|}{8(c-K)}.$$

Even more, one can prove that any biconservative immersion from (M^2, g) in $N^3(c)$ must satisfy $\operatorname{grad} f \neq 0$, and so one obtains the next result.

THEOREM 3.11 ([**64**]). *Let (M^2, g) be an abstract surface and $c \in \mathbb{R}$ an arbitrarily given constant. Assume that $c - K > 0$ and $\operatorname{grad} K \neq 0$ at any point of M, and the level curves of K are circles in M with constant curvature*
$$\kappa = \frac{3|\operatorname{grad} K|}{8(c-K)}.$$
Then, locally, there exists a unique biconservative embedding $\phi : (M^2, g) \to N^3(c)$. Moreover, the mean curvature function is positive and its gradient is different from zero at any point of M.

REMARK 3.12. For a given C, there exists a one parametric family of abstract standard biconservative surfaces indexed by C, and their metrics can be written in a very explicit way.

In the case when the ambient space is the Euclidian sphere, the extrinsic approach proved to be quite difficult, as it implies the gluing of a large number (infinite, in general) of standard biconservative surfaces. When working intrinsically, one first used a gluing procedure, to obtain a closed rotational surface in $\mathbb{R}^3$ (and so complete), with a periodic profile curve, and then, from that surface (or its universal cover $(\mathbb{R}^2, g_{C_1})$, that can be factorized to a non-flat torus), one constructed the desired non-CMC biconservative immersion (see [**62**, Theorem 4.18]).

The hyperbolic case ($c = -1$) is rather similar to the $c = 0$ one. In an extrinsic approach one has to glue two standard biconservative surfaces to obtain complete biconservative surfaces which are not CMC (see [**65**]). In the intrinsic manner, in [**65**] one proposes a new method that can be easily adapted to the other two cases, when $c = 0$ or $c = 1$. The main idea is to rewrite the metric on an abstract standard biconservative surface as

$$g(u, v) = h^2(u)dv^2 + du^2, \quad u > 0, \quad v \in \mathbb{R},$$

where $h(u) > 0$ and $\lim_{u\to 0} h(u) \in \mathbb{R}^*_+$. Then one glues two such abstract standard biconservative surfaces along the axis (Ov) to obtain an abstract simply connected complete surface $(\mathbb{R}^2, g_{C_{-1}})$ that cannot be factorized to a torus. To prove the existence of the non-CMC biconservative immersion from this surface in $\mathbb{H}^3$, one constructs a suitable shape operator and then uses the Fundamental Theorem of Surfaces in space forms. The difficult part of this construction is the fact that $\operatorname{grad} K$ vanishes along the axis (Ov) and, therefore, one cannot directly apply Theorem 3.10.

We remark that all abstract simply connected complete surfaces $(\mathbb{R}^2, g_{C_{-1}})$, $(\mathbb{R}^2, g_{C_0})$ and $(\mathbb{R}^2, g_{C_1})$ are globally conformally equivalent to $\mathbb{R}^2$.

With all these results in mind, two interesting issues can be raised.

PROBLEM 1. Let $\phi : M^2 \to N^3(c)$ be a simply connected complete biconservative surface which is not CMC. Then, up to an isometry of the domain or the codomain, M and ϕ are those given in [**62**], when $c = 0$ or $c = 1$, and in [**65**], when $c = -1$.

PROBLEM 2. Any compact biconservative surface in $N^3(c)$ is CMC.

Very recently, Problem 1, was positively answered in [**66**], and one of the most important steps of the proof consisted in proving that (for the $c = -1$ case) from a given abstract surface $(\mathbb{R}^2, g_{c_{-1}})$ there exists a unique biconservative immersion in $\mathbb{H}^3$. Moreover, it is not CMC. The proof of this step relies, among others, on Theorem 3.11, but here $\operatorname{grad} K$ vanishes along (Ov). This intermediary result in the proof of Problem 1 also holds when $c = 0$ or $c = 1$.

Now, Problem 2 can be easily solved for $c = 0$ or $c = -1$. More precisely, one considers the universal cover $\pi : \tilde{M}^2 \to M^2$ of a compact non-CMC biconservative surface M^2 and thus one obtains a surface $\tilde{\phi} : \tilde{M}^2 \to \mathbb{R}^3$ or $\tilde{\phi} : \tilde{M}^2 \to \mathbb{H}^3$, which is simply connected, complete, non-CMC and biconservative. But then, the immersion $\tilde{\phi}$ would have to be one of those given in [**62**] and [**65**], and, in particular, $\tilde{M}^2 = (\mathbb{R}^2, g_{C_{-1}})$ or $\tilde{M}^2 = (\mathbb{R}^2, g_{C_0})$, for some constants C_{-1} and C_0. But this is a contradiction as $\tilde{M}^2 = (\mathbb{R}^2, g_{C_{-1}})$ or $\tilde{M}^2 = (\mathbb{R}^2, g_{C_0})$ can be factorized only to a cylinder (which is not compact). The case $c = 1$ in Problem 2 cannot be solved as easily as the other two cases, even if the abstract surface $(\mathbb{R}^2, g_{C_1})$ does factorize to a non-flat torus, and the author could not prove whether (some of) the immersions in [**62**] are double-periodic or not.

In fact, Problem 2 was completely solved in [**61**]. Without addressing Problem 1, the authors prove that while in $N^3(c)$, with $c = -1$ or $c = 0$, all compact biconservative surfaces are CMC, in the case $c = 1$, there exists an entire family of compact non-CMC biconservative surfaces. In other words, among the immersions given [**62**] in the case of the sphere, there exists a family of double-periodic ones. To prove the existence of this family of surfaces in $\mathbb{S}^3$, the authors used the known fact that the standard biconservative surface in the sphere is rotational and that the curvature of its profile curve, that lies in a sphere $\mathbb{S}^2$ totally geodesic in $\mathbb{S}^3$, satisfies a certain equation (see Theorem 3.3). They proved, using among other results the Poincare-Bendixon Theorem, that the maximally defined solutions of this equation are defined on the entire real axis $\mathbb{R}$ and are periodic. Now, having these periodic curvature functions and using closure conditions from the theory of elastic curves (see [**2**]), the authors showed that there exist periodic profile curves in $\mathbb{S}^2$ having as curvatures some of these periodic curvature functions. Obviously, the corresponding surface to a periodic profile curve is compact, biconservative and non-CMC.

4. Hypersurfaces in space forms

4.1. Biharmonic hypersurfaces. All results obtained until now for biharmonic hypersurfaces in spheres are partial positive answers to the following two conjectures, which have been proposed in [**3**].

CONJECTURE 1. *Any proper biharmonic hypersurface in $\mathbb{S}^{m+1}$ is either an open part of $\mathbb{S}^m(1/\sqrt{2})$, or an open part of $\mathbb{S}^{m_1}(1/\sqrt{2})\times\mathbb{S}^{m_2}(1/\sqrt{2})$, where $m_1+m_2=m$ and $m_1\neq m_2$.*

CONJECTURE 2. *Any proper biharmonic submanifold in $\mathbb{S}^{m+1}$ is CMC.*

Obviously, Conjecture 2 (for hypersurfaces) is weaker than Conjecture 1, but to directly prove the first conjecture seems to be quite a complicated task. Until now, a proof for the second conjecture, which may be viewed as a step in the proof of Conjecture 1, has not been obtain for any dimension m and without additional hypotheses. Moreover, even with Conjecture 2 proved, the proof of Conjecture 1 will be a real challenge, even in the compact case.

It is worth mentioning that if M is a CMC hypersurface, then it is proper biharmonic if and only if $|A|^2=m$ and $f\neq 0$. Therefore, to study CMC proper biharmonic hypersurfaces and give a positive answer to the first conjecture is to study non-minimal CMC hypersurfaces with $|A|^2=m$ and prove that they can only be open parts of $\mathbb{S}^m(1/\sqrt{2})$, or $\mathbb{S}^{m_1}(1/\sqrt{2})\times\mathbb{S}^{m_2}(1/\sqrt{2})$, where $m_1+m_2=m$ and $m_1\neq m_2$. This challenging problem can be viewed as a generalization, from the minimal to the CMC case, of the classical result of S. S. Chern, M. do Carmo, and S. Kobayashi, that says that any minimal hypersurface with $|A|^2=m$ is an open part of $\mathbb{S}^{m_1}(r_1)\times\mathbb{S}^{m_2}(r_2)$, where $m_1+m_2=m$, $r_1^2+r_2^2=1$, and $r_1=\sqrt{m_1/m}$ (see [**24**]).

We will present, in a quite non-exhaustive way, the most important results to date on biharmonic hypersurfaces in spheres. As presentations of this type of results can be also found in [**71**] and in [**73**], in general, we will focus on the most recent of them.

Conjecture 1 was proved only in the case of surfaces in [**10**]. Conjecture 2 was proved for $m=3$ in [**4**]. Very recently, in [**39**], Conjecture 2 was also proved for $m=4$. The proof of this last result is a long and skilful one and relies on a detailed analysis of the Gauss and Codazzi equations that allow the biharmonic hypothesis. This result actually extends to a non-flat 5-dimensional space form the one obtained by Y. Fu, M.-C. Hong, and X. Zhan in [**36**], which shows that any biharmonic hypersurface in $\mathbb{R}^5$ is CMC and, therefore, minimal.

With geometric or analytical additional hypotheses, the two conjectures have received positive answers in several situations. The geometric hypotheses concern the number of distinct principal curvatures of the hypersurface, or ask the scalar curvature to be constant, or the squared length of the second fundamental form to be bounded by m, etc.

Before listing some of these results, we recall that a CMC proper biharmonic submanifold satisfies $|H| \in (0,1]$. In the case of CMC hypersurfaces, like in the more general case of PMC submanifolds, there is a gap in the admissible range of $|H|$.

THEOREM 4.1 ([**7**]). *Let $\phi : M^m \to \mathbb{S}^{m+1}$ be a CMC proper biharmonic hypersurface with $m > 2$. Then $|H| \in (0,(m-2)/m] \cup \{1\}$. Moreover, $|H| = 1$ if and only if $\phi(M)$ is an open subset of the small hypersphere $\mathbb{S}^m(1/\sqrt{2})$, and $|H| = (m-2)/m$ if and only if $\phi(M)$ is an open subset of the standard product $\mathbb{S}^{m-1}(1/\sqrt{2}) \times \mathbb{S}^1(1/\sqrt{2})$.*

REMARK 4.2. The above result was proved independently in [**82**].

Among the results answering positively to the two conjectures, when adding the hypothesis on the number of distinct principal curvatures, we mention the following theorem.

THEOREM 4.3 ([**3**]). *Let $\phi : M^m \to \mathbb{S}^{m+1}$ be a proper biharmonic hypersurface with at most two distinct principal curvatures everywhere. Then $\phi(M)$ is either an open part of $\mathbb{S}^m(1/\sqrt{2})$, or an open part of $\mathbb{S}^{m_1}(1/\sqrt{2}) \times \mathbb{S}^{m_2}(1/\sqrt{2})$, $m_1+m_2 = m$, $m_1 \neq m_2$. Moreover, if M is complete, then either $\phi(M) = \mathbb{S}^m(1/\sqrt{2})$ and ϕ is an embedding, or $\phi(M) = \mathbb{S}^{m_1}(1/\sqrt{2}) \times \mathbb{S}^{m_2}(1\sqrt{2})$, $m_1 + m_2 = m$, $m_1 \neq m_2$, and ϕ is an embedding when $m_1 \geq 2$ and $m_2 \geq 2$.*

As a direct consequence of the above result, for $m = 2$ we reobtain the proof of Conjecture 1 in [**10**].

The link between biharmonic hypersurfaces and the isoparametric ones was established by T. Ichiyama, J. I. Inoguchi, and H. Urakawa in 2009 and 2010 (see [**42**], [**43**]).

THEOREM 4.4 ([**42**],[**43**]). *Let $\phi : M^m \to \mathbb{S}^{m+1}$ be an isoparametric proper biharmonic hypersurface. Then $\phi(M)$ is either an open part of $\mathbb{S}^m(1/\sqrt{2})$, or an open part of $\mathbb{S}^{m_1}(1/\sqrt{2}) \times \mathbb{S}^{m_2}(1/\sqrt{2})$, $m_1 + m_2 = m$, $m_1 \neq m_2$.*

The proof relies on the expression of principal curvatures for an isoparametric hypersurface which was then used to solve the equation $|A|^2 = m$.

In the case when the number of distinct principal curvatures is less than or equal to 3 we have two results. The first one holds for $m = 3$.

THEOREM 4.5 ([**4**]). *Let $\phi : M^3 \to \mathbb{S}^4$ be a proper biharmonic hypersurface. Then ϕ is CMC.*

The second result is a generalization of Theorem 4.5 to higher dimensions.

THEOREM 4.6 ([**33**]). *Let $\phi : M^m \to \mathbb{S}^{m+1}$ be a proper biharmonic hypersurfaces and assume that it has at most three distinct principal curvatures at any point. Then M is CMC.*

These two results lead to positive answers to Conjecture 1, with supplementary hypotheses. First, assuming that the hypersurface is also complete, Theorem 4.5 implies Conjecture 1.

THEOREM 4.7 ([**4**]). *Let $\phi : M^3 \to \mathbb{S}^4$ be a complete proper biharmonic hypersurface. Then $\phi(M) = \mathbb{S}^3(1/\sqrt{2})$ or $\phi(M) = \mathbb{S}^2(1/\sqrt{2}) \times \mathbb{S}^1(1/\sqrt{2})$.*

Indeed, it is easy to verify that, in general, a CMC proper biharmonic hypersurface satisfies $|A|^2 = m$ and has positive constant scalar curvature. Therefore, by applying Theorem 4.5, one obtains that M^3 is CMC and has constant scalar curvature. Furthermore, since M is complete, using a result of Q. M. Cheng and Q. R. Wan in [**22**], one sees that M is isoparametric and then we conclude with Theorem 4.4.

REMARK 4.8. Theorem 4.7 was actually obtained in [**4**] under the stronger assumption of compactness, because we used a result of S. Chang in [**13**] that shows that any compact CMC hypersurface with constant scalar curvature in $\mathbb{S}^4$ is isoparametric.

Then, using another of S. Chang's results (see [**14**]), Theorem 4.6 implies the next theorem.

THEOREM 4.9 ([**33**]). *Let $\phi : M^m \to \mathbb{S}^{m+1}$ be a compact biharmonic hypersurface and assume that it has three distinct curvatures everywhere. Then M is minimal.*

REMARK 4.10. Here we had to slightly modify the original statement of Theorem 4.9 due to S. Chang's result, that is stated for hypersurfaces with three principal curvatures everywhere (and not for hypersurfaces with at most three principal curvatures everywhere).

Still following the study of proper biharmonic hypersurfaces according to the number of distinct principal curvatures, Y. Fu and M. C. Hong improved Theorem 4.6 for a greater number of distinct principal curvatures (although considering an additional hypothesis on the scalar curvature). More precisely, using Gauss and Codazzi equations and performing some long computations, they proved the following theorem.

THEOREM 4.11 ([**35**]). *Let $\phi : M^m \to \mathbb{S}^{m+1}$ be a proper biharmonic hypersurface with at most six distinct principal curvatures at any point. If M has constant scalar curvature, then M is CMC.*

REMARK 4.12. The above result holds in any space form $N^{m+1}(c)$, as it is actually given in its original form in [**35**].

As we have seen, the CMC hypothesis implies, for a proper biharmonic hypersurface M^m, that $|A|^2 = m$ and its scalar curvature is a positive constant. When M is compact, either hypothesis $|A|^2 =$ constant, or the scalar curvature is a constant, implies that M is CMC.

The fact that $|A|^2 =$ constant implies that M is CMC, and therefore $|A|^2 = m$, is an immediate consequence of two results showing that if the function $|A|^2 - m$ has constant sign, then M is CMC. More precisely, we have the following two results.

THEOREM 4.13 ([**5**],[**71**]). *Let $\phi : M^m \to \mathbb{S}^{m+1}$ be a compact proper biharmonic hypersurface. If $|A|^2 \geq m$, then ϕ is CMC and $|A|^2 = m$.*

THEOREM 4.14 ([**21**]). *Let $\phi : M^m \to \mathbb{S}^{m+1}$ be a compact proper biharmonic hypersurface. If $|A|^2 \leq m$, then ϕ is CMC and $|A|^2 = m$.*

The first of the two results follows easily by integrating Δf^2, which was computed by using the normal part of the biharmonic equation. The second result, on the other side, given by J. H. Chen, is more difficult to prove and requires tensor analysis (among others, a Bochner type formula for $\Delta|\operatorname{grad} f|^2$, the expressions of Δf^2 and Δf^4 for biharmonic hypersurfaces, and a certain inequality involving $|A|^2$ and f^2). Moreover, in [**5**] and [**71**], it was given a proof which is slightly different from the original one, based on the unique continuation property of biharmonic maps.

We mention here that if $|A|^2 = \text{constant}$ and M^m is complete and non-compact, then M^m is CMC, under some additional hypotheses involving the Ricci curvature (see [**5**], [**71**]).

The fact that constant scalar curvature implies that M is CMC has recently been proved by S. Maeta and Y. L. Ou (see [**54**]).

THEOREM 4.15 ([**54**]). *A compact hypersurface $\phi : M^m \to \mathbb{S}^{m+1}$ with constant scalar curvature is biharmonic if and only if it is minimal, or it has non-zero constant mean curvature and $|A|^2 = m$.*

The proof relies, among others, on the same Bochner formula for $\Delta|\operatorname{grad} f|^2$, i.e.,

$$\frac{1}{2}\Delta|\operatorname{grad} f|^2 = \langle \operatorname{grad}\Delta f, \operatorname{grad} f\rangle - \operatorname{Ricci}(\operatorname{grad} f, \operatorname{grad} f) - |\nabla \operatorname{grad} f|^2,$$

and on the general classical inequality

$$|\nabla \operatorname{grad} f|^2 \geq \frac{1}{m}(\Delta f)^2.$$

In conclusion, we can state the following result.

THEOREM 4.16. *Let $\phi : M^m \to \mathbb{S}^{m+1}$ be a compact proper biharmonic hypersurface. Then any of the following hypotheses implies the other two:*

(i) *M is CMC;*
(ii) *$|A|^2$ is a constant;*
(iii) *the scalar curvature s is a constant.*

Using the celebrated theorem on CMC hypersurfaces in space forms obtained by K. Nomizu and B. Smyth in [**67**], the next corollary follows immediately.

COROLLARY 4.17. *Let $\phi : M^m \to \mathbb{S}^{m+1}$ be a compact proper biharmonic hypersurface with constant scalar curvature and $\operatorname{Riem}^M \geq 0$, then $\phi(M) = \mathbb{S}^m(1/\sqrt{2})$, or $\phi(M) = \mathbb{S}^{m_1}(1/\sqrt{2}) \times \mathbb{S}^{m_2}(1/\sqrt{2})$, where $m_1 + m_2 = m$ and $m_1 \neq m_2$.*

REMARK 4.18. With the assumption that M^m is CMC instead of constant scalar curvature, the above corollary holds locally (see [**5**], [**71**]).

Another partial answer to Conjecture 2, this time with additional hypothesis of an analytical type, was given in [**9**], where one proved that a proper biharmonic hypersurface in $\mathbb{S}^{m+1}$ satisfying a certain inequality has the following unique continuation property.

THEOREM 4.19 ([**9**]). *Let $\phi : M^m \to \mathbb{S}^{m+1}$ be a proper biharmonic hypersurface. Assume that there exists a non-negative function h on M such that $|\operatorname{grad}|A|^2| \leq h|\operatorname{grad} f|$ on M. If $\operatorname{grad} f$ vanishes on a non-empty open connected subset of M, then $\operatorname{grad} f = 0$ on M, i.e., M has constant mean curvature.*

The proof relies again on Aronszajn's result (see [**1**]), but here the reduction of the order is not necessary anymore. The main idea is to think $\operatorname{grad} f$ as a vector field tangent to $\mathbb{R}^{m+2}$ along M and identify it with an $\mathbb{R}^{m+2}$-valued function u. This function satisfies the equation $Lu = f \operatorname{grad} |A|^2$, where L is a second order linear elliptic operator. Then, the result follows using the hypothesis and the Aronszajn's result.

The inequality in the theorem is verified when one imposes natural geometric conditions on $|A|^2$ or the scalar curvature of M.

COROLLARY 4.20 ([**9**]). *Let $\phi : M^m \to \mathbb{S}^{m+1}$ be a proper biharmonic hypersurface such that $|A|^2$ is constant or M has constant scalar curvature. Then, either M has constant mean curvature, or the set of points where* $\operatorname{grad} f \neq 0$ *is an open dense subset of M.*

The above corollary is a local version of the results obtained by J. H. Chen, and S. Maeta and Y.-L. Ou, respectively, with the additional hypothesis that f is constant on some open set.

Still with additional hypotheses of an analytic nature (for example, using new Liouville type theorems for superharmonic functions on complete non-compact manifolds, and asking that a certain function on M built by using f to be of class L^p, or asking that $f^{-1} \in L^p(M)$, for some $p \in (0, \infty)$), the second conjecture was proved right by S. Maeta in [**53**] and Y. Luo and S. Maeta in [**52**].

Another interesting result to prove Conjecture 1 was obtained by M. Vieira in [**81**] by using classical techniques from Riemannian geometry.

THEOREM 4.21 ([**81**]). *Let M^m be a compact proper biharmonic hypersurface in a closed hemisphere of $\mathbb{S}^{m+1}$. If $1 - |H|^2$ does not change sign, then M^m is the small hypersphere $\mathbb{S}^m(1/\sqrt{2})$.*

The proof of this theorem relies on a general formula that gives the expression of the bilaplacian of a function f on M, when f is the restriction of some function $\bar{f}$ defined on the ambient manifold N^{m+1}. This formula is given in terms of the covariant derivatives of $\bar{f}$ with respect to N, and the second fundamental form of M in N. A very interesting fact about this formula is the involvement of the tangent and normal parts of the bitension field. Therefore, when M is biharmonic the equation is simpler and can be used in the study of such hypersurfaces. Furthermore, one can consider a given non-zero vector $v \in \mathbb{R}^{m+2}$ which determines the hemisphere containing M in Theorem 4.21, and the function $\bar{f}$ on $\mathbb{S}^{m+1}$ that associates to each point the inner product between its position vector and v. Taking into account the properties of this particular function, the bilaplacian of its restriction to M simplifies even more. Finally, integrating over M, one concludes.

We end this subsection with yet another two positive partial answers to Conjecture 1.

THEOREM 4.22 ([**21**]). *Let $\phi : M^m \to \mathbb{S}^{m+1}$ be a compact proper biharmonic hypersurface. If M^m has non-negative sectional curvature and $m \leq 10$, then ϕ is CMC and $\phi(M)$ is either $\mathbb{S}^m(1/\sqrt{2})$, or $\mathbb{S}^{m_1}(1/\sqrt{2}) \times \mathbb{S}^{m_2}(1/\sqrt{2})$, $m_1 + m_2 = m$, $m_1 \neq m_2$.*

PROOF. We present a simpler proof than the original one in [**21**]. By a long but straightforward computation we get a formula for the Laplacian of the squared norm of the shape operator, which holds for any hypersurface M,

$$(4.1)\quad \frac{1}{2}\Delta|A|^2 = -|\nabla A|^2 - m\operatorname{div}(A(\operatorname{grad} f)) + m^2|\operatorname{grad} f|^2 - \frac{1}{2}\sum_{i,j=1}^{m}(\lambda_i - \lambda_j)^2 R_{ijij},$$

where λ_i are the principal curvature functions of M.

J. H. Chen proved, in [**21**], the following inequality that holds for biconservative hypersurfaces in space forms

$$(4.2)\qquad |\nabla A|^2 \geq \frac{m^2(m+26)}{4(m-1)}|\operatorname{grad} f|^2.$$

Then, we obtain, for any biconservative hypersurface in a space form, that

$$(4.3)\quad \frac{1}{2}\Delta|A|^2 \leq \frac{3(m-10)}{m+26}|\nabla A|^2 - m\operatorname{div}(A(\operatorname{grad} f)) - \frac{1}{2}\sum_{i,j=1}^{m}(\lambda_i - \lambda_j)^2 R_{ijij}.$$

Next, integrating (4.3), we prove that M has at most two distinct principal curvatures at any point and, since M is proper biharmonic, from Theorem 4.3, we conclude. □

The previous theorem deals with compact hypersurfaces, but with the additional hypothesis that the squared norm of the shape operator is constant, we can obtain its local version.

PROPOSITION 4.23. *Let $\phi : M^m \to \mathbb{S}^{m+1}$ be a proper biharmonic hypersurface with non-negative sectional curvature. If the squared norm of the shape operator is constant and satisfies $|A|^2 \leq m$, and $m \leq 6$, then ϕ is CMC and $\phi(M)$ is either an open subset of $\mathbb{S}^m(1/\sqrt{2})$, or an open subset of $\mathbb{S}^{m_1}(1/\sqrt{2}) \times \mathbb{S}^{m_2}(1/\sqrt{2})$, $m_1 + m_2 = m$, $m_1 \neq m_2$.*

PROOF. From the tangent part of the biharmonic equation, we have

$$A(\operatorname{grad} f) = -\frac{m}{2} f \operatorname{grad} f$$

and, also using the normal part, $\operatorname{div}(A(\operatorname{grad} f))$ can be easily written as

$$\operatorname{div}(A(\operatorname{grad} f)) = \frac{m}{2}(m - |A|^2)f^2 - \frac{m}{2}|\operatorname{grad} f|^2.$$

Replacing in (4.1) and using (4.2), we get

$$\frac{1}{2}\Delta|A|^2 \leq \frac{5m-32}{m+26}|\nabla A|^2 - \frac{m^2}{2}(m - |A|^2)f^2 - \frac{1}{2}\sum_{i,j=1}^{m}(\lambda_i - \lambda_j)^2 R_{ijij} \leq 0.$$

Since $|A|^2$ is a constant, it follows that M has at most two distinct principal curvatures at any point, and then, again using Theorem 4.3, we conclude. □

REMARK 4.24. The above result can be also seen as a local version of Theorem 4.14, or of the fact that $|A|^2 = \text{constant}$ implies CMC, with additional hypotheses on the curvature and dimension.

4.2. Biconservative hypersurfaces. The study of biconservative hypersurfaces in space forms is a rather new research direction. In one of the first papers devoted to this study, one obtained the following Simons type formula for the squared norm of S_2, from the more general equation given in [**50**] (see [**27**]).

THEOREM 4.25 ([**27**]). *Let $\phi : M^m \to N^{m+1}(c)$ be a hypersurface in a space form. Then*

$$\begin{aligned}\frac{1}{2}\Delta|S_2|^2 =& 4cm^4f^4 - 4m^3f^3(\operatorname{trace} A^3) - 4m^2f^2|A|^2(cm - |A|^2)\\ &- 8m^4f^2|\operatorname{grad} f|^2 - 4m^2f^2|\nabla A|^2\\ &+ 4m^2f\langle\operatorname{grad} s, \operatorname{grad} f\rangle\\ &- 8m^2\operatorname{div}(f\operatorname{Ricci}(\operatorname{grad} f)) - 2m^2\operatorname{div}\left(|A|^2\operatorname{grad} f^2\right)\\ &+ \frac{m^5}{8}\Delta f^4 - 4cm^2(m-1)\Delta f^2 - 10m^2f\langle\tau_2^\top(\phi), \operatorname{grad} f\rangle\\ &- 4m^2f^2\operatorname{div}\left(\tau_2^\top(\phi)\right) - 2\left|\tau_2^\top(\phi)\right|^2 + 4mf\langle\nabla\tau_2^\top(\phi), A\rangle.\end{aligned} \tag{4.4}$$

Theorem 4.25 leads to the next two results.

THEOREM 4.26 ([**27**]). *Let $\phi : M^m \to N^{m+1}(c)$ be a constant scalar curvature biconservative hypersurface in a space form. Then*

(4.5)

$$\begin{aligned}\frac{3m^2}{2}\Delta f^4 &= 4f^2\{cm^2f^2 - mf(\operatorname{trace} A^3) - |A|^2(cm - |A|^2) - 2m^2|\operatorname{grad} f|^2 - |\nabla A|^2\}\\ &= 2f^2\left\{-\sum_{i,j=1}^m(\lambda_i - \lambda_j)^2R_{ijij} - 4m^2|\operatorname{grad} f|^2 - 2|\nabla A|^2\right\}.\end{aligned}$$

COROLLARY 4.27 ([**27**]). *Let $\phi : M^m \to \mathbb{S}^{m+1}$ be a biharmonic hypersurface with constant scalar curvature. Then the following system holds*

$$\begin{cases}\frac{3m^2}{2}\Delta f^4 = & 4f^2\{m^2f^2 - mf(\operatorname{trace} A^3) - |A|^2(m - |A|^2)\\ & -2m^2|\operatorname{grad} f|^2 - |\nabla A|^2\}\\ \Delta f = & f(m - |A|^2).\end{cases} \tag{4.6}$$

A direct application of equation (4.5) is the following rigidity theorem.

THEOREM 4.28 ([**27**]). *Let $\phi : M^m \to N^{m+1}(c)$ be a compact biconservative hypersurface in a space form $N^{m+1}(c)$, with $c \in \{-1, 0, 1\}$. If M is not minimal, has constant scalar curvature and* $\operatorname{Riem}^M \geq 0$, *then $\phi(M)$ is either*

(i) *$\mathbb{S}^m(r)$, $r > 0$, if $c \in \{-1, 0\}$, i.e., N is either the hyperbolic space $\mathbb{H}^{m+1}$ or the Euclidean space $\mathbb{R}^{m+1}$; or*
(ii) *$\mathbb{S}^m(r)$, $r \in (0, 1)$, or the product $\mathbb{S}^{m_1}(r_1) \times \mathbb{S}^{m_2}(r_2)$, where $r_1^2 + r_2^2 = 1$, $m_1 + m_2 = m$, and $r_1 \neq \sqrt{m_1/m}$, if $c = 1$, i.e., N is the unit Euclidean sphere $\mathbb{S}^{m+1}$.*

REMARK 4.29. In [**23**], compact hypersurfaces M^m in $\mathbb{S}^{m+1}$ with $\operatorname{Riem}^M \geq 0$ and constant scalar curvature $s \geq m(m-1)$ were classified. In Theorem 4.28 there is no restriction on the value of scalar curvature.

REMARK 4.30. Rewriting Equation (4.4) in terms of the shape operator A yields a generalization of the well-known formula for CMC hypersurfaces in [**67**]. Moreover, Theorem 4.28 can be viewed as an alternative to the result of K. Nomizu and B. Smyth about compact CMC hypersurfaces in space forms in [**67**], with the "CMC hypothesis" replaced by a more general one about "the biconservativity and constant scalar curvature".

Imposing the above hypothesis on the curvature of M^m and a restriction on the dimension m, but nothing on the scalar curvature, we come to the same conclusion as in Theorem 4.28.

THEOREM 4.31. *Let $\phi : M^m \to N^{m+1}(c)$ be a compact non-minimal biconservative hypersurface. If* $\operatorname{Riem}^M \geq 0$ *and $m \leq 10$, then ϕ is CMC and $\phi(M)$ is one of the hypersurfaces given by Theorem* 4.28.

PROOF. As we have seen, equation (4.3) holds for biconservative hypersurfaces. Then, for $m \leq 9$, the right hand side term is non-positive and then, integrating over M, we conclude that $\nabla A = 0$.

If $m = 10$, we obtain that M has at most two distinct principal curvatures at any point but, using this property, we cannot deduce that M is CMC, as in the biharmonic case. Actually, when $m = 10$, we get $\sum_{i,j=1}^{m}(\lambda_i - \lambda_j)^2 R_{ijij} = 0$ and equality in J. H. Chen's inequality (4.2), i.e., $|\nabla A|^2 = 100|\operatorname{grad} f|^2$. From these two conditions, following the proof of Theorem 4.6 in [**27**], we get $\Delta f^2 = 0$ on M. Since M is compact, we obtain that f is constant and thus $\nabla A = 0$. □

REMARK 4.32. In the proof of Theorem 4.31, we used the fact that

$$-\frac{1}{2}\sum_{i,j=1}^{m}(\lambda_i - \lambda_j)^2 R_{ijij} = \langle T, A\rangle,$$

where $T(X) = -\operatorname{trace}(RA)(\cdot, X, \cdot)$, and a Ricci commutation formula.

We end this part by presenting a Bochner type formula of independent interest (see (4.7)) that could be used in further studies on biconservative submanifolds. This formula is inspired by a similar one in [**57**]. It involves a 4-tensor Q defined on a Riemannian manifold M:

$$Q(X, Y, Z, W) = \langle Y, Z\rangle\langle X, W\rangle - \langle X, Z\rangle\langle Y, W\rangle,$$

the map

$$\sigma_{24}(X, Y, Z, W) = (X, W, Z, Y)$$

which permutes the second and fourth variables, a symmetric $(1,1)$-tensor field S, and the 1-form θ defined as the contraction $C((Q \circ \sigma_{24}) \otimes g^*, \nabla S \otimes S))$, where g denotes the metric on M and g^* is its dual.

THEOREM 4.33 ([**27**]). *Let M be a Riemannian manifold with the curvature tensor field R and consider a symmetric* $(1,1)$*-tensor field S. Then*

$$\operatorname{div}\theta = \langle T, S\rangle + |\operatorname{div} S|^2 - |\nabla S|^2 + \frac{1}{2}|W|^2, \tag{4.7}$$

where $T(X) = -\operatorname{trace}(RS)(\cdot, X, \cdot)$ *and* $W(X, Y) = (\nabla_X S)Y - (\nabla_Y S)X$.

Unlike the Simons type equations, formula (4.7) extends beyond Codazzi tensors as it involves the antisymmetric part of ∇S.

All the results presented in this section are given for compact biconservative hypersurfaces in $N^n(c)$ and have a certain rigidity. All these hypersurfaces (with additional hypotheses) are given by Theorem 4.28 and satisfy $\nabla A = 0$; in particular, they have at most two distinct principal curvatures at any point.

If we discard the hypothesis on compactness, we get many examples of biconservative hypersurfaces. This fact have been already observed in the section devoted to biconservative surfaces in $N^3(c)$. Then, in the first paper to study biconservative hypersurfaces ([**40**]), Th. Hasanis and Th. Vlachos obtained all these hypersurfaces in $\mathbb{R}^4$, divided in three distinct classes.

In [**80**], N. C. Turgay and A. Upadhyay, studied biconservative hypersurfaces in $N^{m+1}(c)$, obtaining their parametric equation. Then, they explicitly classified these hypersurfaces in $N^4(c)$, $c \neq 0$, that, in addition, have three distinct principal curvatures at any point.

For arbitrary dimensions and for $c = 0$, one obtained classification results in [**79**] and in [**58**]. In [**79**], one proved that biconservative hypersurfaces with three distinct principal curvatures are either certain rotational hypersurfaces, or certain cylinders. Then, in [**58**], we performed a detailed qualitative study of biconservative hypersurfaces that are $SO(p+1) \times SO(q+1)$-invariant in $\mathbb{R}^{p+q+2}$, or $SO(p+1)$-invariant in $\mathbb{R}^{p+2}$. This study was done in the framework of the equivariant differential geometry, by using the profile curve associated to the hypersurface in the orbit space. None of these biconservative hypersurfaces are biharmonic.

5. Open problems

Inspired by the recent result in [**61**], that shows that in $\mathbb{S}^3$ there exists an entire family of compact non-CMC biconservative surfaces, we pose the following problem.

Find examples of compact non-CMC biconservative hypersurfaces in $N^{m+1}(c)$, $m > 2$.

Next, one can see that Remark 4.30 suggests our second open problem.

Find all biconservative hypersurfaces having constant scalar curvature in space forms.

Acknowledgments

The authors would like to thank Bang-Yen Chen, Yu Fu, Matheus Vieira, Álvaro Pámpano, Hiba Bibi, and Simona Nistor for carefully reading the preliminary version of our paper and for their comments and suggestions.

References

[1] N. Aronszajn, *A unique continuation theorem for solutions of elliptic partial differential equations or inequalities of second order*, J. Math. Pures Appl. (9) **36** (1957), 235–249. MR92067

[2] J. Arroyo, O. J. Garay, and J. J. Mencía, *Closed generalized elastic curves in* $S^2(1)$, J. Geom. Phys. **48** (2003), no. 2-3, 339–353, DOI 10.1016/S0393-0440(03)00047-0. MR2007599

[3] A. Balmuş, S. Montaldo, and C. Oniciuc, *Classification results for biharmonic submanifolds in spheres*, Israel J. Math. **168** (2008), 201–220, DOI 10.1007/s11856-008-1064-4. MR2448058

[4] Adina Balmuş, Stefano Montaldo, and Cezar Oniciuc, *Biharmonic hypersurfaces in 4-dimensional space forms*, Math. Nachr. **283** (2010), no. 12, 1696–1705, DOI 10.1002/mana.200710176. MR2560665

[5] Adina Balmuş, Stefano Montaldo, and Cezar Oniciuc, *New results toward the classification of biharmonic submanifolds in* $\mathbb{S}^n$, An. Ştiinţ. Univ. "Ovidius" Constanţa Ser. Mat. **20** (2012), no. 2, 89–114, DOI 10.2478/v10309-012-0043-2. MR2945959

[6] Adina Balmuş, Stefano Montaldo, and Cezar Oniciuc, *Biharmonic PNMC submanifolds in spheres*, Ark. Mat. **51** (2013), no. 2, 197–221, DOI 10.1007/s11512-012-0169-5. MR3090194

[7] A. Balmuş and C. Oniciuc, *Biharmonic submanifolds with parallel mean curvature vector field in spheres*, J. Math. Anal. Appl. **386** (2012), no. 2, 619–630, DOI 10.1016/j.jmaa.2011.08.019. MR2834772

[8] Volker Branding and Cezar Oniciuc, *Unique continuation theorems for biharmonic maps*, Bull. Lond. Math. Soc. **51** (2019), no. 4, 603–621, DOI 10.1112/blms.12240. MR3990379

[9] article author=Bibi, Hiba, author=Loubeau, Eric, author=Oniciuc, Cezar, title=Unique continuation propery for biharmonic hypersurfaces in spheres, journal=Ann. Global Anal. Geom., volume=60, date=2021, number=4, pages=807–827, review= MR4328039, doi=10.1007/s10455-021-09801-5,

[10] R. Caddeo, S. Montaldo, and C. Oniciuc, *Biharmonic submanifolds of* S^3, Internat. J. Math. **12** (2001), no. 8, 867–876, DOI 10.1142/S0129167X01001027. MR1863283

[11] R. Caddeo, S. Montaldo, and C. Oniciuc, *Biharmonic submanifolds in spheres*, Israel J. Math. **130** (2002), 109–123, DOI 10.1007/BF02764073. MR1919374

[12] R. Caddeo, S. Montaldo, C. Oniciuc, and P. Piu, *Surfaces in three-dimensional space forms with divergence-free stress-bienergy tensor*, Ann. Mat. Pura Appl. (4) **193** (2014), no. 2, 529–550, DOI 10.1007/s10231-012-0289-3. MR3180932

[13] Shaoping Chang, *A closed hypersurface with constant scalar and mean curvatures in* S^4 *is isoparametric*, Comm. Anal. Geom. **1** (1993), no. 1, 71–100, DOI 10.4310/CAG.1993.v1.n1.a4. MR1230274

[14] Shaoping Chang, *On closed hypersurfaces of constant scalar curvatures and mean curvatures in* S^{n+1}, Pacific J. Math. **165** (1994), no. 1, 67–76. MR1285564

[15] Bang-Yen Chen, *Total mean curvature and submanifolds of finite type*, Series in Pure Mathematics, vol. 1, World Scientific Publishing Co., Singapore, 1984, DOI 10.1142/0065. MR749575

[16] Bang-Yen Chen, *Some open problems and conjectures on submanifolds of finite type*, Soochow J. Math. **17** (1991), no. 2, 169–188. MR1143504

[17] Bang-Yen Chen, *Total mean curvature and submanifolds of finite type*, 2nd ed., Series in Pure Mathematics, vol. 27, World Scientific Publishing Co. Pte. Ltd., Hackensack, NJ, 2015. With a foreword by Leopold Verstraelen. MR3362186

[18] B.-Y. Chen, *Chen's biharmonic conjecture and submanifolds with parallel normalized mean curvature vector*, Mathematics 2019, 7, 710.

[19] Bang-Yen Chen and Susumu Ishikawa, *Biharmonic surfaces in pseudo-Euclidean spaces*, Mem. Fac. Sci. Kyushu Univ. Ser. A **45** (1991), no. 2, 323–347, DOI 10.2206/kyushumfs.45.323. MR1133117

[20] Bang-yen Chen and Koichi Ogiue, *On totally real submanifolds*, Trans. Amer. Math. Soc. **193** (1974), 257–266, DOI 10.2307/1996914. MR346708

[21] Jian Hua Chen, *Compact 2-harmonic hypersurfaces in* $S^{n+1}(1)$ (Chinese, with Chinese summary), Acta Math. Sinica **36** (1993), no. 3, 341–347. MR1247088

[22] Qing Ming Cheng and Qian Rong Wan, *Hypersurfaces of space forms* $M^4(c)$ *with constant mean curvature*, Geometry and global analysis (Sendai, 1993), Tohoku Univ., Sendai, 1993, pp. 437–442. MR1361210

[23] Shiu Yuen Cheng and Shing Tung Yau, *Hypersurfaces with constant scalar curvature*, Math. Ann. **225** (1977), no. 3, 195–204, DOI 10.1007/BF01425237. MR431043

[24] S. S. Chern, M. do Carmo, and S. Kobayashi, *Minimal submanifolds of a sphere with second fundamental form of constant length*, Functional Analysis and Related Fields (Proc. Conf. for M. Stone, Univ. Chicago, Chicago, Ill., 1968), Springer, New York, 1970, pp. 59–75. MR0273546

[25] James Eells Jr. and J. H. Sampson, *Harmonic mappings of Riemannian manifolds*, Amer. J. Math. **86** (1964), 109–160, DOI 10.2307/2373037. MR164306

[26] D. Fetcu, E. Loubeau, S. Montaldo, and C. Oniciuc, *Biharmonic submanifolds of* $\mathbb{CP}^n$, Math. Z. **266** (2010), no. 3, 505–531, DOI 10.1007/s00209-009-0582-z. MR2719418

[27] Dorel Fetcu, Eric Loubeau, and Cezar Oniciuc, *Bochner-Simons formulas and the rigidity of biharmonic submanifolds*, J. Geom. Anal. **31** (2021), no. 2, 1732–1755, DOI 10.1007/s12220-019-00323-y. MR4215275

[28] Dorel Fetcu, Simona Nistor, and Cezar Oniciuc, *On biconservative surfaces in 3-dimensional space forms*, Comm. Anal. Geom. **24** (2016), no. 5, 1027–1045, DOI 10.4310/CAG.2016.v24.n5.a5. MR3622313

[29] D. Fetcu and C. Oniciuc, *Biharmonic hypersurfaces in Sasakian space forms*, Differential Geom. Appl. **27** (2009), no. 6, 713–722, DOI 10.1016/j.difgeo.2009.03.011. MR2552680

[30] Dorel Fetcu, Cezar Oniciuc, and Ana Lucia Pinheiro, *CMC biconservative surfaces in* $\mathbb{S}^n \times \mathbb{R}$ *and* $\mathbb{H}^n \times \mathbb{R}$, J. Math. Anal. Appl. **425** (2015), no. 1, 588–609, DOI 10.1016/j.jmaa.2014.12.051. MR3299681

[31] Dorel Fetcu, Cezar Oniciuc, and Harold Rosenberg, *Biharmonic submanifolds with parallel mean curvature in* $\mathbb{S}^n \times \mathbb{R}$, J. Geom. Anal. **23** (2013), no. 4, 2158–2176, DOI 10.1007/s12220-012-9323-3. MR3107694

[32] Dorel Fetcu and Ana Lucia Pinheiro, *Biharmonic surfaces with parallel mean curvature in complex space forms*, Kyoto J. Math. **55** (2015), no. 4, 837–855, DOI 10.1215/21562261-3157766. MR3479312

[33] Yu Fu, *Biharmonic hypersurfaces with three distinct principal curvatures in spheres*, Math. Nachr. **288** (2015), no. 7, 763–774, DOI 10.1002/mana.201400101. MR3345102

[34] Yu Fu, *Explicit classification of biconservative surfaces in Lorentz 3-space forms*, Ann. Mat. Pura Appl. (4) **194** (2015), no. 3, 805–822, DOI 10.1007/s10231-014-0399-1. MR3345666

[35] Yu Fu and Min-Chun Hong, *Biharmonic hypersurfaces with constant scalar curvature in space forms*, Pacific J. Math. **294** (2018), no. 2, 329–350, DOI 10.2140/pjm.2018.294.329. MR3770116

[36] Yu Fu, Min-Chun Hong, and Xin Zhan, *On Chen's biharmonic conjecture for hypersurfaces in* $\mathbb{R}^5$, Adv. Math. **383** (2021), Paper No. 107697, 28, DOI 10.1016/j.aim.2021.107697. MR4233281

[37] Yu Fu, Shun Maeta, and Ye-Lin Ou, *Biharmonic hypersurfaces in a product space* $L^m \times \mathbb{R}$, Math. Nachr. **294** (2021), no. 9, 1724–1741, DOI 10.1002/mana.201900457. MR4333889

[38] Yu Fu and Nurettin Cenk Turgay, *Complete classification of biconservative hypersurfaces with diagonalizable shape operator in the Minkowski 4-space*, Internat. J. Math. **27** (2016), no. 5, 1650041, 17, DOI 10.1142/S0129167X16500415. MR3498113

[39] Zhida Guan, Haizhong Li, and Luc Vrancken, *Four dimensional biharmonic hypersurfaces in nonzero space forms have constant mean curvature*, J. Geom. Phys. **160** (2021), Paper No. 103984, 15, DOI 10.1016/j.geomphys.2020.103984. MR4167191

[40] Th. Hasanis and Th. Vlachos, *Hypersurfaces in* E^4 *with harmonic mean curvature vector field*, Math. Nachr. **172** (1995), 145–169, DOI 10.1002/mana.19951720112. MR1330627

[41] David A. Hoffman, *Surfaces of constant mean curvature in manifolds of constant curvature*, J. Differential Geometry **8** (1973), 161–176. MR390973

[42] Toshiyuki Ichiyama, Jun-ichi Inoguchi, and Hajime Urakawa, *Bi-harmonic maps and bi-Yang-Mills fields*, Note Mat. **28** (2009), no. [2008 on verso], suppl. 1, 233–275. MR2640583

[43] Toshiyuki Ichiyama, Jun-ichi Inoguchi, and Hajime Urakawa, *Classifications and isolation phenomena of bi-harmonic maps and bi-Yang-Mills fields*, Note Mat. **30** (2010), no. 2, 15–48, DOI 10.1285/i15900932v30n2p15. MR2943022

[44] Jun-ichi Inoguchi and Toru Sasahara, *Biharmonic hypersurfaces in Riemannian symmetric spaces I*, Hiroshima Math. J. **46** (2016), no. 1, 97–121. MR3482341

[45] Jun-ichi Inoguchi and Toru Sasahara, *Biharmonic hypersurfaces in Riemannian symmetric spaces II*, Hiroshima Math. J. **47** (2017), no. 3, 349–378. MR3719448

[46] Guo Ying Jiang, *2-harmonic maps and their first and second variational formulas* (Chinese), Chinese Ann. Math. Ser. A **7** (1986), no. 4, 389–402. An English summary appears in Chinese Ann. Math. Ser. B **7** (1986), no. 4, 523. MR886529

[47] Guo Ying Jiang, *The conservation law for 2-harmonic maps between Riemannian manifolds* (Chinese), Acta Math. Sinica **30** (1987), no. 2, 220–225. MR891928

[48] Guo Ying Jiang, *Some nonexistence theorems on 2-harmonic and isometric immersions in Euclidean space* (Chinese), Chinese Ann. Math. Ser. A **8** (1987), no. 3, 377–383. An English summary appears in Chinese Ann. Math. Ser. B **8** (1987), no. 3, 389. MR924896

[49] E. Loubeau, S. Montaldo, and C. Oniciuc, *The stress-energy tensor for biharmonic maps*, Math. Z. **259** (2008), no. 3, 503–524, DOI 10.1007/s00209-007-0236-y. MR2395125

[50] Eric Loubeau and Cezar Oniciuc, *Biharmonic surfaces of constant mean curvature*, Pacific J. Math. **271** (2014), no. 1, 213–230, DOI 10.2140/pjm.2014.271.213. MR3259766

[51] Eric Loubeau and Cezar Oniciuc, *Constant mean curvature proper-biharmonic surfaces of constant Gaussian curvature in spheres*, J. Math. Soc. Japan **68** (2016), no. 3, 997–1024, DOI 10.2969/jmsj/06830997. MR3523535

[52] Yong Luo and Shun Maeta, *Biharmonic hypersurfaces in a sphere*, Proc. Amer. Math. Soc. **145** (2017), no. 7, 3109–3116, DOI 10.1090/proc/13320. MR3637957

[53] Shun Maeta, *Biharmonic hypersurfaces with bounded mean curvature*, Proc. Amer. Math. Soc. **145** (2017), no. 4, 1773–1779, DOI 10.1090/proc/13335. MR3601567

[54] Shun Maeta and Ye-Lin Ou, *Some classifications of biharmonic hypersurfaces with constant scalar curvature*, Pacific J. Math. **306** (2020), no. 1, 281–290, DOI 10.2140/pjm.2020.306.281. MR4109916

[55] F. Manfio, N. C. Turgay, and A. Upadhyay, *Biconservative submanifolds in* $\mathbb{S}^n\times\mathbb{R}$ *and* $\mathbb{H}^n\times\mathbb{R}$, J. Geom. Anal. **29** (2019), no. 1, 283–298, DOI 10.1007/s12220-018-9990-9. MR3897013

[56] Yoichiro Miyata, *2-type surfaces of constant curvature in* S^n, Tokyo J. Math. **11** (1988), no. 1, 157–204, DOI 10.3836/tjm/1270134269. MR947954

[57] Ngaiming Mok, Yum Tong Siu, and Sai-Kee Yeung, *Geometric superrigidity*, Invent. Math. **113** (1993), no. 1, 57–83, DOI 10.1007/BF01244302. MR1223224

[58] S. Montaldo, C. Oniciuc, and A. Ratto, *Proper biconservative immersions into the Euclidean space*, Ann. Mat. Pura Appl. (4) **195** (2016), no. 2, 403–422, DOI 10.1007/s10231-014-0469-4. MR3476680

[59] S. Montaldo, C. Oniciuc, and A. Ratto, *Biconservative surfaces*, J. Geom. Anal. **26** (2016), no. 1, 313–329, DOI 10.1007/s12220-014-9551-9. MR3441516

[60] Stefano Montaldo, Irene I. Onnis, and Apoena Passos Passamani, *Biconservative surfaces in BCV-spaces*, Math. Nachr. **290** (2017), no. 16, 2661–2672, DOI 10.1002/mana.201600394. MR3722502

[61] S. Montaldo, A. Pámpano, *On the existence of closed biconservative surfaces in space forms*, Comm. Anal. Geom., to appear.

[62] Simona Nistor, *Complete biconservative surfaces in* $\mathbb{R}^3$ *and* $\mathbb{S}^3$, J. Geom. Phys. **110** (2016), 130–153, DOI 10.1016/j.geomphys.2016.07.013. MR3566105

[63] Simona Nistor, *On biconservative surfaces.* part B, Differential Geom. Appl. **54** (2017), no. part B, 490–502, DOI 10.1016/j.difgeo.2017.08.003. MR3693945

[64] S. Nistor, *Biharmonicity and biconservativity topics in the theory of submanifolds*, PhD Thesis, 2017.

[65] Simona Nistor and Cezar Oniciuc, *Complete biconservative surfaces in the hyperbolic space* $\mathbb{H}^3$, Nonlinear Anal. **198** (2020), 111860, 29, DOI 10.1016/j.na.2020.111860. MR4081860

[66] S. Nistor, C. Oniciuc, *On the uniqueness of complete biconservative surfaces in 3-dimensional space forms*, Ann. Sc. Norm. Super. Pisa Cl. Sci., to appear. DOI 10.2422/2036-2145.202010_037

[67] Katsumi Nomizu and Brian Smyth, *A formula of Simons' type and hypersurfaces with constant mean curvature*, J. Differential Geometry **3** (1969), 367–377. MR266109

[68] Shinji Ohno, Takashi Sakai, and Hajime Urakawa, *Biharmonic homogeneous hypersurfaces in compact symmetric spaces*, Differential Geom. Appl. **43** (2015), 155–179, DOI 10.1016/j.difgeo.2015.09.005. MR3421882

[69] C. Oniciuc, *Biharmonic maps between Riemannian manifolds*, An. Ştiinţ. Univ. Al. I. Cuza Iaşi. Mat. (N.S.) **48** (2002), no. 2, 237–248 (2003). MR2004799

[70] Cezar Oniciuc, *Tangenţă şi proprietăţi de armonicitate* (Romanian), DGDS. Differential Geometry—Dynamical Systems. Monographs, vol. 3, Geometry Balkan Press, Bucharest, 2003. Dissertation, Universitatea "Al. I. Cuza" Iaşi, Iaşi, 2002. MR2159756

[71] C. Oniciuc, *Biharmonic submanifolds in space forms*, Habilitation Thesis, 2012, `www.researchgate.net`, DOI:10.13140/2.1.4980.5605.

[72] Ye-Lin Ou, *Biharmonic hypersurfaces in Riemannian manifolds*, Pacific J. Math. **248** (2010), no. 1, 217–232, DOI 10.2140/pjm.2010.248.217. MR2734173

[73] Ye-Lin Ou and Bang-Yen Chen, *Biharmonic submanifolds and biharmonic maps in Riemannian geometry*, World Scientific Publishing Co. Pte. Ltd., Hackensack, NJ, 2020. MR4265170

[74] Toru Sasahara, *Surfaces in Euclidean 3-space whose normal bundles are tangentially biharmonic*, Arch. Math. (Basel) **99** (2012), no. 3, 281–287, DOI 10.1007/s00013-012-0410-2. MR2969034

[75] Toru Sasahara, *Quasi-biharmonic Lagrangian surfaces in Lorentzian complex space forms*, Ann. Mat. Pura Appl. (4) **192** (2013), no. 2, 191–201, DOI 10.1007/s10231-011-0218-x. MR3035135

[76] Toru Sasahara, *Tangentially biharmonic Lagrangian H-umbilical submanifolds in complex space forms*, Abh. Math. Semin. Univ. Hambg. **85** (2015), no. 2, 107–123, DOI 10.1007/s12188-015-0110-5. MR3416177

[77] Toru Sasahara, *Classification of biharmonic C-parallel Legendrian submanifolds in 7-dimensional Sasakian space forms*, Tohoku Math. J. (2) **71** (2019), no. 1, 157–169, DOI 10.2748/tmj/1552100448. MR3920796

[78] Toru Sasahara, *Classification theorems for biharmonic real hypersurfaces in a complex projective space*, Results Math. **74** (2019), no. 4, Paper No. 136, 10, DOI 10.1007/s00025-019-1062-3. MR3978960

[79] Nurettin Cenk Turgay, *H-hypersurfaces with three distinct principal curvatures in the Euclidean spaces*, Ann. Mat. Pura Appl. (4) **194** (2015), no. 6, 1795–1807, DOI 10.1007/s10231-014-0445-z. MR3419918

[80] Nurettin Cenk Turgay and Abhitosh Upadhyay, *On biconservative hypersurfaces in 4-dimensional Riemannian space forms*, Math. Nachr. **292** (2019), no. 4, 905–921, DOI 10.1002/mana.201700328. MR3937626

[81] M. Vieira, *Biharmonic hypersurfaces in hemispheres*, Preprint 2020, arXiv:2008.08274.

[82] Xian Feng Wang and Lan Wu, *Proper biharmonic submanifolds in a sphere*, Acta Math. Sin. (Engl. Ser.) **28** (2012), no. 1, 205–218, DOI 10.1007/s10114-012-9018-5. MR2863761

[83] Rüya Yeğin Şen and Nurettin Cenk Turgay, *On biconservative surfaces in 4-dimensional Euclidean space*, J. Math. Anal. Appl. **460** (2018), no. 2, 565–581, DOI 10.1016/j.jmaa.2017.12.009. MR3759059

Department of Mathematics and Informatics, Gh. Asachi Technical University of Iasi, Bd. Carol I, 11, 700506 Iasi, Romania
Email address: dorel.fetcu@etti.tuiasi.ro

Faculty of Mathematics, Al. I. Cuza University of Iasi, Bd. Carol I, 11, 700506 Iasi, Romania
Email address: oniciucc@uaic.ro

Contemporary Mathematics
Volume **777**, 2022
https://doi.org/10.1090/conm/777/15629

Recent progress of biharmonic hypersurfaces in space forms

Yu Fu, Dan Yang, and Xin Zhan

Dedicated to the memory of Prof. Tadashi Nagano

ABSTRACT. In the past two decades, biharmonic submanifolds have attracted much attention from mathematicians all over the world. In particular, concerning Chen's conjecture and the generalized Chen's conjecture, many meaningful results have been obtained. In this survey paper, we will restrict our attention to biharmonic hypersurfaces in the real space forms. In the first three parts, we give a short survey on some new developments on biharmonic hypersurfaces in Euclidean spaces, hyperbolic spaces and Euclidean spheres, respectively. In the last section, we outline the proof of Chen's conjecture for hypersurfaces of $\mathbb{R}^5$ [Adv. Math, 383 (2021), Paper No. 107697, 28] and point out the potential importance of the method and the key techniques used in the proof for further studies on biharmonic hypersurfaces in space forms.

1. The original Chen's conjecture

The concept of biharmonic maps between Riemannian manifolds was introduced by Eells and Lemaire [**18, 19**] to study $k-$harmonic maps. The so-called biharmonic map ϕ between an n-dimensional Riemannian manifold (M^n, g) and an m-dimensional Riemannian manifold (N^m, h) is a critical point of the bienergy functional

$$E_2(\phi) = \frac{1}{2}\int_M |\tau(\phi)|^2 dv_g,$$

where $\tau(\phi) = \text{trace}\nabla d\phi$ is the tension field of ϕ that vanishes for a harmonic map. We use the rough Laplacian defined on the set of all sections in $\phi^{-1}(TN)$ as

$$\Delta = -\text{trace}\,(\nabla^\phi)^2 = -\text{trace}\,(\nabla^\phi\nabla^\phi - \nabla^\phi_\nabla).$$

Then the Euler-Lagrange equation associated to the bienergy is stated as

$$\tau_2(\phi) = -\Delta\tau(\phi) - \text{trace}\,R^N(d\phi, \tau(\phi))d\phi = 0, \tag{1.1}$$

where $\tau_2(\phi)$ is the bitension field of ϕ, and $R^N(X,Y) = \nabla_X\nabla_Y - \nabla_Y\nabla_X - \nabla_{[X,Y]}$ is the curvature tensor of N^m (e.g.[**29**],[**30**]).

2020 *Mathematics Subject Classification.* Primary 53C40, 58E20; Secondary 53C42.

Key words and phrases. Minimal submanifolds, biharmonic hypersurfaces, Chen's conjecture.

The first author was supported by Liaoning Provincial Science and Technology Department Project (No.2020-MS-340) and Liaoning BaiQianWan Talents Program. The second author was supported by the NSFC (No.11801246) and the General Project for Department of Liaoning Education (No.LJC201901). The third author was supported by the NSFC (No. 12101083) and the Natural Science Foundation of Jiangsu Province (No. BK20210936).

Based on the equation (1.1), we say that M^n is a biharmonic submanifold of N^m if and only if a fourth order semi-linear elliptic equation holds:

$$\Delta \overrightarrow{H} + \operatorname{trace} R^N(d\phi, \overrightarrow{H})\, d\phi = 0, \tag{1.2}$$

where $\overrightarrow{H}$ is the mean curvature vector field.

Quite evidently, every minimal submanifold is a trivial biharmonic one. A submanifold M^n is called proper biharmonic if M^n is a non-minimal (non-harmonic) biharmonic submanifold. For the Euclidean ambient space, surprisingly, B. Y. Chen [**11**] and G. Y. Jiang [**30**] proved independently that every biharmonic surface in $\mathbb{R}^3$ must be minimal. Later, I. Dimitrić [**16, 17**] confirmed that there exist no proper biharmonic submanifolds for the following special cases:

- Curves in $\mathbb{R}^m$;
- Submanifolds in $\mathbb{R}^m$ with constant mean curvature;
- Submanifolds of finite type in $\mathbb{R}^m$;
- Pseudo-umbilical submanifolds M^n in $\mathbb{R}^m$ satisfying $n \neq 4$;
- Hypersurfaces in $\mathbb{R}^m$ with at most two distinct principal curvatures.

Based on the above results, B. Y. Chen [**11**] in 1991 made the following important conjecture:

Chen's conjecture. *Any biharmonic submanifold in $\mathbb{R}^m$ $(m \geq 3)$ is minimal.*

In the past thirty years, Chen's conjecture has attracted much attentions of many geometers and some important progress have been made ([**1**], [**2**], [**13**], and the recent book [**43**] for a more detailed account). This problem is still far from being solved. As the most fundamental case of Chen's conjecture, one would ask whether Chen's conjecture is true for hypersurfaces of Euclidean space $\mathbb{R}^m$ $(m \geq 5)$. Hence, it is a challenging and important problem to solve the following case:

Chen's conjecture for biharmonic hypersurface. *Every biharmonic hypersurface in $\mathbb{R}^m$ $(m \geq 5)$ is minimal.*

Now we list some recent developments which give affirmative partial answers to the Chen's conjecture for hypersurfaces in the following cases :

- Hypersurfaces in $\mathbb{R}^4$ (Hasanis and Vlachos [**27**] in 1995; see also Defever's result [**15**] with another method);
- $\delta(2)$-ideal and $\delta(3)$-ideal hypersurfaces in $\mathbb{R}^m$ (Chen and Munteanu [**14**] in 2013);
- Weakly convex hypersurfaces in $\mathbb{R}^m$ (Y. Luo [**33**] in 2014);
- Hypersurfaces with at most three distinct principle curvatures in $\mathbb{R}^m$ (Fu [**20, 21**] in 2015);
- Invariant hypersurfaces of cohomogeneity one in $\mathbb{R}^m$ (Montaldo, Oniciuc and Ratto [**36**] in 2016);
- Hypersurfaces with constant scalar curvature in $\mathbb{R}^5$ (Fu [**23**]);
- Generic hypersurfaces with irreducible principal curvature vector in $\mathbb{R}^m$ (Koiso and Urakawa [**31**] in 2018);
- Hypersurfaces with constant scalar curvature and at most six distinct principal curvatures in $\mathbb{R}^m$ (Fu [**24**] in 2018);
- Hypersurfaces in $\mathbb{R}^5$ (Fu, Hong and Zhan [**25**] in 2020).

It is worth pointing out that Fu-Hong-Zhan [**25**] confirmed Chen's conjecture for hypersurfaces in $\mathbb{R}^5$ without any assumption on curvatures or principal curvatures. Also, the proof in [**25**] provided new algebraic equation techniques and ideas for studying biharmonic hypesurfaces. In Section 4, we will give a sketch of the proof.

2. The generalized Chen's conjecture

After realizing the interesting properties and corresponding phenomenon of biharmonic submanifolds in non-flat Riemannian space forms, in 2001 Caddeo, Montaldo and Oniciuc [**9**] proposed the so-called generalized Chen's conjecture as generalization of the original Chen's conjecture.

GENERALIZED CHEN'S CONJECTURE. *There exist no proper biharmonic submanifolds in a Riemannian manifold with non-positive sectional curvature.*

Partial non-existence results on this problem, obtained in the cases listed below, strongly support the Generalized Chen's conjecture:

- Compact submanifolds of non-positive curved spaces (Jiang [**29**] in 1986);
- Curves in $\mathbb{H}^m$ (Caddeo, Montaldo and Oniciuc [**10**] in 2002);
- Submanifolds in $\mathbb{H}^3$ (Caddeo, Montaldo and Oniciuc [**10**] in 2002);
- Pseudo-umbilical submanifolds M^n of $\mathbb{H}^m$ satisfying $n \neq 4$ (Caddeo, Montaldo and Oniciuc [**10**] in 2002);
- Submanifolds of non-positive curved spaces with constant mean curvature (Oniciuc [**39**] in 2002);
- Submanifolds of non-positive curved spaces with finite total mean curvature (Nakauchi and Urakawa [**38**] in 2013).

The most interesting thing for us is whether the generalized Chen's conjecture is true for biharmonic hypersurfaces in a Riemannian manifold with non-positive sectional curvature. Besides the above results, there are also some recent developments on biharmonic hypersurfaces of non-positive curved spaces.

- Surfaces in $\mathbb{H}^3$ (Caddeo, Montaldo and Oniciuc [**10**] in 2002);
- Hypersurfaces in $\mathbb{H}^m$ with at most two distinct principal curvatures (Balmuş, Montaldo and Oniciuc [**3**] in 2008);
- Hypersurfaces in $\mathbb{H}^4$ (Balmuş, Montaldo and Oniciuc [**5**] in 2010);
- Totally umbilical hypersurfaces of Einstein spaces (Ou [**41**] in 2010);
- Hypersurfaces of non-positive Ricci curved spaces with finite total mean curvature (Nakauchi and Urakawa [**37**] in 2011);
- Hypersurfaces of non-positive Ricci curved spaces which either are compact or have constant mean curvature (Oniciuc [**39**] in 2002);
- Hypersurfaces in $\mathbb{H}^m$ with at most three distinct principal curvatures(Fu [**22**] in 2015);
- Hypersurfaces in $\mathbb{H}^5$ with constant scalar curvature (Fu [**23**] in 2015);
- Hypersurfaces with constant scalar curvature and at most six distinct principal curvatures in $\mathbb{H}^m$ (Fu and Hong [**24**] in 2018);
- Hypersurfaces in $\mathbb{H}^5$ (Guan, Li and Vrancken [**26**] in 2020).

Based on the above facts, we have no reason to doubt that the generalized Chen's conjecture is false. However, Ou and Tang in 2012 [**44**] showed that the generalized Chen's conjecture fails to hold via constructing a family of proper biharmonic hypersurfaces in a conformally flat 5-space with non-constant negative curved space.

As the ambient space constructed in Ou-Tang's counter examples have non-constant negative curvature, it would be interesting to know whether the generalized Chen's conjecture holds true for an ambient space of constant negative curvature. In particular, we have:

IMPROVED GENERALIZED CHEN'S CONJECTURE. *Every biharmonic hypersurface in $\mathbb{H}^m$ is minimal.*

3. Biharmonic submanifolds in Euclidean sphere

In this section, we shall pay our attention on the classification of biharmonic hypersurfaces in the Euclidean sphere $\mathbb{S}^{n+1}$. Contrary to the Euclidean space $\mathbb{R}^m$ and the hyperbolic space $\mathbb{H}^m$, there exist proper biharmonic submanifolds in a unit sphere $\mathbb{S}^m$. In fact, two classical examples of proper biharmonic hypersurfaces in spheres were obtained.

EXAMPLE 3.1. The first example of proper biharmonic hypersurface in $\mathbb{S}^{n+1}$ is the generalized Clifford torus

$$\mathbb{S}^k\left(\frac{1}{\sqrt{2}}\right)\times\mathbb{S}^{n-k}\left(\frac{1}{\sqrt{2}}\right),\quad 0\leq k\leq n,\ k\neq\frac{n}{2},$$

which was firstly discovered by G. Y. Jiang in [**28**].

EXAMPLE 3.2. The another example is $\mathbb{S}^n\left(\frac{1}{\sqrt{2}}\right)$–a biharmonic hypersurface in $\mathbb{S}^{n+1}$, which was proved by Caddeo, Montaldo and Oniciuc [**9**].

Balmuş, Montaldo and Oniciuc [**3,4**] made the following two conjectures:

BMO CONJECTURE 1. *Every proper biharmonic submanifold in $\mathbb{S}^m$ has constant mean curvature.*

BMO CONJECTURE 2. *The only proper biharmonic hypersurfaces in $\mathbb{S}^{n+1}$ must be the open parts of $\mathbb{S}^n\left(1/\sqrt{2}\right)$ or $\mathbb{S}^k\left(1/\sqrt{2}\right)\times\mathbb{S}^{n-k}\left(1/\sqrt{2}\right)$ with $0\leq k\leq n$, $k\neq\frac{n}{2}$.*

There are some recent developments on the above two conjectures.

- The only proper biharmonic surface M^2 in $\mathbb{S}^3$ has constant mean curvature, and M^2 is an open part of $\mathbb{S}^2\left(\frac{1}{\sqrt{2}}\right)$ (Caddeo, Montaldo and Oniciuc [**9**] in 2001);
- The only compact proper biharmonic hypersurface M^3 in $\mathbb{S}^4$ has constant mean curvature, and M^3 is either an open part of the hypersphere $\mathbb{S}^3\left(\frac{1}{\sqrt{2}}\right)$ or the generalized Clifford torus $\mathbb{S}^1\left(\frac{1}{\sqrt{2}}\right)\times\mathbb{S}^2\left(\frac{1}{\sqrt{2}}\right)$ (Balmuş, Montaldo and Oniciuc [**5**] in 2010);
- Any proper biharmonic hypersurface M^n in $\mathbb{S}^{n+1}$ with at most two distinct principal curvatures has constant mean curvature, and M^n is either an open part of $\mathbb{S}^n\left(\frac{1}{\sqrt{2}}\right)$ or $\mathbb{S}^{n_1}\left(\frac{1}{\sqrt{2}}\right)\times\mathbb{S}^{n_2}\left(\frac{1}{\sqrt{2}}\right)$ with $n_1+n_2=n$ and $n_1\neq n_2$ (Balmuş, Montaldo and Oniciuc [**3**] in 2008);

- There exist no compact proper biharmonic hypersurface M^n in $\mathbb{S}^{n+1}$ with constant mean curvature and three distinct principal curvatures (Balmuş, Montaldo and Oniciuc [**5**] in 2010);
- There exists no compact proper biharmonic hypersurface M^n in $\mathbb{S}^{n+1}$ with three distinct principal curvatures everywhere (Fu [**22**] in 2015);
- Proper biharmonic hypersurfaces in $\mathbb{S}^5$ with constant scalar curvature must have constant mean curvature (Fu [**23**] in 2015);
- Proper biharmonic hypersurfaces in $\mathbb{S}^{n+1}$ with constant scalar curvature and at most six distinct principal curvatures have constant mean curvature (Fu and Hong [**24**] in 2018);
- Complete biharmonic hypersurface in $\mathbb{S}^{n+1}$ have constant mean curvature under the following assumptions: (i) the mean curvature function $H > 0$, (ii) the squared norm of the shape operator $|A|^2 \leq m$, and (iii) $\ln^{(k)} \frac{e^{(k)}}{H} \in \mathrm{L}^p(M)$ for some $p \in (0, \infty)$ (Luo and Maeta [**34**] in 2017);
- Compact biharmonic hypersurfaces in $\mathbb{S}^{n+1}$ with constant scalar curvature have constant mean curvature (Maeta and Ou in [**35**] in 2020);
- Proper biharmonic hypersurfaces in $\mathbb{S}^{n+1}$ have constant mean curvature assuming that there exists a non-negative function h on M such that $|\mathrm{grad}\,|A|^2| \leq h|\mathrm{grad}\,H|$ on M and $\mathrm{grad}\,H$ vanishes on a non-empty open connected subset of M (Bibi, Loubeau and Oniciuc [**8**] in 2020);
- Any proper biharmonic hypersurface M^4 of $\mathbb{S}^5$ have constant mean curvature (Guan, Li and Vrancken [**26**] in 2020).

4. Biharmonic hypersurfaces in $\mathbb{R}^5$

The case of hypersurfaces for Chen's conjecture is fundamentally important. After Chen in 1985 and Jiang's work in 1987 of $n = 2$ independently, Hasanis and Vlachos's contribution of $n = 3$ in 1995 on Chen's conjecture, very recently, Fu, Hong and Zhan [**25**] proved the following:

THEOREM 4.1 ([**25**]). *Every biharmonic hypersurface in Euclidean space $\mathbb{R}^5$ is minimal.*

This result supports strongly Chen's conjecture for biharmonic hypersurfaces. Moreover, the proof in [**25**] contains some new algebraic techniques of differential equations to deal with biharmonic hypersurfaces in space forms.

Recall that the biharmonic condition (1.2) for a hypersurface in $\mathbb{R}^{n+1}$ is equivalent to two equations:

$$\begin{cases} \Delta H + H \mathrm{trace}\, A^2 = 0, \\ 2A\, \mathrm{grad} H + nH \mathrm{grad}\, H = 0, \end{cases} \tag{4.1}$$

where A is the shape operator of M^n.

For simplicity, we assume that M is oriented, so A and H are defined on the whole M. Assume that the mean curvature H is non-constant and we will derive a contradiction. We put $U = \{p \in M^n : \mathrm{grad}\, H \neq 0 \text{ at } p\}$. Then U is an open subset on M^n. We can also assume that on an open (connected) subset of U, denoted again by U, the multiplicities of the distinct principal curvatures are constant and $H \neq 0$ at any point. Therefore, on U, the principal curvatures are smooth functions and A can be diagonalized by smooth vector fields.

The second equation of (4.1) shows that $\operatorname{grad} H$ is an eigenvector of the shape operator A with the corresponding principal curvature $\lambda_1 = -\frac{nH}{2}$. Note that $\sum_{i=1}^n \lambda_i = nH$ and $\lambda_1 = -\frac{nH}{2}$. Then we have $\sum_{i=2}^n \lambda_i = -3\lambda_1$. As $\operatorname{grad} H = \sum_{i=1}^n e_i(H)e_i$ and e_1 is parallel to the direction of $\operatorname{grad} H$, we have that $e_1(H) \neq 0$, at any point of U, $e_i(H) = 0$, on U, $2 \leq i \leq n$ and hence

$$e_1(\lambda_1) \neq 0, \quad e_i(\lambda_1) = 0, \quad 2 \leq i \leq n. \tag{4.2}$$

In this paper, by $e_1(\lambda_1) \neq 0$ we mean that the function $e_1(\lambda_1)$ is different from 0 at any point of U, and by $e_i(\lambda_1) = 0$ we mean that the function vanishes at any point of U.

Set

$$\nabla_{e_i} e_j = \sum_{k=1}^n \omega_{ij}^k e_k, \ 1 \leq i, j \leq n.$$

Using the Gauss and Codazzi equations, biharmonic equations (4.1) can be reduced into a system of $2n - 1$ differential equations as follows:

LEMMA 4.2 ([**24**]). *The principal curvature functions λ_i and the coefficients of connection ω_{ii}^1 $(i = 2, \ldots, n)$ satisfy on U the following differential equations:*

$$e_1 e_1(\lambda_1) = e_1(\lambda_1)\Big(\sum_{i=2}^n \omega_{ii}^1\Big) + \lambda_1 S, \tag{4.3}$$

$$e_1(\lambda_i) = \lambda_i \omega_{ii}^1 - \lambda_1 \omega_{ii}^1, \tag{4.4}$$

$$e_1(\omega_{ii}^1) = (\omega_{ii}^1)^2 + \lambda_1 \lambda_i, \tag{4.5}$$

where $\lambda_1 = -nH/2$, S is the squared norm of the second fundamental form h of M and $e_1 = \operatorname{grad} H/|\operatorname{grad} H|$.

REMARK 4.3. Notice that (4.3)-(4.5) is an over-determined system of equations. To find the complete solution of system of differential equations (4.3)-(4.5) is very crucial for solving Chen's conjecture.

Let us consider biharmonic hypersurfaces in $\mathbb{R}^5$. Set

$$f_k = (\omega_{22}^1)^k + (\omega_{33}^1)^k + (\omega_{44}^1)^k, \text{ for } k = 1, \ldots, 5.$$

For simplicity, we write

$$\lambda = \lambda_1, \quad f_1 = T, \quad T' = e_1(T), \quad T'' = e_1 e_1(T),$$
$$T''' = e_1 e_1 e_1(T), \quad T'''' = e_1 e_1 e_1 e_1(T).$$

The proof of Theorem 4.1 can be outlined in two major steps as follows:

STEP 1. The principal curvatures λ_i satisfy $e_j(\lambda_i) = 0$ on U with $1 \leq i \leq 4$ and $2 \leq j \leq 4$.

First, we have following algebraic relations involving f_k.

LEMMA 4.4 ([**25**]). *With the notions f_k, the following two relations hold*

$$f_1^4 - 6f_1^2 f_2 + 3f_2^2 + 8f_1 f_3 - 6f_4 = 0, \tag{4.6}$$

$$f_1^5 - 5f_1^3 f_2 + 5f_1^2 f_3 + 5f_2 f_3 - 6f_5 = 0. \tag{4.7}$$

Moreover, after a direct computation we find that f_k can be expressed in the terms of λ, T and their higher-order derivative forms. This enables us to solve the Step 1 for hypersurfaces in $\mathbb{R}^5$.

LEMMA 4.5 ([**25**]). *f_1, f_2, f_3, f_4 and f_5 can be written as*

$$\begin{cases} f_1 = T, \\ f_2 = T' + 3\lambda^2, \\ f_3 = \frac{1}{2}T'' - \lambda^2 T + 6\lambda\lambda', \\ f_4 = \frac{1}{6}T''' - \frac{4}{3}\lambda^2 T' - \frac{5}{3}\lambda\lambda' T + 2\lambda'^2 + 4\lambda\lambda'' - 2\lambda^4, \\ f_5 = \frac{1}{24}T'''' - \frac{5}{6}\lambda^2 T'' - \frac{25}{12}\lambda\lambda' T' - \frac{1}{12}(13\lambda\lambda'' + \lambda'^2 - 12\lambda^4)T \\ \quad +2\lambda\lambda''' + \frac{5}{3}\lambda'\lambda'' - \frac{26}{3}\lambda^3\lambda'. \end{cases} \tag{4.8}$$

Substituting (4.8) into (4.6) and (4.7), we can obtain two entirely different differential equations with respect to T:

$$\begin{aligned} & - T''' + 4TT'' + 3T'^2 + (-6T^2 + 26\lambda^2)T' + (T^4 - 26\lambda^2 T^2 \\ & + 58\lambda\lambda' T) + 39\lambda^4 - 24\lambda\lambda'' - 12\lambda'^2 = 0, \end{aligned} \tag{4.9}$$

$$\begin{aligned} & - T'''' + 10T'T'' + (10T^2 + 50\lambda^2)T'' - (20T^3 + 20\lambda^2 T \\ & - 170\lambda\lambda')T' + (4T^5 - 80\lambda^2 T^3 + 120\lambda\lambda' T^2 - 84\lambda^4 T + 26\lambda\lambda'' T \\ & + 2\lambda'^2 T) + 568\lambda^3\lambda' - 48\lambda\lambda''' - 40\lambda'\lambda'' = 0. \end{aligned} \tag{4.10}$$

We can eliminate the terms of T'''', T''', T'', T' term by term and derive a non-trivial polynomial equation of T with the coefficients depending only on λ and its higher derivative forms. Hence

LEMMA 4.6 ([**25**]). *The function T satisfies $e_j(T) = 0$ on U with $2 \leq j \leq 4$.*

According to Lemmas 4.5 and 4.6, (4.8) implies that f_k for $k = 1, \dots, 5$ satisfy $e_i(f_k) = 0$ for $2 \leq i \leq 4$. Furthermore, taking into account the determinant of the coefficient matrix, we differentiate both sides of equations $f_k = \sum_{i=2}^{4} (\omega_{ii}^1)^k$ for $k = 1, 2, 3$ with respect to e_i $(2 \leq i \leq 4)$. According to Cramer's rule in linear algebra, one gets

$$e_i(\omega_{22}^1) = e_i(\omega_{33}^1) = e_i(\omega_{44}^1) = 0.$$

Moreover, taking into account into the principal curvature functions λ_i, one completes the proof of the Step 1.

STEP 2. To derive a contradiction with the assumption that H is non-constant.

We first recall some basic relations concerning the coefficients of connection and principal curvature functions (c.f.[**24**]).

LEMMA 4.7 ([**24**], [**25**]). *For three distinct principal curvatures λ_2, λ_3 and λ_4, we have the following relations on U:*

$$\omega_{23}^4(\lambda_3 - \lambda_4) = \omega_{32}^4(\lambda_2 - \lambda_4) = \omega_{43}^2(\lambda_3 - \lambda_2), \tag{4.11}$$

$$\omega_{23}^4\omega_{32}^4 + \omega_{34}^2\omega_{43}^2 + \omega_{24}^3\omega_{42}^3 = 0, \tag{4.12}$$

$$\omega_{23}^4(\omega_{33}^1 - \omega_{44}^1) = \omega_{32}^4(\omega_{22}^1 - \omega_{44}^1) = \omega_{43}^2(\omega_{33}^1 - \omega_{22}^1). \tag{4.13}$$

LEMMA 4.8 ([**24**], [**25**]). *The following relations hold on U*

$$\omega_{22}^1\omega_{33}^1 - 2\omega_{23}^4\omega_{32}^4 = -\lambda_2\lambda_3, \tag{4.14}$$

$$\omega_{22}^1\omega_{44}^1 - 2\omega_{24}^3\omega_{42}^3 = -\lambda_2\lambda_4, \tag{4.15}$$

$$\omega_{33}^1\omega_{44}^1 - 2\omega_{34}^2\omega_{43}^2 = -\lambda_3\lambda_4. \tag{4.16}$$

To finish the proof, we analyse the situation when the functions $\omega_{23}^4, \omega_{32}^4$, and ω_{43}^2 vanish or not. One situation is when all of them vanish on U. Then, we have the cases when at least one does not vanish. In the case when exactly one does not vanish, by restricting U, we can assume that it does not vanish at any point of an open subset denoted again by U. When exactly two functions do not vanish, the interesting case is that when both do not vanish at any point of the same open subset, otherwise we are basically in the previous situation. And so on. In fact, we only need to consider two case:

CASE A. $\omega_{23}^4 \neq 0$, $\omega_{32}^4 \neq 0$, and $\omega_{43}^2 \neq 0$ on U. Since ω_{ii}^1 and λ_i are smooth functions satisfying $e_i(\omega_{jj}^1) = e_i(\lambda_j) = 0$ on U for $2 \leq i, j \leq 4$, it follows that (4.11) and (4.13) reduce to

$$\frac{\omega_{33}^1 - \omega_{44}^1}{\lambda_3 - \lambda_4} = \frac{\omega_{33}^1 - \omega_{22}^1}{\lambda_3 - \lambda_2} = \frac{\omega_{44}^1 - \omega_{22}^1}{\lambda_4 - \lambda_2} := \alpha$$

for a smooth function α satisfying $e_i(\alpha) = 0$ for $2 \leq i \leq 4$. Hence there exists another smooth function β satisfying $e_i(\beta) = 0$ such that

$$\omega_{ii}^1 = \alpha\lambda_i + \beta \tag{4.17}$$

for $i = 2, 3, 4$. Differentiating with respect to e_1 on both sides of equation (4.17), using (4.4) and (4.5) we get

$$e_1(\alpha) = \lambda(\alpha^2 + 1) + \alpha\beta, \tag{4.18}$$

$$e_1(\beta) = \beta(\alpha\lambda + \beta). \tag{4.19}$$

Taking into account (4.12) and (4.14)-(4.17), we find

$$(1 + \alpha^2)S = 10(1 + \alpha^2)\lambda^2 - 12\alpha\beta\lambda + 6\beta^2. \tag{4.20}$$

Furthermore, by (4.17), (4.4) and (4.5), the equations $-3\lambda = \lambda_2 + \lambda_3 + \lambda_4$ and (4.3) turn into

$$-3e_1(\lambda) = \alpha(2\lambda_1^2 + S) - 6\beta\lambda, \tag{4.21}$$

$$e_1e_1(\lambda) = 3(-\alpha\lambda + \beta)e_1(\lambda) + \lambda S. \tag{4.22}$$

From (4.18)-(4.22), we can eliminate the terms of $e_1e_1(\lambda), e_1(\lambda), S, \beta$, and obtain

$$\left\{484844765184\alpha^{65} + \cdots + 846526464\alpha^7\right\}\lambda^{16} = 0.$$

Since $\lambda \neq 0$, the above equation is a non-trivial polynomial equation concerning α with constant coefficients. This implies that α must be a constant. By utilizing (4.18)-(4.21) again, we may obtain a contradiction.

CASE B. $\omega_{23}^4 = \omega_{32}^4 = \omega_{43}^2 = 0$ on U. In this case, it follows from (4.14)-(4.16) that

$$\omega_{22}^1\omega_{33}^1 = -\lambda_2\lambda_3, \tag{4.23}$$

$$\omega_{22}^1\omega_{44}^1 = -\lambda_2\lambda_4, \tag{4.24}$$

$$\omega_{33}^1\omega_{44}^1 = -\lambda_3\lambda_4. \tag{4.25}$$

CASE B.1. All principal curvatures $\lambda_2, \lambda_3, \lambda_4$ are nonzero on U.

Eliminating $\omega_{22}^1, \omega_{44}^1, \lambda_2, \lambda_3$ from the equations (4.23)-(4.25), we find $(\omega_{33}^1)^2 + \lambda_3^2 = 0$. This leads to $\omega_{33}^1 = \lambda_3 = 0$, which is impossible.

CASE B.2. At least one of three principal curvatures $\lambda_2, \lambda_3, \lambda_4$ is zero on U. Without loss of generality, we assume $\lambda_4 = 0$. The crucial idea is to deduce the following relations:

$$\omega_{22}^1 \omega_{33}^1 = L, \tag{4.26}$$

$$M_1(\omega_{22}^1)^2 + N_1(\omega_{33}^1)^2 = K_1, \tag{4.27}$$

$$M_2(\omega_{22}^1)^2 + N_2(\omega_{33}^1)^2 = K_2, \tag{4.28}$$

where $L = 3\lambda\lambda_2 + \lambda_2^2$ and

$$\begin{aligned}
&M_1 = 2\lambda_2 - 2\lambda, \quad N_1 = -8\lambda - 2\lambda_2,\\
&K_1 = -17\lambda\lambda_2^2 - 51\lambda^2\lambda_2 - 63\lambda^3,\\
&M_2 = 8\lambda\lambda_2^2 + 114\lambda^2\lambda_2 + 178\lambda^3,\\
&N_2 = 44\lambda\lambda_2^2 + 132\lambda^2\lambda_2 + 124\lambda^3,\\
&K_2 = 1449\lambda^5 + 2343\lambda^4\lambda_2 + 1636\lambda^3\lambda_2^2 + 543\lambda^2\lambda_2^3 + 65\lambda\lambda_2^4.
\end{aligned}$$

By (4.26)-(4.28), some similar analysis to the one in the proof of Case A also yields a contraction. This completes the proof of Step 2.

REMARK 4.9. In Case B.2, the proof is extremely similar to the case of hypersurfaces with three distinct principal curvatures in $\mathbb{R}^5$, which was considered in [**20**].

REMARK 4.10. In dimension $n > 4$, Step 1 and Step 2 are not easily verified generally, which are the main difficulties in the study of Chen's conjecture. However, once one knows more about those coefficients of f_k and the geometric structure of biharmonic hypersurfaces, one will probably prove Chen's conjecture for higher dimension by our method.

REMARK 4.11. Recently, Guan-Li-Vrancken [**26**] used the method in [**25**] to prove that every biharmonic hypersurface in the 5-dimensional nonzero space form has constant mean curvature. All the information shows that, the algebraic equations' technique is useful and crucial for studying biharmonic hypersurfaces.

References

[1] Kazuo Akutagawa and Shun Maeta, *Biharmonic properly immersed submanifolds in Euclidean spaces*, Geom. Dedicata **164** (2013), 351–355, DOI 10.1007/s10711-012-9778-1. MR3054632

[2] Luis J. Alías, S. Carolina García-Martínez, and Marco Rigoli, *Biharmonic hypersurfaces in complete Riemannian manifolds*, Pacific J. Math. **263** (2013), no. 1, 1–12, DOI 10.2140/pjm.2013.263.1. MR3069073

[3] A. Balmuş, S. Montaldo, and C. Oniciuc, *Classification results for biharmonic submanifolds in spheres*, Israel J. Math. **168** (2008), 201–220, DOI 10.1007/s11856-008-1064-4. MR2448058

[4] Adina Balmuş, Stefano Montaldo, and Cezar Oniciuc, *Classification results and new examples of proper biharmonic submanifolds in spheres*, Note Mat. **28** (2009), no. [2008 on verso], suppl. 1, 49–61. MR2640575

[5] Adina Balmuş, Stefano Montaldo, and Cezar Oniciuc, *Biharmonic hypersurfaces in 4-dimensional space forms*, Math. Nachr. **283** (2010), no. 12, 1696–1705, DOI 10.1002/mana.200710176. MR2560665

[6] Adina Balmuş, Stefano Montaldo, and Cezar Oniciuc, *Biharmonic PNMC submanifolds in spheres*, Ark. Mat. **51** (2013), no. 2, 197–221, DOI 10.1007/s11512-012-0169-5. MR3090194

[7] A. Balmuş and C. Oniciuc, *Biharmonic submanifolds with parallel mean curvature vector field in spheres*, J. Math. Anal. Appl. **386** (2012), no. 2, 619–630, DOI 10.1016/j.jmaa.2011.08.019. MR2834772

[8] H. Bibi, E. Loubeau, and C. Oniciuc, *Unique continuation property for Biharmonic hypersurfaces in spheres*, arXiv:2007.06527.

[9] R. Caddeo, S. Montaldo, and C. Oniciuc, *Biharmonic submanifolds of* S^3, Internat. J. Math. **12** (2001), no. 8, 867–876, DOI 10.1142/S0129167X01001027. MR1863283

[10] R. Caddeo, S. Montaldo, and C. Oniciuc, *Biharmonic submanifolds in spheres*, Israel J. Math. **130** (2002), 109–123, DOI 10.1007/BF02764073. MR1919374

[11] Bang-Yen Chen, *Some open problems and conjectures on submanifolds of finite type*, Soochow J. Math. **17** (1991), no. 2, 169–188. MR1143504

[12] Bang-Yen Chen, *Total mean curvature and submanifolds of finite type*, 2nd ed., Series in Pure Mathematics, vol. 27, World Scientific Publishing Co. Pte. Ltd., Hackensack, NJ, 2015. With a foreword by Leopold Verstraelen. MR3362186

[13] B. Y. Chen, *Chen's biharmonic conjecture and submanifolds with parallel normalized mean curvature vector*, Mathematics 7 (2019) no. 8: 710. https://doi.org/10.3390/math7080710.

[14] Bang-Yen Chen and Marian Ioan Munteanu, *Biharmonic ideal hypersurfaces in Euclidean spaces*, Differential Geom. Appl. **31** (2013), no. 1, 1–16, DOI 10.1016/j.difgeo.2012.10.008. MR3010073

[15] Filip Defever, *Hypersurfaces of* $\mathbf{E}^4$ *with harmonic mean curvature vector*, Math. Nachr. **196** (1998), 61–69, DOI 10.1002/mana.19981960104. MR1657990

[16] Ivko M. Dimitric, *Quadric representation and submanifolds of finite type*, ProQuest LLC, Ann Arbor, MI, 1989. Thesis (Ph.D.)–Michigan State University. MR2638104

[17] Ivko Dimitrić, *Submanifolds of* E^m *with harmonic mean curvature vector*, Bull. Inst. Math. Acad. Sinica **20** (1992), no. 1, 53–65. MR1166218

[18] James Eells Jr. and J. H. Sampson, *Harmonic mappings of Riemannian manifolds*, Amer. J. Math. **86** (1964), 109–160, DOI 10.2307/2373037. MR164306

[19] James Eells and Luc Lemaire, *Selected topics in harmonic maps*, CBMS Regional Conference Series in Mathematics, vol. 50, Published for the Conference Board of the Mathematical Sciences, Washington, DC; by the American Mathematical Society, Providence, RI, 1983, DOI 10.1090/cbms/050. MR703510

[20] Yu Fu, *Biharmonic hypersurfaces with three distinct principal curvatures in Euclidean 5-space*, J. Geom. Phys. **75** (2014), 113–119, DOI 10.1016/j.geomphys.2013.09.004. MR3126938

[21] Yu Fu, *Biharmonic hypersurfaces with three distinct principal curvatures in Euclidean space*, Tohoku Math. J. (2) **67** (2015), no. 3, 465–479, DOI 10.2748/tmj/1446818561. MR3420554

[22] Yu Fu, *Biharmonic hypersurfaces with three distinct principal curvatures in spheres*, Math. Nachr. **288** (2015), no. 7, 763–774, DOI 10.1002/mana.201400101. MR3345102

[23] Yu Fu, *On biharmonic hypersurfaces with constant scalar curvatures in* $\mathbb{S}^5$, Proc. Amer. Math. Soc. **143** (2015), no. 12, 5399–5409, DOI 10.1090/proc/12677. MR3411155

[24] Yu Fu and Min-Chun Hong, *Biharmonic hypersurfaces with constant scalar curvature in space forms*, Pacific J. Math. **294** (2018), no. 2, 329–350, DOI 10.2140/pjm.2018.294.329. MR3770116

[25] Yu Fu, Min-Chun Hong, and Xin Zhan, *On Chen's biharmonic conjecture for hypersurfaces in* $\mathbb{R}^5$, Adv. Math. **383** (2021), Paper No. 107697, 28, DOI 10.1016/j.aim.2021.107697. MR4233281

[26] Zhida Guan, Haizhong Li, and Luc Vrancken, *Four dimensional biharmonic hypersurfaces in nonzero space forms have constant mean curvature*, J. Geom. Phys. **160** (2021), Paper No. 103984, 15, DOI 10.1016/j.geomphys.2020.103984. MR4167191

[27] Th. Hasanis and Th. Vlachos, *Hypersurfaces in* E^4 *with harmonic mean curvature vector field*, Math. Nachr. **172** (1995), 145–169, DOI 10.1002/mana.19951720112. MR1330627

[28] Guo Ying Jiang, *2-harmonic isometric immersions between Riemannian manifolds* (Chinese), Chinese Ann. Math. Ser. A **7** (1986), no. 2, 130–144. An English summary appears in Chinese Ann. Math. Ser. B **7** (1986), no. 2, 255. MR858581

[29] Guo Ying Jiang, *2-harmonic maps and their first and second variational formulas* (Chinese), Chinese Ann. Math. Ser. A **7** (1986), no. 4, 389–402. An English summary appears in Chinese Ann. Math. Ser. B **7** (1986), no. 4, 523. MR886529

[30] Guo Ying Jiang, *Some nonexistence theorems on 2-harmonic and isometric immersions in Euclidean space* (Chinese), Chinese Ann. Math. Ser. A **8** (1987), no. 3, 377–383. An English summary appears in Chinese Ann. Math. Ser. B **8** (1987), no. 3, 389. MR924896

[31] Norihito Koiso and Hajime Urakawa, *Biharmonic submanifolds in a Riemannian manifold*, Osaka J. Math. **55** (2018), no. 2, 325–346. MR3787748

[32] E. Loubeau, S. Montaldo, and C. Oniciuc, *The stress-energy tensor for biharmonic maps*, Math. Z. **259** (2008), no. 3, 503–524, DOI 10.1007/s00209-007-0236-y. MR2395125

[33] Yong Luo, *Weakly convex biharmonic hypersurfaces in nonpositive curvature space forms are minimal*, Results Math. **65** (2014), no. 1-2, 49–56, DOI 10.1007/s00025-013-0328-4. MR3162427

[34] Yong Luo and Shun Maeta, *Biharmonic hypersurfaces in a sphere*, Proc. Amer. Math. Soc. **145** (2017), no. 7, 3109–3116, DOI 10.1090/proc/13320. MR3637957

[35] Shun Maeta and Ye-Lin Ou, *Some classifications of biharmonic hypersurfaces with constant scalar curvature*, Pacific J. Math. **306** (2020), no. 1, 281–290, DOI 10.2140/pjm.2020.306.281. MR4109916

[36] S. Montaldo, C. Oniciuc, and A. Ratto, *On cohomogeneity one biharmonic hypersurfaces into the Euclidean space*, J. Geom. Phys. **106** (2016), 305–313, DOI 10.1016/j.geomphys.2016.04.012. MR3508923

[37] Nobumitsu Nakauchi and Hajime Urakawa, *Biharmonic hypersurfaces in a Riemannian manifold with non-positive Ricci curvature*, Ann. Global Anal. Geom. **40** (2011), no. 2, 125–131, DOI 10.1007/s10455-011-9249-1. MR2811621

[38] Nobumitsu Nakauchi and Hajime Urakawa, *Biharmonic submanifolds in a Riemannian manifold with non-positive curvature*, Results Math. **63** (2013), no. 1-2, 467–474, DOI 10.1007/s00025-011-0209-7. MR3009698

[39] C. Oniciuc, *Biharmonic maps between Riemannian manifolds*, An. Ştiinţ. Univ. Al. I. Cuza Iaşi. Mat. (N.S.) **48** (2002), no. 2, 237–248 (2003). MR2004799

[40] C. Oniciuc, *Biharmonic submanifolds in space forms*, Habilitation Thesis, (2012), 149p.

[41] Ye-Lin Ou, *Biharmonic hypersurfaces in Riemannian manifolds*, Pacific J. Math. **248** (2010), no. 1, 217–232, DOI 10.2140/pjm.2010.248.217. MR2734173

[42] Ye-Lin Ou, *Some recent progress of biharmonic submanifolds*, Recent advances in the geometry of submanifolds—dedicated to the memory of Franki Dillen (1963–2013), Contemp. Math., vol. 674, Amer. Math. Soc., Providence, RI, 2016, pp. 127–139, DOI 10.1090/conm/674. MR3565250

[43] Ye-Lin Ou and Bang-Yen Chen, *Biharmonic submanifolds and biharmonic maps in Riemannian geometry*, World Scientific Publishing Co. Pte. Ltd., Hackensack, NJ, [2020] ©2020. MR4265170

[44] Ye-Lin Ou and Liang Tang, *On the generalized Chen's conjecture on biharmonic submanifolds*, Michigan Math. J. **61** (2012), no. 3, 531–542, DOI 10.1307/mmj/1347040257. MR2975260

Research Center of Applied Mathematics, Dongbei University of Finance and Economics, Dalian 116025, People's Republic of China

Email address: yufu@dufe.edu.cn

School of Mathematics, Liaoning University, Shenyang 110044, People's Republic of China

Email address: dlutyangdan@126.com

School of Mathematics and Statistics, Changshu Institute of Technology, Suzhou 215500, People's Republic of China

Email address: zhanxin_math@163.com

Contemporary Mathematics
Volume **777**, 2022
https://doi.org/10.1090/conm/777/15631

A commutativity condition for subsets in quandles — a generalization of antipodal subsets

Akira Kubo, Mika Nagashiki, Takayuki Okuda, and Hiroshi Tamaru

Dedicated to the memory of Professor Tadashi Nagano

Abstract. In this paper, we introduce some commutativity condition for subsets in quandles, which we call the s-commutativity. Note that quandles can be regarded as a generalization of symmetric spaces, and the notion of s-commutative subsets is a generalization of antipodal subsets. We study maximal s-commutative subsets in quandles, and show that they have some nice properties. As one example, any maximal s-commutative subsets in quandles are subquandles. We also determine maximal s-commutative subsets in spheres, projective spaces, and dihedral quandles. In these quandles, maximal s-commutative subsets turn out to be unique up to automorphisms.

1. Introduction

A quandle is a set equipped with a binary operator, whose axioms are corresponding to the Reidemeister moves of knots. Although the notion of quandles is originated in knot theory ([**5**]), it now plays important roles in many branches of mathematics. Quandles have been studied actively from various perspectives and viewpoints. As one aspect, quandles can be regarded as a generalization of symmetric spaces. The definition of quandles can be formulated as follows.

Definition 1.1. Let Q be a set, and s be a map from Q into $\mathrm{Map}(Q, Q)$, that is, a map $s_x : Q \to Q$ is equipped for each $x \in Q$. Then a pair (Q, s) is called a *quandle* if

- $(Q1)$ for any $x \in Q$, $s_x(x) = x$,
- $(Q2)$ for any $x \in Q$, s_x is bijective,
- $(Q3)$ for any $x, y \in Q$, $s_x \circ s_y = s_{s_x(y)} \circ s_x$.

As has already been mentioned in [**5**], symmetric spaces are quandles by taking s_x as the point symmetry at x (for symmetric spaces, see Subsection 2.4). It is then

2020 *Mathematics Subject Classification.* Primary 53C35; Secondary 57K12.
Key words and phrases. Symmetric spaces, quandles, antipodal subsets, poles.
This work was partly supported by Osaka City University Advanced Mathematical Institute (MEXT Joint Usage/Research Center on Mathematics and Theoretical Physics JPMXP0619217849). The first author was supported by Hiroshima University Research Grant (support of young scientists). The third author was supported by JSPS KAKENHI Grant Number 20K14310. The fourth author was supported by JSPS KAKENHI Grant Number 19K21831.

natural to study quandles by referring to the theory of symmetric spaces. There have been such studies, for which we refer to [**3**, **4**, **8**] and references therein.

This paper also studies quandles from the viewpoint of symmetric spaces. In the theory of Riemannian symmetric spaces, important notions include not only Riemannian geometric properties (such as curvatures), but also ones derived from the point symmetries. A typical and most important example is the notion of antipodal subsets, introduced by Chen and Nagano ([**2**]). The definition can easily be transferred to quandles as follows.

DEFINITION 1.2. A subset X in a quandle (Q, s) is said to be *antipodal* if $s_x(y) = y$ holds for any $x, y \in X$.

This notion generalizes the antipodal points in the unit sphere S^n. For each $x \in S^n$, the point $-x$ is an antipodal point of x. The subset $\{\pm x\}$ is an antipodal subset in S^n, which is maximal with respect to inclusion relation. Furthermore, every maximal antipodal subset in S^n is of this form (see Section 3 for details and more examples). For other spaces, it would probably be surprising that the classifications of maximal antipodal subsets are involved in general, and it is still open for some symmetric spaces (see Remark 3.12).

As one of important applications of antipodal subsets, the notion of 2-numbers has been introduced ([**2**]). The *2-number* of a compact Riemannian symmetric space is defined by the supremum of the cardinalities of antipodal subsets, which is in fact finite. The 2-number is related to topological information of the symmetric space, such as the Euler characteristic ([**2**]) and the $\mathbb{Z}_2$-Betti number ([**7**]). For further results and applications, we refer to a survey article [**1**] and references therein.

In a possible structure theory of quandles, it would be reasonable to expect to have a nice class of subsets, which reflect some properties of the ambient quandles. In this paper, we introduce the notion of s-commutative subsets in quandles, which would form a nice class of subsets. The definition is given as follows (see Section 4 for details).

DEFINITION 1.3. A subset X in a quandle (Q, s) is said to be *s-commutative* if $s_x \circ s_y = s_y \circ s_x$ holds for any $x, y \in X$.

Note that the s-commutativity condition itself can also been seen in [**2**, Proposition 3.4], but the notion of s-commutative subsets would be new. One of the purposes of this paper is to study some fundamental properties of s-commutative subsets. Recall that antipodal subsets play important roles in the study of symmetric spaces. The notion of s-commutative subsets is a generalization of antipodal subsets.

PROPOSITION 1.4. *Every antipodal subset in a quandle (Q, s) is s-commutative.*

For s-commutative subsets in quandles, it is natural to consider maximal ones with respect to inclusion relation. Recall that a subset X in (Q, s) is called a subquandle if X is normalized by $s_x^{\pm 1}$ for every $x \in X$.

THEOREM 1.5. *Every maximal s-commutative subset in a quandle (Q, s) is a subquandle.*

There are notions of the direct products and the interaction-free unions of quandles. We also study, in Section 5, the behaviour of antipodal subsets and s-commutative subsets under these operations.

In Section 6, we determine maximal s-commutative subsets in some symmetric spaces and quandles. The first example is the n-dimensional unit sphere S^n with $n \in \mathbb{Z}_{>0}$. Denote by $\{e_1, \ldots, e_{n+1}\}$ the standard basis of $\mathbb{R}^{n+1}$, and by $\mathrm{SO}(n+1)$ the special orthogonal group of degree $n+1$.

PROPOSITION 1.6. *The subset $\{\pm e_1, \ldots, \pm e_{n+1}\}$ is maximal s-commutative in the unit sphere S^n, which is unique up to the congruence by $\mathrm{SO}(n+1)$.*

Since the above subset is not antipodal, this example clarifies a difference between antipodal subsets and s-commutative subsets. The second example is the n-dimensional real projective space $P(\mathbb{R}^{n+1})$. We denote by $\mathbb{R}x \in P(\mathbb{R}^{n+1})$ the line spanned by $x \in \mathbb{R}^{n+1} \setminus \{0\}$.

PROPOSITION 1.7. *The subset $\{\mathbb{R}e_1, \ldots, \mathbb{R}e_{n+1}\}$ is maximal s-commutative in $P(\mathbb{R}^{n+1})$ if $n > 1$. If $n = 1$, then $\{\mathbb{R}e_1, \mathbb{R}e_2, \mathbb{R}(e_1 + e_2), \mathbb{R}(e_1 - e_2)\}$ is a maximal s-commutative subset in $P(\mathbb{R}^2)$. In both cases, a maximal s-commutative subset in $P(\mathbb{R}^{n+1})$ is unique up to the congruence by $\mathrm{SO}(n+1)$.*

It is natural but interesting that the shape of a maximal s-commutative subset in $P(\mathbb{R}^{n+1})$ depends on n. Note that $\{\mathbb{R}e_1, \mathbb{R}e_2, \mathbb{R}(e_1 + e_2), \mathbb{R}(e_1 - e_2)\}$ is s-commutative in $P(\mathbb{R}^2)$, but not in $P(\mathbb{R}^3)$. This yields that the s-commutativity is an extrinsic property. The third example is the dihedral quandle R_n. Recall that $R_n := (\mathbb{Z}/n\mathbb{Z}, s)$ with $s_x(y) := 2x - y$ (see Example 2.5 for details).

PROPOSITION 1.8. *For the dihedral quandle R_n, the following subsets are maximal antipodal and maximal s-commutative, respectively, and all of them are unique up to automorphisms:*

condition	*maximal antipodal*	*maximal s-commutative*
n is odd	$\{0\}$	$\{0\}$
$n = 4l - 2$ $(l \in \mathbb{Z}_{>0})$	$\{0, 2l - 1\}$	$\{0, 2l - 1\}$
$n = 4l$ $(l \in \mathbb{Z}_{>0})$	$\{0, 2l\}$	$\{0, l, 2l, 3l\}$

By the results of the above three propositions, one can see that antipodal subsets and s-commutative subsets are sometimes the same, but sometimes different, both in the finite and infinite cardinality cases. In Remark 6.6, we also mention a difference between poles and antipodal subsets in finite quandles.

Note that the study on s-commutative subsets is new and undeveloped. There are many open problems related to symmetric spaces and quandles. In particular, it would be interesting to determine maximal s-commutative subsets in symmetric spaces and quandles, and study the uniqueness properties. Note that $\mathrm{SO}(n+1)$ (and also the orthogonal group $\mathrm{O}(n+1)$) acts on S^n and $P(\mathbb{R}^{n+1})$ as automorphisms of quandles. Therefore, for the above cases S^n, $P(\mathbb{R}^{n+1})$, and R_n, maximal s-commutative subsets are unique up to automorphisms. However, this would be not true in general.

Our studies on quandles from the viewpoint of symmetric spaces have been influenced strongly by Professor Tadashi Nagano. In particular, the fourth author obtained his PhD degree under the supervision of Professor Tadashi Nagano, and learned many things about symmetric spaces. Without his guidance, our studies would not have been possible. The authors are also grateful to Hiroyuki Tasaki and Makiko Sumi Tanaka for their kind supports and helpful discussions.

2. Preliminaries

In this section, we recall some basic notions on quandles. In particular, the direct product and the interaction-free union of quandles are studied. We also recall some examples of quandles, including symmetric spaces, whose subsets will be studied in the following sections. Throughout this paper, quandles are denoted by $Q = (Q, s)$ as in Definition 1.1, and s is called the quandle structure on Q.

2.1. Quandles. In this subsection, we recall some basic notions on quandles. First of all, we recall the notion of subquandles.

DEFINITION 2.1. Let (Q, s) be a quandle. A subset X of Q is called a *subquandle* if for any $x \in X$ it satisfies $s_x^{\pm 1}(X) \subset X$.

Note that X is a subquandle if and only if $s_x(X) = X$ for any $x \in X$. It is easy to see that every subquandle is naturally a quandle. We also note that the condition $s_x(X) \subset X$ is not sufficient to define subquandles, since s_x is not necessary involutive.

DEFINITION 2.2. Let (M, s^M) and (N, s^N) be quandles. Then a map $f : M \to N$ is called a *homomorphism* if it satisfies

$$s^N_{f(x)} \circ f = f \circ s^M_x \quad (\forall x \in M).$$

Note that the inverse map of a bijective homomorphism is also a homomorphism. Hence, a bijective homomorphism is called an *isomorphism.* The notions of automorphisms and the automorphism groups are defined naturally.

DEFINITION 2.3. For a quandle $Q = (Q, s)$, a bijective homomorphism from (Q, s) to (Q, s) itself is called an *automorphism.* The group consisting of all automorphisms is called the *automorphism group* of (Q, s), and denoted by $\mathrm{Aut}(Q)$ or $\mathrm{Aut}(Q, s)$.

A quandle Q is said to be *homogeneous* if $\mathrm{Aut}(Q)$ acts transitively on Q. Note that $s_x \in \mathrm{Aut}(Q)$ holds for any $x \in Q$.

EXAMPLE 2.4. Let M be a set. Then one obtains a quandle structure s on M by putting $s_x := \mathrm{id}_M$ for every $x \in M$. This (M, s) is called a *trivial quandle.* A trivial quandle (M, s) is homogeneous, since any bijection from M to M is an automorphism.

The next example is the dihedral quandle, which would be the simplest non-trivial quandles. We denote by $\mathbb{Z}/n\mathbb{Z}$ the cyclic group of order n.

EXAMPLE 2.5. Let us fix $n \in \mathbb{Z}_{>0}$, and put $R_n := \mathbb{Z}/n\mathbb{Z}$. Then we have a quandle structure s on R_n by putting

$$s_x : R_n \to R_n : y \mapsto 2x - y,$$

for each $x \in R_n$. This (R_n, s) is called the dihedral quandle of order n. The dihedral quandle (R_n, s) is homogeneous, since the following f is an automorphism:

$$f : R_n \to R_n : x \mapsto x + 1.$$

We note that, in terms of the above automorphism f, the cyclic group $\mathbb{Z}/n\mathbb{Z}$ acts on (R_n, s) as automorphisms.

2.2. Direct products of quandles. In this subsection, we consider a family of quandles $\{Q_\lambda = (Q_\lambda, s^\lambda)\}_{\lambda\in\Lambda}$, and study the direct product quandle $\prod_{\lambda\in\Lambda} Q_\lambda$. As a set, it is the direct product set of $\{Q_\lambda\}_{\lambda\in\Lambda}$, whose elements will be denoted as $(x_\lambda)_\lambda$. The next proposition has been known, and can be proved directly.

PROPOSITION 2.6. *Let $Q := \prod_{\lambda\in\Lambda} Q_\lambda$ be the direct product set. Then the map $s^Q : Q \to \mathrm{Map}(Q,Q)$ defined by the following is a quandle structure on Q:*

$$s^Q_{(x_\lambda)_\lambda} : Q \to Q : (y_\lambda)_\lambda \mapsto (s^\lambda_{x_\lambda}(y_\lambda))_\lambda.$$

The obtained quandle is called the *direct product quandle* of a family of quandles $\{Q_\lambda\}_{\lambda\in\Lambda}$, and denoted by $\prod_{\lambda\in\Lambda} Q_\lambda$. In the remaining of this subsection, we show some properties of the direct product quandles. We start from the following.

PROPOSITION 2.7. *For any $\eta \in \Lambda$, the natural projection $p_\eta : \prod_{\lambda\in\Lambda} Q_\lambda \to Q_\eta$ is a quandle homomorphism.*

PROOF. We denote by s^Q the quandle structure on the direct product quandle $Q := \prod_{\lambda\in\Lambda} Q_\lambda$. Take any $(x_\lambda)_\lambda, (y_\lambda)_\lambda \in Q$, and any $\eta \in \Lambda$. Then we have

$$(p_\eta \circ s^Q_{(x_\lambda)_\lambda})((y_\lambda)_\lambda) = p_\eta((s^\lambda_{x_\lambda}(y_\lambda))_\lambda) = s^\eta_{x_\eta}(y_\eta) = (s^\eta_{p_\eta((x_\lambda)_\lambda)} \circ p_\eta)((y_\lambda)_\lambda),$$

which proves that p_η is a quandle homomorphism. □

The next proposition gives a key property of direct product quandles. Namely, a family of quandle homomorphisms into Q_λ gives rise to a quandle homomorphism into $\prod_{\lambda\in\Lambda} Q_\lambda$.

PROPOSITION 2.8. *Let Z be a quandle, and $\pi_\lambda : Z \to Q_\lambda$ be a quandle homomorphism for each $\lambda \in \Lambda$. Then the map*

$$h : Z \to \prod_{\lambda\in\Lambda} Q_\lambda : x \mapsto (\pi_\lambda(x))_\lambda$$

is the unique quandle homomorphism satisfying $p_\lambda \circ h = \pi_\lambda$ for any $\lambda \in \Lambda$.

PROOF. Denote by s^Z and s^Q the quandle structures on Z and $Q := \prod_{\lambda\in\Lambda} Q_\lambda$, respectively. Then, since each π_λ is a quandle homomorphism, we have

$$\begin{aligned}(h \circ s^Z_x)(y) = h(s^Z_x(y)) &= (\pi_\lambda(s^Z_x(y)))_\lambda = ((s^\lambda_{\pi_\lambda(x)} \circ \pi_\lambda)(y))_\lambda \\ &= s^Q_{(\pi_\lambda(x))_\lambda}(\pi_\lambda(y))_\lambda = (s^Q_{h(x)} \circ h)(y)\end{aligned}$$

for any $x, y \in Z$. This shows that h is a quandle homomorphism. It is clear that h satisfies $p_\lambda \circ h = \pi_\lambda$ for any $\lambda \in \Lambda$. The uniqueness follows easily. □

By applying Proposition 2.8, we obtain a property of quandle automorphisms on $\prod_{\lambda\in\Lambda} Q_\lambda$. Namely, the direct product group $\prod_{\lambda\in\Lambda} \mathrm{Aut}(Q_\lambda)$ of the automorphism groups of Q_λ can be understood as a subgroup of $\mathrm{Aut}(\prod_{\lambda\in\Lambda} Q_\lambda)$.

PROPOSITION 2.9. *For each $(f_\lambda)_\lambda \in \prod_{\lambda\in\Lambda} \mathrm{Aut}(Q_\lambda)$, the following map $\prod_{\lambda\in\Lambda} f_\lambda$ is an automorphism on the quandle $\prod_{\lambda\in\Lambda} Q_\lambda$:*

$$\prod_{\lambda\in\Lambda} f_\lambda : \prod_{\lambda\in\Lambda} Q_\lambda \to \prod_{\lambda\in\Lambda} Q_\lambda : (x_\lambda)_\lambda \mapsto (f_\lambda(x_\lambda))_\lambda.$$

PROOF. Put $Z := \prod_{\lambda\in\Lambda} Q_\lambda$, and $\pi_\lambda := f_\lambda \circ p_\lambda$ for each $\lambda \in \Lambda$. Since $\pi_\lambda : Z \to Q_\lambda$ is a quandle homomorphism, it follows from Proposition 2.8 that

$$h : Z \to \prod_{\lambda\in\Lambda} Q_\lambda : x \mapsto (\pi_\lambda(x))_\lambda$$

is a quandle homomorphism. Then so is $\prod_{\lambda\in\Lambda} f_\lambda$, since

$$h((x_\lambda)_\lambda) = ((f_\lambda \circ p_\lambda)((x_\lambda)_\lambda))_\lambda = (f_\lambda(x_\lambda))_\lambda = (\prod_{\lambda\in\Lambda} f_\lambda)((x_\lambda)_\lambda).$$

Furthermore, this is bijective, since $\prod_{\lambda\in\Lambda}(f_\lambda^{-1})$ is the inverse map. □

As another application of Proposition 2.8, one can see that the quandle structure on $\prod_{\lambda\in\Lambda} Q_\lambda$ is canonical, in the following sense.

PROPOSITION 2.10. *Let s' be a quandle structure on the direct product set $\prod_{\lambda\in\Lambda} Q_\lambda$, and suppose that the projection p_λ is a quandle homomorphism onto Q_λ for any $\lambda \in \Lambda$. Then $(\prod_{\lambda\in\Lambda} Q_\lambda, s')$ coincides with the direct product quandle of $\{Q_\lambda\}_{\lambda\in\Lambda}$.*

PROOF. We apply Proposition 2.8 for the quandle $Z := (\prod_{\lambda\in\Lambda} Q_\lambda, s')$ and the quandle homomorphisms $p_\lambda : Z \to Q_\lambda$. Then the obtained quandle homomorphism $h : Z \to \prod_{\lambda\in\Lambda} Q_\lambda$ is the identity map, since

$$h((x_\lambda)_\lambda) = (p_\lambda((x_\lambda)_\lambda))_\lambda = (x_\lambda)_\lambda$$

for any $(x_\lambda)_\lambda \in Z$. Since $h = \mathrm{id}$ is a quandle homomorphism, we conclude that s' coincides with the quandle structure s^Q on the direct product quandle $\prod_{\lambda\in\Lambda} Q_\lambda$. This completes the proof. □

It should be noted that, by Proposition 2.8, the direct product quandle $\prod_{\lambda\in\Lambda} Q_\lambda$ is the product of the family $\{Q_\lambda = (M_\lambda, s^\lambda)\}_{\lambda\in\Lambda}$ in the category of quandles and quandle homomorphisms. This means that it has the following universal property.

> **Universal property of direct products:** Let us fix any quandle Z and any quandle homomorphism $\pi_\lambda : Z \to Q_\lambda$ for each $\lambda \in \Lambda$. Then there exists a unique quandle homomorphism $h : Z \to \prod_{\lambda\in\Lambda} Q_\lambda$ such that $p_\lambda \circ h = \pi_\lambda$ for any $\lambda \in \Lambda$.

2.3. Interaction-free unions of quandles. Same as in the previous subsection, we consider a family of quandles $\{Q_\lambda = (Q_\lambda, s^\lambda)\}_{\lambda\in\Lambda}$. In this subsection, we study the interaction-free union quandle $\bigsqcup^{\mathrm{free}}_{\lambda\in\Lambda} Q_\lambda$, defined by the following.

PROPOSITION 2.11. *Let $Q := \bigsqcup_{\lambda\in\Lambda} Q_\lambda$ be the disjoint union as a set. Then the map $s^{\mathrm{free}} : Q \to \mathrm{Map}(Q, Q)$ defined by the following is a quandle structure on Q: for each $x \in Q_\eta$,*

$$s_x^{\mathrm{free}}(y) := \begin{cases} s_x^\eta(y) & (\textit{if } y \in Q_\eta), \\ y & (\textit{otherwise}). \end{cases}$$

PROOF. Let $x \in Q_\eta$. Then the quandle structure s^{free} can be written as

$$s_x^{\mathrm{free}}|_{Q_\eta} = s_x^\eta, \qquad s_x^{\mathrm{free}}|_{Q_{\eta'}} = \mathrm{id} \ (\text{for } \eta' \neq \eta).$$

Then one can check Conditions (Q1), (Q2), and (Q3) of quandles easily. □

We denote by $\bigsqcup^{\text{free}}_{\lambda\in\Lambda} Q_\lambda := (\bigsqcup_{\lambda\in\Lambda} Q_\lambda, s^{\text{free}})$, which will be called the *interaction-free union* of the family of quandles $\{Q_\lambda\}_{\lambda\in\Lambda}$ in this paper. Similar to the case of direct product quandles, it satisfies the following.

PROPOSITION 2.12. *The natural inclusion map $\iota_\eta : Q_\eta \to \bigsqcup^{\text{free}}_{\lambda\in\Lambda} Q_\lambda$ is a quandle homomorphism for any $\eta \in \Lambda$.*

PROOF. Take any $x, y \in Q_\eta$. Then we have

$$(\iota_\eta \circ s^\eta_x)(y) = s^\eta_x(y) = s^{\text{free}}_x(y) = (s^{\text{free}}_{\iota_\eta(x)} \circ \iota_\eta)(y),$$

which completes the proof. □

We here give a property on quandle homomorphisms. Namely, if a family of quandle homomorphisms from Q_λ satisfies some condition, then it gives rise to a quandle homomorphism from $\bigsqcup^{\text{free}}_{\lambda\in\Lambda} Q_\lambda$.

PROPOSITION 2.13. *Let (Z, s^Z) be a quandle, and $\iota'_\lambda : Q_\lambda \to Z$ be a quandle homomorphism for each $\lambda \in \Lambda$. Assume that it satisfies*

$$(s^Z_{\iota'_\eta(x)} \circ \iota'_\xi)(y) = \iota'_\xi(y) \quad (\forall x \in Q_\eta,\ \forall y \in Q_\xi \textit{ with } \eta \neq \xi).$$

Then the map $h : \bigsqcup^{\text{free}}_{\lambda\in\Lambda} Q_\lambda \to Z$ defined by

$$h(x) := \iota'_\eta(x) \quad (\textit{if } x \in Q_\eta)$$

is the unique quandle homomorphism satisfying $h \circ \iota_\eta = \iota'_\eta$ for any $\eta \in \Lambda$.

PROOF. It is easy to see that $h \circ \iota_\eta = \iota'_\eta$ for any $\eta \in \Lambda$, and that the map h with this property is unique. Hence we have only to show that h is a quandle homomorphism. Take any $\eta, \xi \in \Lambda$, $x \in Q_\eta$, and $y \in Q_\xi$. We claim that

$$(s^Z_{h(x)} \circ h)(y) = (h \circ s^{\text{free}}_x)(y).$$

If $\eta = \xi$, then the claim follows from the property of $\iota'_\xi : Q_\xi \to Z$, which is a quandle homomorphism. In fact, we have

$$(s^Z_{h(x)} \circ h)(y) = (s^Z_{\iota'_\eta(x)} \circ \iota'_\eta)(y) = (\iota'_\eta \circ s^\eta_x)(y) = (h \circ s^{\text{free}}_x)(y).$$

If $\eta \neq \xi$, then the claim follows from the definition of s^{free} and the assumption. One can see that

$$(s^Z_{h(x)} \circ h)(y) = (s^Z_{\iota'_\eta(x)} \circ \iota'_\xi)(y) = \iota'_\xi(y) = h(y) = (h \circ s^{\text{free}}_x)(y).$$

This proves the claim, which completes the proof of the proposition. □

The above proposition gives us a relationship between quandle automorphism groups. Let us consider the automorphism group $\mathrm{Aut}(\bigsqcup^{\text{free}}_{\lambda\in\Lambda} Q_\lambda)$ of the interaction-free union quandle, and the direct product group $\prod_{\lambda\in\Lambda} \mathrm{Aut}(Q_\lambda)$ of the automorphism groups of Q_λ. The next proposition means that the latter can be understood as a subgroup of the former.

PROPOSITION 2.14. *Let $(f_\lambda)_\lambda \in \prod_{\lambda\in\Lambda} \mathrm{Aut}(Q_\lambda)$. Then the map $\bigsqcup_{\lambda\in\Lambda} f_\lambda$ defined by the following is an automorphism on the quandle $\bigsqcup^{\text{free}}_{\lambda\in\Lambda} Q_\lambda$:*

$$\bigsqcup_{\lambda\in\Lambda} f_\lambda : \bigsqcup^{\text{free}}_{\lambda\in\Lambda} Q_\lambda \to \bigsqcup^{\text{free}}_{\lambda\in\Lambda} Q_\lambda : x_\eta \mapsto f_\eta(x_\eta) \quad (\textit{for } x_\eta \in Q_\eta).$$

PROOF. We consider $Z := \bigsqcup^{\mathrm{free}}_{\lambda\in\Lambda} Q_\lambda$ and $\iota'_\lambda := \iota_\lambda \circ f_\lambda$. Then they satisfy the assumption of Proposition 2.13. We thus have a quandle homomorphism h, which satisfies for any $x_\eta \in Q_\eta$ that

$$h(x_\eta) = \iota'_\eta(x_\eta) = (\iota_\eta \circ f_\eta)(x_\eta) = f_\eta(x_\eta) = (\bigsqcup_{\lambda\in\Lambda} f_\lambda)(x_\eta).$$

This yields that $\bigsqcup_{\lambda\in\Lambda} f_\lambda$ is a homomorphism. One also knows that it is bijective, since $\bigsqcup_{\lambda\in\Lambda}(f_\lambda^{-1})$ is the inverse map. $\square$

We saw some properties of interaction-free union quandles, which are similar to the case of direct product quandles. However, we cannot say that s^{free} is a canonical quandle structure on $\bigsqcup_{\lambda\in\Lambda} Q_\lambda$, in the following sense.

REMARK 2.15. A quandle structure s' on $Q := \bigsqcup_{\lambda\in\Lambda} Q_\lambda$, satisfying that each inclusion map $\iota_\lambda : Q_\lambda \to (Q, s')$ is a quandle homomorphism, is not unique in general. A simple example is given by the dihedral quandle R_4. Let us consider

$$R_4 = \{0, 1, 2, 3\}, \quad Q_1 := \{0, 2\}, \quad Q_2 := \{1, 3\}.$$

Then Q_1 and Q_2 are trivial quandles. Therefore, $R_4 = Q_1 \bigsqcup Q_2$ as a set, both inclusion maps are quandle homomorphisms, but the quandle structure on R_4 is different from s^{free}.

We also note that, in the category of quandles and quandle homomorphisms, the interaction-free union $\bigsqcup^{\mathrm{free}}_{\lambda\in\Lambda} Q_\lambda$ is not the coproduct of $\{Q_\lambda\}_{\lambda\in\Lambda}$ in general.

2.4. Symmetric spaces. Symmetric spaces form an important subclass of quandles. As is well-known, there are several equivalent definitions of symmetric spaces. For the following definition, we refer to the book by Loos ([**6**]).

DEFINITION 2.16. A smooth manifold M equipped with a family of diffeomorphisms $\{s_x : M \to M\}_{x\in M}$ indexed by M is called a *symmetric space* if

(1) The map $M \times M \to M : (x, y) \mapsto s_x(y)$ is smooth.
(2) For each $x \in M$, the point x is an isolated fixed point of $s_x : M \to M$.
(3) $s_x^2 = \mathrm{id}_M$ for each $x \in M$.
(4) $s_x \circ s_y = s_y \circ s_{s_y(x)}$ for each $x, y \in M$.

The diffeomorphism $s_x : M \to M$ is called the *point symmetry* at $x \in M$. Note that Loos ([**6**]) uses a binary operator $x \cdot y$, but the above definition is a paraphrase in terms of the correspondence $s_x(y) = x \cdot y$.

REMARK 2.17. According to this definition, one can easily see that every symmetric space is a quandle. We identify a family of diffeomorphisms $\{s_x : M \to M\}_{x\in M}$ with a quandle structure $s : M \to \mathrm{Map}(M, M)$ naturally.

In the following we give fundamental examples of symmetric spaces, such as spheres and real projective spaces. For both examples, we use the standard inner-product $\langle \cdot, \cdot \rangle$ on $\mathbb{R}^{n+1}$, and the reflection $r_V : \mathbb{R}^{n+1} \to \mathbb{R}^{n+1}$ with respect to a subspace V. Recall that r_V is defined by

$$r_V(v + w) = v - w \quad (\text{for any } v \in V \text{ and } w \in V^\perp),$$

where $V^\perp$ is the orthogonal compliment of V in $\mathbb{R}^{n+1}$ with respect to $\langle , \rangle$.

EXAMPLE 2.18. Let $n \in \mathbb{Z}_{>0}$. We realize the n-dimensional sphere as

$$S^n := \{x \in \mathbb{R}^{n+1} \mid \langle x, x \rangle = 1\}.$$

For each $x \in S^n$, we define $s_x := r_{\mathbb{R}x}$, the reflection with respect to the line $\mathbb{R}x$. One also can write s_x as

$$s_x : S^n \to S^n : y \mapsto -y + 2\langle x, y \rangle x.$$

Then S^n is a compact symmetric space with point symmetries $s := \{s_x\}_{x \in S^n}$. One can easily see that the orthogonal group $\mathrm{O}(n+1)$ can be considered as a subgroup of $\mathrm{Aut}(S^n, s)$. Since the action of $\mathrm{O}(n+1)$ on S^n is transitive, spheres are homogeneous quandles.

Note that $\mathrm{Aut}(S^n, s)$ is the automorphism group of the quandle (S^n, s) in the sense of Definition 2.3. There is another notion of the automorphism groups of symmetric spaces, but we do not use it in this paper.

EXAMPLE 2.19. Let $n \in \mathbb{Z}_{>0}$. We realize the n-dimensional real projective space by

$$P(\mathbb{R}^{n+1}) := \{\text{one-dimensional subspaces in } \mathbb{R}^{n+1}\}.$$

For each $\ell \in P(\mathbb{R}^{n+1})$, in terms of the reflection r_ℓ, let us define

$$s_\ell : P(\mathbb{R}^{n+1}) \to P(\mathbb{R}^{n+1}) : \ell' \mapsto r_\ell(\ell').$$

Then $P(\mathbb{R}^{n+1})$ is a compact symmetric space with respect to the point symmetries $s := \{s_\ell\}_{\ell \in P(\mathbb{R}^{n+1})}$. One can easily see that the orthogonal group $\mathrm{O}(n+1)$ acts on $P(\mathbb{R}^{n+1})$ as automorphisms, and the projective orthogonal group $\mathrm{O}(n+1)/\{\pm I_{n+1}\}$ can be considered as a subgroup of $\mathrm{Aut}(P(\mathbb{R}^{n+1}), s)$. Since the action of $\mathrm{O}(n+1)$ on $P(\mathbb{R}^{n+1})$ is transitive, real projective spaces are homogeneous quandles.

3. Poles and antipodal subsets

There are notions of poles and antipodal subsets in the theory of (Riemannian) symmetric spaces. In this section, we directly transfer these notions to quandles. Throughout this section, $Q = (Q, s)$ denotes a quandle, and $\mathrm{Aut}(Q)$ denotes the automorphism group of Q.

3.1. Poles. In this subsection, we introduce the notion of poles in quandles, and describe some examples. We start from the notion of pole pairs in a quandle (Q, s), which gives a binary relation on Q.

DEFINITION 3.1. Let $x, y \in Q$. Then a pair (x, y) is called a *pole pair* in Q if it satisfies $s_x = s_y$.

By definition, (x, x) is always a pole pair, which is said to be *trivial.* We are interested in studying non-trivial pole pairs in given quandles.

REMARK 3.2. The notion of poles has been defined for symmetric spaces by Chen and Nagano ([**2**]). In fact, the condition of the original definition is different, but $s_x = s_y$ is one of equivalent conditions for (x, y) to be a pole pair ([**2**, Proposition 2.9]).

In terms of the notion of pole pairs, one can define the notion of pole subsets of quandles.

DEFINITION 3.3. A subset X in Q is called a *pole subset* if (x, y) is a pole pair for any $x, y \in X$.

It is clear that any subset of a pole subset is also a pole subset. Therefore it is natural to consider the maximal ones with respect to inclusion relation. We denote the collection of all maximal pole subsets in Q by

$$\mathcal{MS}(Q; \mathrm{pole}).$$

PROPOSITION 3.4. *Let $f \in \mathrm{Aut}(Q)$. If X is a pole or maximal pole subset in Q, then $f(X)$ is also a pole or maximal pole subset in Q, respectively. In particular, the automorphism group* $\mathrm{Aut}(Q)$ *acts on* $\mathcal{MS}(Q; \mathrm{pole})$.

PROOF. By the definition of quandle automorphisms, we have

$$f \circ s_x \circ f^{-1} = s_{f(x)} \quad (\forall x \in Q).$$

Therefore, (x, y) is a pole pair if and only if so is $(f(x), f(y))$. Then one can easily show the first assertion, that is, if X is a pole subset then so is $f(X)$.

In order to show the assertion on maximality, assume that X is a maximal pole subset in Q. Then $f(X)$ is a pole subset. Take a pole subset B with $f(X) \subset B$. Since $f^{-1}(B)$ is a pole subset containing X, one has $X = f^{-1}(B)$, that is $f(X) = B$. This completes the proof of the maximality of $f(X)$ as a pole subset. □

By this proposition, it is natural to consider the classification of maximal pole subsets, up to automorphisms. This problem is essentially equivalent to determine the orbit space of the action of $\mathrm{Aut}(Q)$,

$$\mathrm{Aut}(Q) \backslash \mathcal{MS}(Q; \mathrm{pole}).$$

We here note that a maximal pole subset is determined by one element in the following sense. For each $x \in Q$, we denote by

$$P(Q; x) := \{y \in Q \mid (x, y) \text{ is a pole pair in } Q\}.$$

Then $P(Q; x)$ is a maximal pole subset in Q, since the concept of pole pairs defines an equivalent relation on Q. Conversely, a maximal pole subset in Q containing x coincides with $P(Q; x)$. By this argument, for a homogeneous quandle Q, the above orbit space is always a singleton, that is, a maximal pole subset in Q is unique up to automorphisms.

Although we know the uniqueness of pole subsets in homogeneous quandles, it is a different problem to give an explicit expression. In the remaining of this subsection, we describe maximal pole subsets in the quandles given in the previous section. Note that the maximal pole subsets in the spheres S^n and the real projective spaces $P(\mathbb{R}^{n+1})$ have been well-known. We here recall them for the reader's convenience.

EXAMPLE 3.5. Let us consider the n-dimensional sphere S^n as in Example 2.18. Then a pair of points (x, y) on S^n is a pole pair if and only if $x = \pm y$. Therefore every maximal pole subset in S^n is of the form $\{\pm x\}$.

PROOF. Take $x, y \in S^n$. Recall that s_x and s_y are the reflections with respect to the lines $\mathbb{R}x$ and $\mathbb{R}y$, respectively. Therefore, (x, y) is a pole pair if and only if $\mathbb{R}x = \mathbb{R}y$, that is, $x = \pm y$. □

We next describe pole subsets in the real projective spaces $P(\mathbb{R}^{n+1})$. Recall that the point symmetry s_ℓ at $\ell \in P(\mathbb{R}^{n+1})$ is defined by the reflection $r_\ell : \mathbb{R}^{n+1} \to \mathbb{R}^{n+1}$ with respect to ℓ.

EXAMPLE 3.6. Let us consider the n-dimensional real projective space $P(\mathbb{R}^{n+1})$ as in Example 2.19. If $n > 1$, then $P(\mathbb{R}^{n+1})$ does not admit non-trivial pole pairs. If $n = 1$, then every maximal pole subset in $P(\mathbb{R}^2)$ is of the form $\{\ell_1, \ell_2\}$, where ℓ_1 is perpendicular to ℓ_2.

PROOF. Take $\ell_1, \ell_2 \in P(\mathbb{R}^{n+1})$. First of all, we claim that (ℓ_1, ℓ_2) is a pole pair if and only if the reflections satisfy $r_{\ell_1} = \pm r_{\ell_2}$. The "if"-part is easy to check. In order to show the converse, assume that (ℓ_1, ℓ_2) is a pole pair. By definition, one has $r_{\ell_1}(\ell) = r_{\ell_2}(\ell)$ for every $\ell \in P(\mathbb{R}^{n+1})$. We consider

$$f := r_{\ell_2}^{-1} \circ r_{\ell_1} \in \mathrm{O}(n+1).$$

Since it is orthogonal and normalizes $\mathbb{R}e_i$ for each $i \in \{1, \ldots, n+1\}$, there exists $\varepsilon_i \in \{\pm 1\}$ such that $f(e_i) = \varepsilon_i e_i$. Furthermore, since f normalizes $\mathbb{R}(e_i + e_j)$, one can see that $\varepsilon_i = \varepsilon_j$. This shows $f = \pm\,\mathrm{id}$, which completes the proof of the claim.

Note that each r_ℓ has eigenvalues 1 and -1 with multiplicities 1 and n, respectively. Therefore, if $n > 1$, then one has $r_{\ell_1} \neq -r_{\ell_2}$. This yields that, in this case, (ℓ_1, ℓ_2) is a pole pair if and only if $r_{\ell_1} = r_{\ell_2}$, that is, $\ell_1 = \ell_2$. This completes the proof of the case of $n > 1$. For the case of $n = 1$, one can see that $r_{\ell_1} = r_{\ell_2}$ is equivalent to $\ell_1 = \ell_2$, and $r_{\ell_1} = -r_{\ell_2}$ is equivalent to $\ell_1 \perp \ell_2$. □

Finally in this subsection, we describe pole subsets of the dihedral quandles R_n. It would be interesting that it depends on the parity of n.

EXAMPLE 3.7. Let us consider the dihedral quandle R_n of order n as in Example 2.5. Then we have the following:

(1) If n is odd, then R_n does not admit non-trivial pole pairs.
(2) If n is even, say $n = 2l$ with $l \in \mathbb{Z}_{>0}$, then (x, y) is a pole pair if and only if $x \in \{y, y+l\}$. Therefore every maximal pole subset in R_n is of the form $\{x, x+l\}$.

PROOF. Recall that $s_x(z) = 2x - z$ for $R_n = \mathbb{Z}/n\mathbb{Z}$. Hence, (x, y) is a pole pair in R_n if and only if

$$2x = 2y \quad (\text{as in } \mathbb{Z}/n\mathbb{Z}).$$

If n is odd, then $2(x - y) = 0$ is equivalent to $x - y = 0$, which completes the proof of (1). If $n = 2l$ with $l \in \mathbb{Z}_{>0}$, then $2(x - y) = 0$ is equivalent to $x - y \in \{0, l\}$. Therefore, in this case, a maximal pole subset in R_n containing x is of the form $\{x, x + l\}$. □

3.2. Antipodal subsets. In this subsection, we introduce and study antipodal subsets in quandles Q. Similarly to the studies on poles in the previous subsection, we start from a binary relation on Q.

DEFINITION 3.8. Let $x, y \in Q$. Then (x, y) is called an *antipodal pair* on Q if it satisfies $s_x(y) = y$ and $s_y(x) = x$.

We then define the notion of antipodal subsets. The definition would be simple and natural, but this notion turns out to be quite interesting for many cases.

DEFINITION 3.9. A subset X in Q is called an *antipodal subset* if (x, y) is an antipodal pair in Q for any $x, y \in X$.

It should be noted that any antipodal subset is a subquandle, which is intrinsically a trivial quandle. Namely, one can rephrase antipodal subsets as subquandles which are trivial quandles.

REMARK 3.10. As mentioned in the introduction, the notion of antipodal subsets has also been defined for symmetric spaces by Chen and Nagano ([**2**]). Note that, for connected Riemannian symmetric spaces, $s_x(y) = y$ and $s_y(x) = x$ are equivalent. In general they are not equivalent for quandles.

As in the case of pole subsets, it is easy to see that any subset of an antipodal subset is also antipodal. Therefore it is natural to consider the maximal ones with respect to inclusion relation. We denote the collection of all maximal antipodal subsets in Q by

$$\mathcal{MS}(Q; \text{antipodal}).$$

PROPOSITION 3.11. *Let $f \in \mathrm{Aut}(Q)$. If X is an antipodal or maximal antipodal subset in Q, then $f(X)$ is also an antipodal or maximal antipodal subset in Q, respectively. In particular,* $\mathrm{Aut}(Q)$ *acts on* $\mathcal{MS}(Q; \text{antipodal})$.

PROOF. Recall that $f \circ s_x \circ f^{-1} = s_{f(x)}$ holds for every $x \in Q$. Therefore, a pair (x, y) is antipodal if and only if so is $(f(x), f(y))$. Then the assertions of this proposition can be proved by the same argument as for Proposition 3.4. □

Similarly to the case of poles, it is natural to consider the classification of maximal antipodal subsets up to automorphisms. This problem is essentially equivalent to determine the orbit space

$$\mathrm{Aut}(Q)\backslash\mathcal{MS}(Q; \text{antipodal}).$$

However, in contrast to maximal pole subsets, maximal antipodal subsets are not necessarily unique even in homogeneous quandles and symmetric spaces.

REMARK 3.12. The study on maximal antipodal subsets in symmetric spaces has been initiated in [**2**]. Among others, it has been known that the uniqueness holds for several symmetric spaces, such as compact Hermitian symmetric spaces ([**2**]) and symmetric R-spaces ([**9**]). However the uniqueness does not hold in general, in which cases the classifications of all maximal antipodal subsets are sometimes involved. For example, the classification has not been completed for the oriented real Grassmannians $G_k(\mathbb{R}^n)^\sim$ of oriented k-planes in $\mathbb{R}^n$ with higher k. For more details and recent results, we refer to [**10–12**] and references therein.

We here recall the classification of maximal antipodal subsets in the spheres S^n and the real projective spaces $P(\mathbb{R}^{n+1})$. In both cases, a maximal antipodal subset is unique up to automorphisms. In fact we show a priori stronger statements, that is, it is unique up to the congruence by $\mathrm{SO}(n+1)$.

EXAMPLE 3.13. Let S^n be the n-dimensional sphere as in Example 2.18. Then every antipodal pair in S^n is a pole pair. Therefore, every antipodal subset in S^n is a pole subset, and a maximal antipodal subset in S^n is unique up to the congruence by $\mathrm{SO}(n+1)$.

PROOF. Let (x, y) be an antipodal pair in S^n. Since s_x is the reflection with respect to the line $\mathbb{R}x$, it follows from $s_x(y) = y$ that $y \in \mathbb{R}x$. This yields that (x, y) is a pole pair in S^n. Therefore, every maximal antipodal subset in S^n is of the form $\{\pm x\}$, which is unique up to the congruence by $\mathrm{SO}(n+1)$. □

We next study the real projective space $P(\mathbb{R}^{n+1})$. In this case, antipodal subsets are not necessarily pole subsets.

EXAMPLE 3.14. Let us consider the n-dimensional real projective space $P(\mathbb{R}^{n+1})$ as in Example 2.19. A subset X in $P(\mathbb{R}^{n+1})$ is maximal antipodal if and only if there exists an orthonormal basis $\{x_1, \ldots, x_{n+1}\}$ of $\mathbb{R}^{n+1}$ such that

$$X = \{\mathbb{R}x_1, \ldots, \mathbb{R}x_{n+1}\}.$$

In particular, a maximal antipodal subset in $P(\mathbb{R}^{n+1})$ is unique up to the congruence by $\mathrm{SO}(n+1)$.

PROOF. For each $\ell \in P(\mathbb{R}^{n+1})$, recall that s_ℓ is given as the reflection with respect to the line ℓ. Therefore, for $\ell_1, \ell_2 \in P(\mathbb{R}^{n+1})$, they satisfy $s_{\ell_1}(\ell_2) = \ell_2$ if and only if

$$\ell_2 = \ell_1 \text{ or } \ell_2 \perp \ell_1.$$

By applying this condition, one can directly prove the assertion. In fact, the "if"-part of the assertion is easy. In order to show the converse implication, let X be a maximal antipodal subset in $P(\mathbb{R}^{n+1})$. Take $\ell_1 \in X$. By definition of antipodal subsets, one has

$$X \subset \mathrm{Fix}(s_{\ell_1}, P(\mathbb{R}^{n+1})),$$

where the right-hand side denotes the fixed point set of s_{ℓ_1}. Then, according to the above condition, we have

$$\mathrm{Fix}(s_{\ell_1}, P(\mathbb{R}^{n+1})) = \{\ell_1\} \cup \{\ell \in P(\mathbb{R}^{n+1}) \mid \ell \perp \ell_1\}.$$

The second component of the right-hand side is $P(\ell_1^\perp) \cong P(\mathbb{R}^n)$, and $X - \{\ell_1\}$ is a maximal antipodal subset in it. Therefore, by an induction on the dimension, we obtain $X = \{\ell_1, \ldots, \ell_{n+1}\}$, consisting of $n+1$ orthogonal lines. The uniqueness follows from the fact that all orthonormal bases of $\mathbb{R}^{n+1}$ are congruent by $\mathrm{SO}(n+1)$ up to sign. □

The above proof was essentially given by Chen and Nagano (see Proposition 2.10 and Example 2.11 in [**2**]). Note that they proved a proposition in more general settings, and our proof describes some details for the case of $P(\mathbb{R}^{n+1})$.

EXAMPLE 3.15. Let R_n be the dihedral quandle R_n of order n as in Example 2.5. Then every antipodal pair in R_n is a pole pair. Therefore, every antipodal subset in R_n is a pole subset, and a maximal antipodal subset in R_n is unique up to automorphisms.

PROOF. Let (x, y) be an antipodal pair in R_n. Then it satisfies

$$y = s_x(y) = 2x - y \quad (\text{in } R_n).$$

Therefore, it follows from an argument in the proof of Example 3.7 that (x, y) is a pole pair. The uniqueness follows from the fact that the cyclic group $\mathbb{Z}/n\mathbb{Z}$ acts on R_n as automorphisms. □

4. s-commutative subsets

Let $Q = (Q, s)$ be a quandle. In this section, we introduce the concept of s-commutative subsets in Q, which is a generalization of antipodal subsets. We also prove that any maximal s-commutative subset in Q is a subquandle.

4.1. s-commutative subsets. In this subsection, we introduce the concept of s-commutative subsets, and study some basic properties. Similarly to the cases of poles and antipodal subsets, we start from s-commutative pairs of points.

DEFINITION 4.1. Let $x, y \in Q$. Then (x, y) is called an *s-commutative pair* in Q if it satisfies $s_x \circ s_y = s_y \circ s_x$.

The following gives a characterization of s-commutative pairs in terms of pole pairs, which is simple but useful.

LEMMA 4.2. *Let $x, y \in Q$. Then the following three conditions are mutually equivalent:*

(1) *(x, y) is an s-commutative pair in Q.*
(2) *$(x, s_y(x))$ is a pole pair in Q.*
(3) *$(s_x(y), y)$ is a pole pair in Q.*

PROOF. By the axiom of quandles, we have

$$s_x \circ s_y \circ s_x^{-1} = s_{s_x(y)}. \tag{4.1}$$

Therefore, s_x and s_y are commutative if and only if $s_y = s_{s_x(y)}$ holds. This proves the equivalence of (1) and (3). By changing the roles of x and y, one can easily see that (1) and (2) are equivalent. □

Note that every pole pair is an antipodal pair. The next proposition gives a relationship between antipodal pairs and s-commutative pairs. Recall that a pole pair (x, y) is said to be trivial if $x = y$ holds.

PROPOSITION 4.3. *Every antipodal pair in Q is an s-commutative pair in Q. The converse also holds if Q does not admit non-trivial pole pairs.*

PROOF. Let (x, y) be an antipodal pair in Q. Then we have $s_y(x) = x$. In particular, $(x, s_y(x)) = (x, x)$ is a pole pair. It then follows from Lemma 4.2 that (x, y) is an s-commutative pair in Q, which proves the first assertion.

We show the second assertion. Let (x, y) be an s-commutative pair in Q. Then, by Lemma 4.2, both of $(x, s_y(x))$ and $(s_x(y), y)$ are pole pairs in Q. Since Q does not admit non-trivial pole pairs by assumption, we have $x = s_y(x)$ and $s_x(y) = y$, which yields that (x, y) is antipodal. □

In terms of the notion of s-commutative pairs, one can naturally define s-commutative subsets in quandles.

DEFINITION 4.4. A subset X in Q is said to be *s-commutative* if (x, y) is an s-commutative pair in Q for any $x, y \in X$.

According to Proposition 4.3, one can easily obtain the following relationship between antipodal subsets and s-commutative subsets.

PROPOSITION 4.5. *Every antipodal subset in Q is an s-commutative subset in Q. The converse also holds if Q does not admit non-trivial pole pairs.*

One can apply this proposition to the real projective space $P(\mathbb{R}^{n+1})$ with $n \geq 2$. In such cases, a subset is s-commutative if and only if it is antipodal. However, as we mentioned in the introduction, an s-commutative subset in a quandle Q is not necessarily antipodal in general.

4.2. Maximal s-commutative subsets. As with the pole subsets and antipodal subsets, any subset of an s-commutative subset is also s-commutative. Therefore it is natural to consider the maximal ones with respect to inclusion relation. In this subsection, we study maximal s-commutative subsets in quandles. We denote the collection of all maximal s-commutative subsets in Q by

$$\mathcal{MS}(Q; s\text{-commutative}).$$

PROPOSITION 4.6. *Let $f \in \mathrm{Aut}(Q)$. If X is an s-commutative or maximal s-commutative subset in Q, then $f(X)$ is also s-commutative or maximal s-commutative in Q, respectively. In particular, the automorphism group $\mathrm{Aut}(Q)$ acts on $\mathcal{MS}(Q; s\text{-commutative})$.*

PROOF. One can prove the proposition similarly to the proofs of Propositions 3.4 and 3.11. □

As with the cases of maximal pole subsets and maximal antipodal subsets, the classification of maximal s-commutative subsets in Q up to automorphisms is essentially equivalent to determine the orbit space

$$\mathrm{Aut}(Q)\backslash\mathcal{MS}(Q; s\text{-commutative}).$$

For example, this orbit space is a singleton if and only if a maximal s-commutative subset in Q is unique up to automorphisms. In fact, such phenomena do occur for many symmetric spaces and quandles.

The following is the main theorem of this section. Recall that a subset X in Q is a subquandle if and only if it satisfies $s_x(X) = X$ for any $x \in X$.

THEOREM 4.7. *Let X be a maximal s-commutative subset in Q. Then X is a subquandle in Q.*

PROOF. Take any $x \in X$. First of all, we claim that $X \cup s_x(X)$ is an s-commutative subset in Q. Note that X is s-commutative by assumption, and so is $s_x(X)$ by Proposition 4.6. Take any $y, z \in X$, and show that $(y, s_x(z))$ is an s-commutative pair. Since (x, z) is an s-commutative pair, $(z, s_x(z))$ is a pole pair by Lemma 4.2. Then we have

$$s_y \circ s_{s_x(z)} = s_y \circ s_z = s_z \circ s_y = s_{s_x(z)} \circ s_y,$$

which completes the proof of the claim.

One knows $X \subset X \cup s_x(X)$. Since X is maximal s-commutative and $X \cup s_x(X)$ is s-commutative, we have $X = X \cup s_x(X)$. In particular, it satisfies

$$s_x(X) \subset X.$$

Since $s_x(X)$ is maximal s-commutative, one can prove $s_x(X) = X$ by the same argument. □

5. Subsets in direct products and interaction-free unions of quandles

In this section, we consider the direct product quandle $\prod_{\lambda \in \Lambda} Q_\lambda$ and interaction-free union quandle $\bigsqcup^{\mathrm{free}}_{\lambda \in \Lambda} Q_\lambda$ of a family of quandles $\{Q_\lambda\}_{\lambda \in \Lambda}$. We study pole, antipodal, and s-commutative subsets in these quandles. The results of this section yield that, reduction arguments to each component Q_λ work for many cases, but not always.

5.1. Subsets in direct products of quandles. In this subsection, we consider the direct product quandle $\prod_{\lambda \in \Lambda} Q_\lambda$ of a family of quandles $\{Q_\lambda\}_{\lambda \in \Lambda}$, defined in Subsection 2.2. For simplicity of the notation we denote it just by $\prod Q_\lambda$. First of all, we study properties of pairs of points.

PROPOSITION 5.1. *Let $(x_\lambda)_\lambda, (y_\lambda)_\lambda \in \prod Q_\lambda$. Then, $((x_\lambda)_\lambda, (y_\lambda)_\lambda)$ is a pole, antipodal, or s-commutative pair in $\prod Q_\lambda$ if and only if (x_η, y_η) is a pole, antipodal, or s-commutative pair in Q_η for every $\eta \in \Lambda$, respectively.*

PROOF. Denote the quandle structures on Q_λ and $\prod Q_\lambda$ by s^λ and s, respectively. Then, by definition, one knows

$$s_{(x_\lambda)_\lambda}((z_\lambda)_\lambda) = (s^\lambda_{x_\lambda}(z_\lambda))_\lambda.$$

Therefore, $s_{(x_\lambda)_\lambda} = s_{(y_\lambda)_\lambda}$ holds if and only if it satisfies $s^\eta_{x_\eta} = s^\eta_{y_\eta}$ for every $\eta \in \Lambda$. This proves the assertion on pole pairs. Similarly one can show the other assertions on antipodal pairs and s-commutative pairs. □

By applying this properties of pairs, one can obtain properties of subsets. Recall that $p_\eta : \prod Q_\lambda \to Q_\eta$ denotes the projection for $\eta \in \Lambda$. For a family of subsets $\{A_\lambda \subset Q_\lambda\}_{\lambda \in \Lambda}$, one can naturally define the subset $\prod A_\lambda := \prod_{\lambda \in \Lambda} A_\lambda$ in $\prod Q_\lambda$.

PROPOSITION 5.2. *Let A be a subset in $\prod Q_\lambda$. Then, A is a pole, antipodal, or s-commutative subset in $\prod Q_\lambda$ if and only if $p_\eta(A)$ is a pole, antipodal, or s-commutative subset in Q_η for every $\eta \in \Lambda$, respectively. In particular, if A_λ is pole, antipodal, or s-commutative in Q_λ for every $\lambda \in \Lambda$, then $\prod A_\lambda$ is pole, antipodal, or s-commutative in $\prod Q_\lambda$, respectively.*

PROOF. The first assertion is a direct consequence of Proposition 5.1. The second one follows easily from $p_\eta(\prod A_\lambda) = A_\eta$ and the first assertion. □

Note that it is not true in general that every pole, antipodal, or s-commutative subset in $\prod Q_\lambda$ is of the form $\prod A_\lambda$. However, this is in fact true for the maximal cases.

PROPOSITION 5.3. *Let A be a subset in $\prod Q_\lambda$. Then, A is a maximal pole, maximal antipodal, or maximal s-commutative subset in $\prod Q_\lambda$ if and only if there exists a maximal pole, maximal antipodal, or maximal s-commutative subset A_λ in Q_λ for every $\lambda \in \Lambda$, respectively, such that $A = \prod A_\lambda$.*

PROOF. The proof follows from Proposition 5.2. We only prove the case of s-commutative subsets, since the other cases are completely the same.

First of all, assume that A_λ is maximal s-commutative in Q_λ for each $\lambda \in \Lambda$, and prove that $\prod A_\lambda$ is maximal s-commutative in $\prod Q_\lambda$. One knows $\prod A_\lambda$ is s-commutative. Let us take an s-commutative subset A' in $\prod Q_\lambda$ with $\prod A_\lambda \subset A'$. Then one has $A_\eta \subset p_\eta(A')$ for each $\eta \in \Lambda$. Since A_η is maximal s-commutative and $p_\eta(A')$ is s-commutative, we have $A_\eta = p_\eta(A')$. This yields $\prod A_\lambda = A'$, which shows the maximality of $\prod A_\lambda$.

In order to show the converse, let A be a maximal s-commutative subset in $\prod Q_\lambda$. Put $A_\lambda := p_\lambda(A)$. Then it satisfies $A \subset \prod A_\lambda$ by definition. Note that each A_λ is s-commutative, and hence so is $\prod A_\lambda$. Therefore the maximality of A as an s-commutative subset yields that $A = \prod A_\lambda$. One can also show that each A_λ is maximal s-commutative in Q_λ, which follows from the maximality of A and a similar argument. This completes the proof. □

This proposition can be rephrased in terms of bijective correspondences as follows. Recall that, for a quandle Q, we denote by

$$\mathcal{MS}(Q; \text{pole}), \quad \mathcal{MS}(Q; \text{antipodal}), \quad \mathcal{MS}(Q; s\text{-commutative}),$$

the set of all maximal pole, maximal antipodal, and maximal s-commutative subsets in Q, respectively.

THEOREM 5.4. *Let A_λ be a subset in Q_λ for each $\lambda \in \Lambda$, and consider the correspondence to a subset $\prod A_\lambda$ in $\prod Q_\lambda$. Then, this defines the following bijective maps, which are equivariant with respect to the natural actions by $\prod \mathrm{Aut}(Q_\lambda)$:*

(1) $\prod \mathcal{MS}(Q_\lambda; \text{pole}) \to \mathcal{MS}(\prod Q_\lambda; \text{pole})$.
(2) $\prod \mathcal{MS}(Q_\lambda; \text{antipodal}) \to \mathcal{MS}(\prod Q_\lambda; \text{antipodal})$.
(3) $\prod \mathcal{MS}(Q_\lambda; s\text{-commutative}) \to \mathcal{MS}(\prod Q_\lambda; s\text{-commutative})$.

Note that, by Proposition 2.9, the direct product group $\prod \mathrm{Aut}(Q_\lambda)$ acts naturally on the direct product quandle $\prod Q_\lambda$ as automorphisms.

5.2. Subsets in interaction-free unions of quandles. Let us consider the interaction-free union $\bigsqcup^{\text{free}}_{\lambda\in\Lambda} Q_\lambda$ of a family of quandles $\{Q_\lambda\}_{\lambda\in\Lambda}$, defined in Subsection 2.3. As in the previous subsection, we denote it by $\bigsqcup^{\text{free}} Q_\lambda$ for simplicity, and start with studying properties of pairs of points.

PROPOSITION 5.5. *Let $x \in Q_\lambda$ and $y \in Q_\eta$ with $\lambda, \eta \in \Lambda$.*

(1) *Assume $\lambda = \eta$. Then, (x, y) is a pole, antipodal, or s-commutative pair in $\bigsqcup^{\text{free}} Q_\lambda$ if and only if (x, y) is a pole, antipodal, or s-commutative pair in Q_λ, respectively.*
(2) *Assume $\lambda \neq \eta$. Then, (x, y) is a pole pair in $\bigsqcup^{\text{free}} Q_\lambda$ if and only if it satisfies $s^\lambda_x = \mathrm{id}_{Q_\lambda}$ and $s^\eta_y = \mathrm{id}_{Q_\eta}$. On the other hand, (x, y) is always an antipodal and s-commutative pair in $\bigsqcup^{\text{free}} Q_\lambda$.*

PROOF. Recall that $s^{\text{free}}_x = s^\eta_x$ on Q_η if $x \in Q_\eta$, and s^{free}_x is the identity map on Q_η if $x \notin Q_\eta$. Then all the assertions follow from this definition of s^{free}. For example, assume that (x, y) is a pole pair in $\bigsqcup^{\text{free}} Q_\lambda$ with $x \in Q_\lambda$, $y \in Q_\eta$, and $\lambda \neq \eta$. Then one can show that

$$s^\lambda_x = s^{\text{free}}_x |_{Q_\lambda} = s^{\text{free}}_y |_{Q_\lambda} = \mathrm{id}_{Q_\lambda}.$$

Other assertions can be proved easily. □

One can obtain properties of subsets A in $\bigsqcup^{\text{free}} Q_\lambda$, by using the properties of pairs. In the case of direct products, we have used the projections p_η for $\eta \in \Lambda$. For the case of interaction-free unions, the intersections $A \cap Q_\eta$ play similar roles.

PROPOSITION 5.6. *Let A be a subset in $\bigsqcup^{\text{free}} Q_\lambda$. If A is a pole, antipodal, or s-commutative subset in $\bigsqcup^{\text{free}} Q_\lambda$, then $A \cap Q_\eta$ is a pole, antipodal, or s-commutative subset in Q_η for every $\eta \in \Lambda$, respectively. The converse statements also hold for antipodal and s-commutative subsets.*

PROOF. The first and second assertions follow from (1) and (2) of Proposition 5.5, respectively. □

REMARK 5.7. One can easily show from Proposition 5.6 that, if A_λ is an antipodal or s-commutative subset in Q_λ for every $\lambda \in \Lambda$, then $\bigsqcup A_\lambda$ is an antipodal or s-commutative subset in $\bigsqcup^{\text{free}} Q_\lambda$, respectively. However, the same statement for pole subsets is not true in general. Namely, the union $\bigsqcup A_\lambda$ is not necessarily a pole subset even if all A_λ are pole. A simple example can be given by $R_3 \bigsqcup^{\text{free}} R_3$, the interaction-free union of two copies of R_3. Let us denote its element as $(a)_i$ with $a \in R_3$ and $i \in \{1, 2\}$. Then $\{(0)_1, (0)_2\}$ is the union of pole subsets, but not pole in $R_3 \bigsqcup^{\text{free}} R_3$.

We then study maximal subsets. Note that there are differences between properties of maximal pole subsets and those of maximal antipodal or s-commutative subsets. The following describes maximal pole subsets.

PROPOSITION 5.8. *Let A be a nonempty subset in $\bigsqcup^{\text{free}} Q_\lambda$. Then, A is a maximal pole subset in $\bigsqcup^{\text{free}} Q_\lambda$ if and only if one of the following holds:*

(i) *there exists $\eta \in \Lambda$ such that A is a maximal pole subset in Q_η, and there exists $x \in A$ satisfying $s_x^\eta \neq \mathrm{id}_{Q_\eta}$, or*
(ii) $A = \bigsqcup_{\lambda \in \Lambda} \{x \in Q_\lambda \mid s_x^\lambda = \mathrm{id}_{Q_\lambda}\}$.

PROOF. First of all, assume that A satisfies (i). Since A is a pole subset in Q_η, it is also pole in $\bigsqcup^{\text{free}} Q_\lambda$ by the definition of s^{free}. It also satisfies the maximality, since there exists $x \in A$ such that $s_x^\eta \neq \mathrm{id}_{Q_\eta}$. In fact, if (x, y) is a pole pair, then $y \in Q_\eta$ by Proposition 5.5, and hence $y \in A$ by the maximality of A in Q_η.

We next consider the case that A is of the form (ii). Since $A \neq \emptyset$ by assumption, there exists $x \in A$. Note that it satisfies $s_x^{\text{free}} = \mathrm{id}$. Therefore, (x, y) is a pole pair if and only if $s_y^{\text{free}} = \mathrm{id}$, which is equivalent to $y \in A$. This proves that A is a maximal pole subset in $\bigsqcup^{\text{free}} Q_\lambda$.

One can prove the converse implication by a similar argument. Let A be a maximal pole subset in $\bigsqcup^{\text{free}} Q_\lambda$. Since $A \neq \emptyset$, there exists $x \in A \cap Q_\eta$ for some $\eta \in \Lambda$. If it satisfies $s_x^\eta \neq \mathrm{id}_{Q_\eta}$, then one can show that A is of the form of (i). If it satisfies $s_x^\eta = \mathrm{id}_{Q_\eta}$, then A must be of the form of (ii). □

The properties of maximal antipodal and s-commutative subsets in $\bigsqcup^{\text{free}} Q_\lambda$ are described in the next proposition. Namely, a reduction argument to each component Q_λ works for these subsets.

PROPOSITION 5.9. *Let A be a subset in $\bigsqcup^{\text{free}} Q_\lambda$. Then, A is a maximal antipodal or s-commutative subset in $\bigsqcup^{\text{free}} Q_\lambda$ if and only if there exists a maximal antipodal or maximal s-commutative subset A_λ in Q_λ for every $\lambda \in \Lambda$, respectively, such that $A = \bigsqcup A_\lambda$.*

PROOF. The proof follows from Proposition 5.6. We only give a sketch of the proof for the s-commutative case. Let A be a maximal s-commutative subset in

$\bigsqcup^{\mathrm{free}} Q_\lambda$, and put $A_\lambda := A \cap Q_\lambda$ for each $\lambda \in \Lambda$. Then A_λ is s-commutative. If A'_λ is an s-commutative subset in Q_λ with $A_\lambda \subset A'_\lambda$, then one can see that

$$A = \bigsqcup A_\lambda \subset \bigsqcup A'_\lambda.$$

Note that $\bigsqcup A'_\lambda$ is s-commutative, and A has the maximality. Hence it satisfies $A_\lambda = A'_\lambda$, which yields that each A_λ is maximal s-commutative. The converse implication can be proved in a similar way. □

This proposition can be rephrased in terms of bijective correspondences as follows. Recall that a family of maximal antipodal subsets $\{A_\lambda \subset Q_\lambda\}_{\lambda\in\Lambda}$ is regarded as an element in $\prod \mathcal{MS}(Q_\lambda; \text{antipodal})$.

THEOREM 5.10. *Let A_λ be a subset in Q_λ for each $\lambda \in \Lambda$, and consider the correspondence to a subset $\bigsqcup A_\lambda$ in $\bigsqcup^{\mathrm{free}} Q_\lambda$. Then, this defines the following bijective maps, which are equivariant with respect to the natural actions by $\prod \mathrm{Aut}(Q_\lambda)$:*

(1) $\prod \mathcal{MS}(Q_\lambda; \text{antipodal}) \to \mathcal{MS}(\bigsqcup^{\mathrm{free}} Q_\lambda; \text{antipodal})$.
(2) $\prod \mathcal{MS}(Q_\lambda; s\text{-commutative}) \to \mathcal{MS}(\bigsqcup^{\mathrm{free}} Q_\lambda; s\text{-commutative})$.

We refer to Proposition 2.14 for the action of $\prod \mathrm{Aut}(Q_\lambda)$ on $\bigsqcup^{\mathrm{free}} Q_\lambda$. Note that a similar statement on maximal pole subsets cannot be true.

6. Examples

In this section, we determine maximal s-commutative subsets in the spheres S^n, the real projective spaces $P(\mathbb{R}^{n+1})$, and the dihedral quandles R_n. In these quandles, a maximal s-commutative subset is turned out to be unique up to automorphisms.

First of all, we study maximal s-commutative subsets in the spheres S^n, and prove Proposition 1.6. For this purpose, it is enough to show the next proposition.

PROPOSITION 6.1. *Let S^n be the n-dimensional sphere as in Example 2.18. Then we have the following:*

(1) *Let $x, y \in S^n$. Then a pair (x, y) is an s-commutative pair if and only if $x = \pm y$ or $\langle x, y\rangle = 0$.*
(2) *A subset $X \subset S^n$ is maximal s-commutative if and only if there exists an orthonormal basis $\{x_1, \ldots, x_{n+1}\}$ of $\mathbb{R}^{n+1}$ such that*

$$X = \{\pm x_1, \ldots, \pm x_{n+1}\}.$$

(3) *A maximal s-commutative subset in S^n is unique up to the congruence by $\mathrm{SO}(n+1)$.*

PROOF. Recall that (x, y) is an s-commutative pair if and only if $(x, s_y(x))$ is a pole pair by Lemma 4.2. Since every pole subset in the sphere S^n is of the form $\{\pm x\}$, the above condition is equivalent to

$$s_y(x) = \pm x.$$

In the case of $s_y(x) = x$, the pair (x, y) is antipodal, which is equivalent to $y = \pm x$. In the case of $s_y(x) = -x$, one can easily obtain $\langle x, y\rangle = 0$, which completes the proof of (1).

We show (2). The "if"-part follows from (1) easily. In order to show the "only if"-part, let X be a maximal s-commutative subset in S^n. Take $x_1 \in X$. Since any element in X is s-commutative with x_1, it follows from (1) that

$$X \subset \{\pm x_1\} \cup \{y \in S^n \mid \langle y, x_1 \rangle = 0\}.$$

Note that $X \cup \{-x_1\}$ is s-commutative, since $s_{x_1} = s_{-x_1}$. Hence the maximality of X yields that $\{\pm x_1\} \subset X$. Furthermore, since X is maximal s-commutative in S^n, we can show that $X - \{\pm x_1\}$ is maximal s-commutative in

$$\{y \in S^n \mid \langle y, x_1 \rangle = 0\} \cong S^{n-1}.$$

By an inductive argument, one can complete the proof of (2).

The assertion (3) is a direct consequence of (2). □

REMARK 6.2. Recall that every maximal s-commutative subset is a subquandle by Theorem 4.7. Therefore, $X := \{\pm x_1, \ldots, \pm x_{n+1}\}$ is a subquandle of S^n, where $\{x_1, \ldots, x_{n+1}\}$ is an orthonormal basis of $\mathbb{R}^{n+1}$. The quandle structure is given by

$$s_{x_i} = s_{-x_i}, \quad s_{x_i}(\pm x_j) = \begin{cases} \pm x_j & (\text{if } i = j), \\ \mp x_j & (\text{if } i \neq j). \end{cases}$$

If $n = 1$, then the quandle $\{\pm x_1, \pm x_2\}$ is isomorphic to the dihedral quandle R_4 of order 4. One can also see that, for every $n \in \mathbb{Z}_{>0}$, the above quandle X is homogeneous and disconnected. For more details of this quandle, we refer to [**3**].

We next study maximal s-commutative subsets in the real projective spaces $P(\mathbb{R}^{n+1})$. Proposition 1.7 follows from the next proposition. Recall that $\mathbb{R}x \in P(\mathbb{R}^{n+1})$ denotes the line spanned by $x \in \mathbb{R}^{n+1}$.

PROPOSITION 6.3. *Let $P(\mathbb{R}^{n+1})$ be the real projective space as in Example 2.19. Then we have the following:*

(1) *Let $n > 1$. Then a subset $X \subset P(\mathbb{R}^{n+1})$ is maximal s-commutative if and only if it is maximal antipodal, that is, there exists an orthonormal basis $\{x_1, \ldots, x_{n+1}\}$ of $\mathbb{R}^{n+1}$ such that*
$$X = \{\mathbb{R}x_1, \ldots, \mathbb{R}x_{n+1}\}.$$
(2) *Let $n = 1$. Then a subset $X \subset P(\mathbb{R}^2)$ is maximal s-commutative if and only if there exists an orthonormal basis $\{x_1, x_2\}$ of $\mathbb{R}^2$ such that*
$$X = \{\mathbb{R}x_1,\ \mathbb{R}x_2,\ \mathbb{R}(x_1 + x_2),\ \mathbb{R}(x_1 - x_2)\}.$$
(3) *For any $n \in \mathbb{Z}_{>0}$, a maximal s-commutative subset in $P(\mathbb{R}^{n+1})$ is unique up to the congruence by* $\mathrm{SO}(n+1)$.

PROOF. The case of $n > 1$ is easy. In this case, recall that $P(\mathbb{R}^{n+1})$ does not admit non-trivial pole pairs by Example 3.6. Therefore, Proposition 4.5 yields that $X \subset P(\mathbb{R}^{n+1})$ is maximal s-commutative if and only if it is maximal antipodal. Hence one can complete the proof of (1) by Example 3.14.

We then consider the case of $n = 1$. For simplicity of the notation, we denote by $\angle(\ell_1, \ell_2) \in [0, \pi/2]$ the angle between $\ell_1, \ell_2 \in P(\mathbb{R}^2)$. We claim that (ℓ_1, ℓ_2) is an s-commutative pair in $P(\mathbb{R}^2)$ if and only if

$$\angle(\ell_1, \ell_2) \in \{0, \pi/4, \pi/2\}.$$

It follows from Lemma 4.2 that (ℓ_1, ℓ_2) is an s-commutative pair if and only if $(\ell_1, s_{\ell_2}(\ell_1))$ is a pole pair. By Example 3.6, the latter condition is equivalent to

$$\angle(\ell_1, s_{\ell_2}(\ell_1)) \in \{0, \pi/2\}.$$

One can see that $\angle(\ell_1, s_{\ell_2}(\ell_1)) = 0$ is equivalent to $\angle(\ell_1, \ell_2) \in \{0, \pi/2\}$. Furthermore, $\angle(\ell_1, s_{\ell_2}(\ell_1)) = \pi/2$ is equivalent to $\angle(\ell_1, \ell_2) = \pi/4$. This completes the proof of the claim. By using this claim, the assertion (2) can be proved directly.

One can then show (3) by the shapes of maximal s-commutative subsets described in (1) and (2). □

Finally in this section, we study maximal s-commutative subsets in the dihedral quandles R_n, and prove Proposition 1.8. This follows from the next proposition. Recall Example 2.5, where the dihedral quandle is defined by $R_n := \mathbb{Z}/n\mathbb{Z}$ with quandle structure $s_x(y) = 2x - y$.

PROPOSITION 6.4. *Let R_n be the dihedral quandle of order n. Then we have the following:*

(1) *Let $x, y \in R_n$. Then a pair (x, y) is s-commutative if and only if it satisfies $4(x - y) = 0$ as in $\mathbb{Z}/n\mathbb{Z}$.*
(2) *If n is odd, then every maximal s-commutative subset is of the form $\{x\}$, which is a maximal pole subset.*
(3) *If $n = 4l - 2$ with $l \in \mathbb{Z}_{>0}$, then every maximal s-commutative subset is of the form $\{x, x + 2l - 1\}$, which is a maximal pole subset.*
(4) *If $n = 4l$ with $l \in \mathbb{Z}_{>0}$, then every maximal s-commutative subset is of the form $\{x, x + l, x + 2l, x + 3l\}$, which is not antipodal.*
(5) *For any $n \in \mathbb{Z}_{>0}$, a maximal s-commutative subset in R_n is unique up to the congruence by* $\mathrm{Aut}(R_n, s)$.

PROOF. Let $x, y \in R_n = \mathbb{Z}/n\mathbb{Z}$. In the following, we calculate everything as elements of the cyclic group $\mathbb{Z}/n\mathbb{Z}$. By the definition of the quandle structure of R_n, we have

$$s_x \circ s_y(z) = s_x(2y - z) = 2x - (2y - z) = 2x - 2y + z.$$

Therefore, a pair (x, y) is s-commutative if and only if

$$2x - 2y = 2y - 2x.$$

This completes the proof of (1). Then, by using this condition, one can prove (2), (3), and (4) as follows.

If n is odd, then $4(x - y) = 0$ is equivalent to $x - y = 0$. Therefore, in this case, every maximal s-commutative subset consists of a single point. This is also a maximal pole subset by Example 3.7, which completes the proof of (2).

If $n = 4l - 2$, then $4(x - y) = 0$ is equivalent to $x - y \in \{0, 2l - 1\}$. In this case, every maximal s-commutative subset is of the form $\{x, x + 2l - 1\}$. This is a maximal pole subset by Example 3.7, which completes the proof of (3).

If $n = 4l$, then $4(x - y) = 0$ is equivalent to $x - y \in \{0, l, 2l, 3l\}$. In this case, every maximal s-commutative subset consists of four points as desired. This is not antipodal by Example 3.15, which proves (4).

The uniqueness claimed in (5) is also easy to see, since the cyclic $\mathbb{Z}/n\mathbb{Z}$-action on R_n is an automorphism. □

REMARK 6.5. Recall that maximal s-commutative subsets are subquandles. In the case of R_{4l-2} with $l \in \mathbb{Z}_{>0}$, the maximal s-commutative subset is isomorphic to R_2, which is a trivial quandle. In the case of R_{4l} with $l \in \mathbb{Z}_{>0}$, the maximal s-commutative subset is isomorphic to R_4. One can see the pictures for the cases of $n \in \{6, 7, 8\}$ in Figure 1.

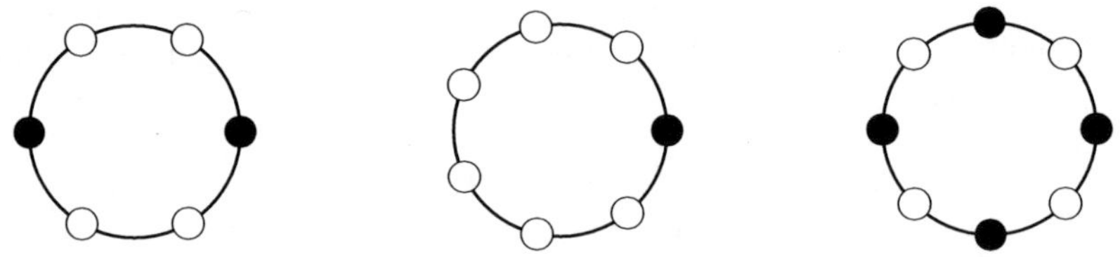

FIGURE 1. Maximal s-commutative subsets in R_n with $n \in \{6, 7, 8\}$

REMARK 6.6. Recall that every pole subset is antipodal, and every antipodal subset is s-commutative. These three notions are different for symmetric spaces. Recall that a maximal antipodal subset in the real projective space $P(\mathbb{R}^{n+1})$ is not pole, and a maximal s-commutative subset in the sphere S^n is not antipodal. We here note that these three notions are also different for finite quandles. By the above proposition, a maximal s-commutative subset in R_{4l} with $l \in \mathbb{Z}_{>0}$ is not antipodal. One can see the difference between pole subsets and antipodal subsets as follows. Consider the interaction-free union $X = \{a, 0, 1, 2\}$ of the trivial quandle $\{a\}$ and the dihedral quandle $R_3 = \{0, 1, 2\}$. Then the subset $\{a, 0\}$ is (maximal) antipodal, but not a pole subset, since $s_a^{\mathrm{free}} = \mathrm{id}_X \neq s_0^{\mathrm{free}}$.

References

[1] Bang-Yen Chen, *Two-numbers and their applications—a survey*, Bull. Belg. Math. Soc. Simon Stevin **25** (2018), no. 4, 565–596. MR3896273

[2] Bang-Yen Chen and Tadashi Nagano, *A Riemannian geometric invariant and its applications to a problem of Borel and Serre*, Trans. Amer. Math. Soc. **308** (1988), no. 1, 273–297, DOI 10.2307/2000963. MR946443

[3] K. Furuki and H. Tamaru, *Flat homogeneous quandle and vertex-transitive graphs*, preprint.

[4] Yoshitaka Ishihara and Hiroshi Tamaru, *Flat connected finite quandles*, Proc. Amer. Math. Soc. **144** (2016), no. 11, 4959–4971, DOI 10.1090/proc/13095. MR3544543

[5] David Joyce, *A classifying invariant of knots, the knot quandle*, J. Pure Appl. Algebra **23** (1982), no. 1, 37–65, DOI 10.1016/0022-4049(82)90077-9. MR638121

[6] Ottmar Loos, *Symmetric spaces. I: General theory*, W. A. Benjamin, Inc., New York-Amsterdam, 1969. MR0239005

[7] Masaru Takeuchi, *Two-number of symmetric R-spaces*, Nagoya Math. J. **115** (1989), 43–46, DOI 10.1017/S0027763000001513. MR1018081

[8] Hiroshi Tamaru, *Two-point homogeneous quandles with prime cardinality*, J. Math. Soc. Japan **65** (2013), no. 4, 1117–1134, DOI 10.2969/jmsj/06541117. MR3127819

[9] Makiko Sumi Tanaka and Hiroyuki Tasaki, *Antipodal sets of symmetric R-spaces*, Osaka J. Math. **50** (2013), no. 1, 161–169. MR3080635

[10] Makiko Sumi Tanaka and Hiroyuki Tasaki, *Maximal antipodal subgroups of some compact classical Lie groups*, J. Lie Theory **27** (2017), no. 3, 801–829. MR3611097

[11] Makiko Sumi Tanaka and Hiroyuki Tasaki, *Maximal antipodal sets of compact classical symmetric spaces and their cardinalities I*, Differential Geom. Appl. **73** (2020), 101682, 32, DOI 10.1016/j.difgeo.2020.101682. MR4149631

[12] Hiroyuki Tasaki, *Antipodal sets in oriented real Grassmann manifolds*, Internat. J. Math. **24** (2013), no. 8, 1350061, 28, DOI 10.1142/S0129167X13500614. MR3103877

Department of Food Sciences and Biotechnology, Faculty of Life Sciences, Hiroshima Institute of Technology, 2-1-1 Miyake, Saeki-ku, Hiroshima, 731-5193 Japan
Email address: a.kubo.3r@cc.it-hiroshima.ac.jp

Nariwa Junior High School, Takahashi-City, Okayama, 716-0121 Japan

Graduate School of Advanced Science and Engineering, Hiroshima University, 1-3-1 Kagamiyama, Higashi-Hiroshima, 739-8526 Japan
Email address: okudatak@hiroshima-u.ac.jp

Department of Mathematics, Graduate School of Science, Osaka City University, Sugimoto, Sumiyoshi-ku, Osaka, 558-8585 Japan
Email address: tamaru@sci.osaka-cu.ac.jp

Contemporary Mathematics
Volume **777**, 2022
https://doi.org/10.1090/conm/777/15625

Spectral gaps of the Laplacian on differential forms

Helton Leal and Zhiqin Lu

Abstract. In this short article, we explore some basic results associated to the Generalized Weyl criterion for the essential spectrum of the Laplacian on Riemannian manifolds. We use the language of Gromov-Hausdorff convergence to recall a spectral gap theorem. Finally, we make the necessary adjustments to extend our main results, and construct a class of complete noncompact manifolds with an arbitrarily large number of gaps in the spectrum of the Hodge Laplacian acting on differential forms.

1. Introduction

Let X be a complete noncompact Riemannian manifold of dimension n and denote by Δ the Laplace-Beltrami operator acting on smooth functions with compact support $\mathcal{C}_0^\infty(X)$. It is well-known that the unique self-adjoint extension of Δ on $L^2(X)$ is a nonnegative definite and densely defined linear operator.

The spectrum of Δ, denoted by $\sigma(\Delta)$, consists of all $\lambda \in \mathbb{C}$ for which $\Delta - \lambda I$ fails to be invertible. The essential spectrum of Δ, $\sigma_{\text{ess}}(\Delta)$, consists of the cluster points in the spectrum and of isolated eigenvalues of infinite multiplicity. The pure point spectrum is defined by

$$\sigma_{\text{pp}}(\Delta) = \sigma(\Delta) \backslash \sigma_{\text{ess}}(\Delta).$$

A consequence of the Hodge Decomposition Theorem is that, on a compact manifold,

$$\sigma(\Delta) = \sigma_{\text{pp}}(\Delta).$$

In the case of a noncompact manifold, however, the spectral structure is generally more complex than in the compact case.

Nonetheless, while it is impossibe to precisely compute the pure point spectrum for most compact manifolds, it is possible to locate the essential spectrum of the Laplacian for a large class of complete, noncompact Riemannian manifolds.

Historically, there are many results that exhibit a large class of noncompact manifolds whose essential spectrum is a connected subset of the real line. See, for example [**7, 9–11, 15, 23**]. Likewise, one can find large sets of manifolds for which the essential spectrum has an arbitrarily large number of "gaps" (that is, the number of connected components of $\mathbb{R} \backslash \sigma_{\text{ess}}(X)$ can be arbitrarily large).

Key words and phrases. Essential spectrum, spectral gap, Laplacian, Gromov-Hausdorff convergence.

The second author was partially supported by the NSF grant DMS-19-08513.

In [**2**], we explore this set of manifolds, in collaboration with N. Charalambous, by first considering spectral continuity and then observing the evolution of the spectrum of a sequence of manifolds under Gromov-Hausdorff convergence. Using these ideas, we prove the existence of gaps in the essential spectrum of a periodic manifold. Our result is, in essence, a new proof to that by Schoen and Tran [**22**], as well as Post [**19**], Lledó and Post [**16**] who used Floquet theory; Khrabustovskyi [**14**] who presented more refined results; and Exner and Post [**12**] who exhibit limiting results.

It is a more complicated task to obtain similar results for the spectrum of the Hodge Laplacian acting on k-forms on a manifold, due to the stronger connection of the operator to the curvature.

In this note, we generalize the ideas present in [**2**] to the k-form spectrum, and construct a class of complete noncompact manifolds with an arbitrary number of gaps in the spectrum of the Hodge-Laplacian on k-forms.

2. The spectrum of the Laplacian on k-forms

Given k-forms $\omega = a_{i_1\cdots i_k}\omega_{i_1} \wedge \cdots \wedge \omega_{i_k}$ and $\eta = b_{i_1\cdots i_k}\omega_{i_1} \wedge \cdots \wedge \omega_{i_k}$, where $(\omega_1 \cdots, \omega_n)$ is a orthonormal co-frame, the L^2 *inner product* in $\Lambda^k M$ is defined as

$$(\omega, \eta) = k! \int_M a_I b_I \, dV,$$

where dV is the volume form, $\{e_1, \cdots, e_n\}$ is a global orthonormal frame and $\{\omega_1, \cdots, \omega_n\}$ its dual frame, and I is the corresponding multi-index.

Let $d : \Lambda^k M \to \Lambda^{k+1} M$ be the exterior differential. The adjoint δ of d, which is called the *codifferential operator*, is $\delta : \Lambda^{k+1} M \to \Lambda^k M$ satisfying

$$(d\omega, \eta) = (\omega, \delta\eta)$$

for all $\omega \in \Lambda^k M, \eta \in \Lambda^{k+1} M$ with compact support. It is worth noting that, contrary to the differential operator d, its adjoint operator δ depends on the metric g on the manifold. We have

$$(\delta\eta)_{i_1\cdots i_k} = -(k+1)\nabla_s \eta_{s i_1 \cdots i_k}.$$

The *Hodge Laplacian*, also known as the *Laplace–de Rham* operator, is defined as

$$\Delta_k = \Delta := d\delta + \delta d = (d + \delta)^2. \tag{1}$$

Similar to the case of the Laplace-Beltrami operator, it is well-known that the Hodge Laplacian extends to a self-adjoint, nonnegative operator densely defined over $L^2(\Lambda^k M)$. In particular, when $k = 0$, the Hodge Laplacian coincides with the Laplace-Beltrami operator acting on functions.

The Weitzenböck formula (see [**18**], Theorem 9.4.1) gives that

$$\Delta\omega = \nabla^*\nabla\omega + E(\mathcal{R}),$$

where $\mathcal{R}$ is the curvature operator and $E(\mathcal{R})$ is an algebraic operator of $\mathcal{R}$. Here

$$\nabla^*\nabla = -\sum_i \nabla^2_{E_i},$$

where $\nabla^2_{E_i} = \nabla_{E_i}\nabla_{E_i} - \nabla_{\nabla_{E_i} E_i}$ for an orthonormal frame $\{E_i\}$, is called the connection Laplacian. The results of this paper also hold for connection Laplacian.

We denote $\sigma(k, \Delta)$ the spectrum of the Hodge Laplacian Δ_k. In order to use the language of Gromov-Hausdorff convergence (see [**4**]), we abuse notation and use $\sigma(k, \Delta)$ to refer to the pointed metric space

$$(\sigma(k, \Delta) \cup \{-1\}, -1).$$

Similarly, we use $\sigma_{\text{ess}}(k, \Delta)$ for the pointed metric space

$$(\sigma_{\text{ess}}(k, \Delta) \cup \{-1\}, -1).$$

Note that these definitions imply

$$\sigma(-1, \Delta) = \sigma(n+1, \Delta) = \sigma_{\text{ess}}(-1, \Delta) = \sigma_{\text{ess}}(n+1, \Delta) = \emptyset,$$

meaning that the set consists of a single point metric space $\{-1\}$.

We will omit the degree k of the differential form when there is no risk of confusion.

As mentioned above, directly computing the essential spectrum of the Laplacian on forms has been a complicated task, even for the case of 1-forms, due to their stronger connection to the curvature, and because of the lack of good test forms on the manifold.

In view of the above obstacle, it would be more efficient to study the evolution of the spectrum under various deformations of the manifolds. The first natural case to consider is the evolution of eigenvalues under the continuous deformation its Riemannian metric. J. Dodziuk proved the following result

THEOREM 2.1 ([**8**]). *Let X be a compact manifold and let g_t be a family of Riemannian metrics on X. Assume that*

$$g_t \to g$$

in the C^0 topology. Then the spectrum (eigenvalues) of g_t converges to the spectrum of g (as pointed metric spaces in the Gromov-Hausdorff sense).

A remarkable feature of the above theorem is that it doesn't depend on the curvature of the family of deformed manifolds. For the application in this paper, we need to use a generalized version Dodziuk's result. In the paper [**4**] (also see [**20**] for related results), N. Charalambous and the second author generalized spectral continuity to the case when the quadratic forms of two self-adjoint operators are ε-close.

Let $\mathcal{H}$ be a Hilbert space with two inner products $(\cdot,\cdot)_0$ and $(\cdot,\cdot)_1$. Consider two densely defined nonnegative operators H_0 and H_1 on $\mathcal{H}$ that are self-adjoint with respect to the inner products $(\cdot,\cdot)_0$ and $(\cdot,\cdot)_1$ respectively. Let Q_0, Q_1 be their respective quadratic forms and denote the two norms on $\mathcal{H}$ by $\|\cdot\|_0$ and $\|\cdot\|_1$. Note that both Q_0 and Q_1 are nonnegative.

Denote the domain of the Friedrichs extension of H_0 and H_1 by $\mathfrak{Dom}(H_0)$ and $\mathfrak{Dom}(H_1)$ respectively. We assume that there exists a dense subspace $\mathcal{C} \subset \mathcal{H}$ such that $\mathcal{C}$ is contained in $\mathfrak{Dom}(H_0) \cap \mathfrak{Dom}(H_1)$. In the case of the Laplacian operators $H_0 = \Delta_{g_0}$ and $H_1 = \Delta_{g_1}$ associated to two different metrics g_0 and g_1, $\mathcal{C}$ will be the space of smooth functions/forms with compact support.

DEFINITION 1 ([**4**]). We say that the operators H_0, H_1 are ε-close, if there exists a positive constant $0 < \varepsilon < 1$ such that for all $u \in \mathcal{C}$ the following two inequalities hold

$$(1-\varepsilon)\,\|u\|_0^2 \le \|u\|_1^2 \le (1+\varepsilon)\,\|u\|_0^2; \tag{2}$$

$$(1-\varepsilon)\,Q_0(u,u) \le Q_1(u,u) \le (1+\varepsilon)\,Q_0(u,u). \tag{3}$$

We note that if H_0, H_1 are ε-close, then for any $u, v \in \mathcal{C}$

$$|(u,v)_1 - (u,v)_0| \le \varepsilon(\|u\|_0\,\|v\|_0); \tag{4}$$

$$|Q_1(u,v) - Q_0(u,v)| \le \varepsilon\,[Q_0(u,u)\,Q_0(v,v)]^{1/2}. \tag{5}$$

It has been shown in [**4**] that two ε-close operators have nearby spectra. This result has an important application in the context of the Hodge Laplacian on k-forms over a Riemannian manifold with two ε-close metrics over it. In particular, it allows for the proof of the following theorem which holds even in the noncompact case.

THEOREM 2.2 ([**4**]). *Let X be an orientable manifold, and let g_0, g_1 be two smooth complete Riemannian metrics on X that are ε-close for some $0 < \varepsilon < 1/2$. Fix $A > 0$. Then for any $\lambda \in \sigma(k, \Delta_1) \cap [0, A]$,*

$$\operatorname{dist}(\lambda, \sigma(k, \Delta_0)) < c(A,n)\,\varepsilon^{\frac{1}{3}}$$

for some constant $c(A)$ depending only on A. A similar result holds for the essential spectra of the operators. In particular,

$$d_{\mathfrak{h}}(\sigma(k, \Delta_1), \sigma(k, \Delta_0)) = o(1),$$

where $o(1) \to 0$, as $\varepsilon \to 0$.

To clarify the notation in the above theorem, $d_{\mathfrak{h}}$ denotes the pointed Gromov-Hausdorff distance between the spectra as subsets of the real line with a common fixed point -1. $\sigma(k, \Delta_i)$ denotes the spectrum of nonnegative definite Hodge Laplacian Δ_i acting on k-forms which corresponds to the metric g_i for $i = 0, 1$.

In contrast, in the setting of a family of compact Riemannian manifolds which is convergent in the Gromov-Hausdorff sense, we have the following important results. The first result is due to K. Fukaya

THEOREM 2.3 ([**13**]). *Let X_t be a family of compact Riemannian manifolds which is Gromov-Hausdorff convergent to a compact metric space X. We assume that X is not a point. Assume that the curvatures of the manifolds X_t are uniformly bounded. Then the eigenvalues of X_t converge to those of X.*

The above result was later generalized by J. Cheeger and T. H. Colding

THEOREM 2.4 ([**6**]). *Let X_t be a family of compact Riemannian manifolds which is Gromov-Hausdorff convergent to a compact metric space X. We assume that X is not a point. Assume that the Ricci curvatures of the manifolds X_t are uniformly bounded below. Then the eigenvalues of X_t converge to those of X.*

There is no known common generalization of Theorems 2.1, 2.2. 2.3 and 2.4. In [**2**] we studied a special case, which allowed us to find manifolds with gaps in their L^2 essential spectrum.

Let H be a self-adjoint operator on a Hilbert space $\mathcal{H}$. The norm and inner product in $\mathcal{H}$ are noted by $\|\cdot\|$ and $(\cdot,\cdot)$, respectively. Let $\sigma(H)$, $\sigma_{\mathrm{ess}}(H)$ be the spectrum and the essential spectrum of H, respectively. Let $\mathfrak{Dom}(H)$ be the domain of H. The Generalized Weyl criterion states the following.

THEOREM 2.5 (Charalambous-Lu [**3**]). *Let f be a bounded positive continuous function over $[0,\infty)$. A nonnegative real number λ belongs to the spectrum $\sigma(H)$ if, and only if, there exists a sequence $\{\psi_n\}_{n\in\mathbb{N}}$ such that*

(1) $\|\psi\|$, $\forall n \in \mathbb{N}$;
(2) $(f(H)(H-\lambda)\psi_n, (H-\lambda)\psi_n) \to \infty$ *as* $n \to \infty$; *and*
(3) $(\psi_n, (H-\lambda)\psi_n) \to 0$ *as* $n \to \infty$.

Moreover, λ belongs to $\sigma_{\text{ess}}(H)$ of H if, and only if, in addition to the above properties

(4) $\psi_n \to 0$, *weakly as* $n \to \infty$ *in* $\mathcal{H}$.

One direct consequence is the following.

THEOREM 2.6 (Charalambous-Lu [**4**]). *Let M be a complete Riemannian manifold. Suppose that $\lambda > 0$ belongs to the essential spectrum of the Laplacian on k-forms, $\sigma_{\text{ess}}(k,\Delta)$. Then one of the following holds:*

(1) $\lambda \in \sigma_{\text{ess}}(k-1,\Delta)$ *or*
(2) $\lambda \in \sigma_{\text{ess}}(k+1,\Delta)$.

3. Basic construction

We recreate the construction in [**2**]. Let $(X_1,g_1),(X_2,g_2)$ be two complete Riemannian manifolds. Let $x_1 \in X_1$ and $x_2 \in X_2$ be two fixed points on the manifolds respectively.

Let

$$N = S^{n-1} \times (-2,2)$$

be the product manifold equipped with the metric $g_N = \varepsilon^2 g_0$, where g_0 is the standard product metric.

For any $\varepsilon > 0$, we construct the manifold X_ε by gluing the three manifolds X_1, X_2, N in the following way. Let $f_1 : S^{n-1} \times (-2,-1) \to X_1$ be the function

$$f_1(\theta,t) = \exp_{x_1}(-t\varepsilon\theta),$$

where $\exp_{x_1}$ is the exponential map from $T_{x_1}X_1 \to X_1$. In particular $\exp_{x_1}(0) = x_1$. Similarly, let $f_2 : S^{n-1} \times (1,2) \to X_2$ be the function

$$f_2(\theta,t) = \exp_{x_2}(t\varepsilon\theta),$$

where $\exp_{x_2}$ is the exponential map from $T_{x_2}X_2 \to X_2$. In particular $\exp_{x_2}(0) = x_2$. It is clear that f_i $(i = 1,2)$ are diffeomorphisms between their domains and ranges. Let X_ε denote the composite manifold defined by (X_1, X_2, N, f_1, f_2), such that

$$X_\varepsilon = (X_1 \backslash B_{x_1}(\varepsilon)) \cup (X_2 \backslash B_{x_2}(\varepsilon)) \cup N/\sim, \tag{6}$$

where we identify f_i with their images respectively for $i = 1,2$. Roughly speaking, X_ε is constructed from X_1, X_2 by removing two balls of radius ε and adding a neck connecting them.

Abusing notation, we will identify g_i with $f_i^*(g_i)$ for $i = 1,2$ on the sets where they are defined.

We construct the metric g_ε on X_ε as follows. For $\delta \geq \varepsilon > 0$, let $\rho_0^\delta, \rho_1^\delta, \rho_2^\delta$ be a partition of unity for X_ε in the following sense. Let

$$\text{supp}(\rho_0^\delta) \subset \{p \in X_\varepsilon \mid \text{dist}(p, S^{n-1} \times \{0\}) < 2\delta\},$$

and assume that $\rho_0^\delta \equiv 1$ on $\{p \in X_\varepsilon \mid \mathrm{dist}(p, S^{n-1} \times \{0\}) < \delta\}$. Moreover, assume that $|\nabla \rho_0^\delta| \le C/\delta$. Then $X_\varepsilon \backslash \mathrm{supp}(\rho_0^\delta)$ has two connected components X_1^δ and X_2^δ. Let

$$\rho_i^\delta = 1 - \rho_0^\delta$$

for $i = 1, 2$ be the functions on X_1^δ and X_2^δ, respectively. Then

$$\rho_0^\delta + \rho_1^\delta + \rho_2^\delta = 1$$

on X_ε.

We define

$$g_\varepsilon = \rho_0^\varepsilon g_N + \rho_1^\varepsilon g_1 + \rho_2^\varepsilon g_2.$$

Then we have

PROPOSITION 3.1 ([**2**]). *Let X_1, X_2 be two compact Riemannian manifolds. Using the above notations, we have*

(1) *$(X_\varepsilon, g_\varepsilon)$ is Gromov-Hausdorff convergent to the metric space X_0, which is the union $X_1 \cup X_2$ with x_1 identified with x_2 ;*
(2) *Let x_0 be a reference point in the middle of the neck N. The pointed Riemannian manifolds $(X_\varepsilon, \varepsilon^{-2} g_\varepsilon, x_0)$ are Gromov-Hausdorff convergent to $(S^{n-1} \times (-1, 1), x_0)$.*

We call $(X_\varepsilon, g_\varepsilon)$ a *smoothing* of X_0. Apparently, the curvature of X_ε are not bounded as $\varepsilon \to 0$. Nevertheless, The key property of the family of manifolds X_ε is that it has uniform local Sobolev constants. We shall use this fact to prove the spectrum continuity when the Ricci curvature of X_ε doesn't have a lower bound.

LEMMA 1 ([**2**]). *The (local) Sobolev constants for both $(X_\varepsilon, g_\varepsilon)$ and $(X_\varepsilon, \varepsilon^{-2} g_\varepsilon)$ are uniformly bounded.*

4. Proof of the main theorems

We begin this section with the following

LEMMA 2. *Let Δ_k be the Hodge Laplacian on k-forms of $(N, \varepsilon^2 g_0)$ with the Friedrichs extension, where g is the product metric. Then the first eigenvalue λ_1 of Δ_k diverges to $+\infty$ as $\varepsilon \to 0$.*

PROOF. We consider the first eigenvalue of (N, g_0) with the standard product metric. It is well known that if $k = 0$, then the first eigenvalue has a positive lower bound $\sigma > 0$. By scaling, for the metric, $\varepsilon^2 g_0$, the first eigenvalue is bounded below by $\sigma \varepsilon^{-2}$, which diverges to ∞ when $\varepsilon \to 0$. By duality, the same is true for $k = n$.

Similarly, we can work out the cases when $k \ne 0, n$. Since g_0 is the product metric, we have the following decomposition for $\Delta = \Delta_k$:

$$\Delta_N = \Delta_{S^{n-1}} \otimes 1 + 1 \otimes \Delta_{\mathbb{R}}.$$

Let $\lambda_1^k(N)$ denote the first eigenvalue of the manifold N on k-forms. Let ω be the first eigenform. Then we can write ω as

$$\omega = \omega_1 \wedge f(t)dt + g(t)\omega_2,$$

where $f(t), g(t)$ are functions such that $f(\pm 2) = g(\pm 2) = 0$ (by the Friedrichs extension) , and ω_1, ω_2 are $(k-1)$ and k eigenforms of S^{n-2} respectively.

If $g \not\equiv 0$, then

$$\lambda_1^k(S^{n-1} \times \mathbb{R}) \ge \lambda_1^0((-2, 2)) > 0$$

if $k \neq 0, n$, and the conclusion of the lemma proved. If $g \equiv 0$, then f must not be identically zero. As a result, we have

$$\lambda_1^k(S^{n-1} \times \mathbb{R}) \geq \lambda_1^1((-2, 2)).$$

Since $f(\pm 2) = 0$, we have $\lambda_1^1((-2, 2)) > 0$, thus completes the proof of the lemma. $\square$

Using the above, we shall prove the main technical result of this paper in the following.

THEOREM 4.1. *Let X_1, X_2 be two compact Riemannian manifolds and take $\lambda \notin \operatorname{Spec}(X_1) \cup \operatorname{Spec}(X_2)$. Consider the manifold $(X_\varepsilon, g_\varepsilon)$ defined above. Set $2\sigma = \operatorname{dist}(\lambda, \operatorname{Spec}(X_1) \cup \operatorname{Spec}(X_2))$ and take $\lambda' \in (\lambda - \sigma, \lambda + \sigma)$. Then, for any $\varepsilon > 0$ small enough, $\lambda' \notin \operatorname{Spec}(X_\varepsilon)$.*

PROOF. Let λ' be an eigenvalue of X_ε and let ω be an eigenform such that $\|\omega\|_{L^2} = 1$. Write

$$\omega = \rho_0^\delta \omega + \rho_1^\delta \omega + \rho_2^\delta \omega := \omega_0 + \omega_1 + \omega_2.$$

it turns out that

$$(\Delta\omega_0, \omega) + (\Delta\omega_1, \omega) + (\Delta\omega_2, \omega) = \lambda'. \tag{7}$$

Note that in fact, $\omega_i = \omega_i^{\delta,\varepsilon}$ depending on both ε, δ. For fixed $\delta > 0$, we let $\varepsilon \to 0$. Since $(\Delta\omega_i, \omega)/\|\omega\|_{L^2}^2$ is bounded for $i = 1, 2$, then using Lemma 1, by the elliptic regularity, we have $\mathcal{C}^{2,\alpha}$-estimate. Therefore, we have a sequence limit

$$\lim_{\varepsilon_j \to 0} \omega_i^{\delta,\varepsilon_j} = \omega_i^{\delta,0}$$

as $\varepsilon_j \to 0$ and $i = 1, 2$.

If any of $\omega_1^{\delta,0}$, or $\omega_2^{\delta,0}$ is not zero, say $\omega_1^{\delta,0} \neq 0$. Then we have

$$\Delta\omega_1^{\delta,0} = \lambda' \omega_1^{\delta,0}.$$

Letting $\delta \to 0$, the form $\omega_1^{\delta,0}$ will be convergent to an eigenform on $X_1 \backslash \{x_1\}$. By the removable singularity theorem, it would be extended to an eigenform of X_1, contradicting to the fact that λ' is away from $\operatorname{Spec}(X_1)$.

It remains to prove that it is not possible that both $\omega_1^{\delta,0}$ and $\omega_2^{\delta,0}$ are zero. Assume otherwise, then by (7), $(\Delta\omega_0, \omega)/\|\omega\|_{L^2}^2$ must be bounded when $\varepsilon \to 0$. But this would imply that

$$(\Delta\omega_0, \omega_0)/\|\omega_0\|_{L^2}^2$$

is bounded, which contradicts to Lemma 2. This completes the proof of the theorem. $\square$

Using the above technical lemma, we can prove the following

THEOREM 4.2. *There exists a complete non-compact Riemannian manifold M whose essential spectrum of the Hodge Laplacian on k-forms has an arbitrarily large number of connected components.*

PROOF. Let $X = S^n$ be the n-dimensional sphere and let N, S be the north pole and the south pole of X, respectively. Define a sequence of compact manifolds $X_i = X$, with the corresponding north and south poles N_i, S_i. Define the metric space Y by glueing the north pole of X_i to the south pole of X_{i+1} for any $i \geq 0$. Then the spectrum of Y is the same as the spectrum of $X = S^n$, which is discrete.

Using Theorem 4.1, after a smoothing of Y to Y_ε, then the essential spectrum of Y_ε would have arbitrarily large number of connected components. □

Using a similar argument, we can prove

COROLLARY 1. *There exists a complete non-compact Riemannian manifold M whose spectrum of the connection Laplacian on k-forms has an arbitrarily large number of connected components.*

References

[1] C. Anné, G. Carron, and O. Post, *Gaps in the differential forms spectrum on cyclic coverings*, Math. Z. **262** (2009), no. 1, 57–90, DOI 10.1007/s00209-008-0363-0. MR2491601

[2] N. Charalambous, H. Leal, and Z. Lu, *Spectral gaps on complete Riemannian manifolds*, Geometry of submanifolds, Contemp. Math., vol. 756, Amer. Math. Soc., Providence, RI, 2020, pp. 57–67, DOI 10.1090/conm/756/15196. MR4186938

[3] N. Charalambous and Z. Lu, *On the spectrum of the Laplacian*, Math. Ann. **359** (2014), no. 1-2, 211–238, DOI 10.1007/s00208-013-1000-8. MR3201899

[4] N. Charalambous and Z. Lu, *The spectrum of continuously perturbed operators and the Laplacian on forms*, Differential Geom. Appl. **65** (2019), 227–240, DOI 10.1016/j.difgeo.2019.05.002. MR3948872

[5] N. Charalambous and Z. Lu, *The spectrum of the Laplacian on forms over flat manifolds*, Math. Z. **296** (2020), no. 1-2, 1–12, DOI 10.1007/s00209-019-02407-5. MR4140728

[6] J. Cheeger and T. H. Colding, *On the structure of spaces with Ricci curvature bounded below. III*, J. Differential Geom. **54** (2000), no. 1, 37–74. MR1815411

[7] Z. H. Chen and Z. Q. Lu, *Essential spectrum of complete Riemannian manifolds*, Sci. China Ser. A **35** (1992), no. 3, 276–282. MR1183713

[8] J. Dodziuk, *Eigenvalues of the Laplacian on forms*, Proc. Amer. Math. Soc. **85** (1982), no. 3, 437–443, DOI 10.2307/2043863. MR656119

[9] H. Donnelly, *Exhaustion functions and the spectrum of Riemannian manifolds*, Indiana Univ. Math. J. **46** (1997), no. 2, 505–527, DOI 10.1512/iumj.1997.46.1338. MR1481601

[10] J. F. Escobar, *On the spectrum of the Laplacian on complete Riemannian manifolds*, Comm. Partial Differential Equations **11** (1986), no. 1, 63–85, DOI 10.1080/03605308608820418. MR814547

[11] J. F. Escobar and A. Freire, *The spectrum of the Laplacian of manifolds of positive curvature*, Duke Math. J. **65** (1992), no. 1, 1–21, DOI 10.1215/S0012-7094-92-06501-X. MR1148983

[12] P. Exner and O. Post, *Convergence of spectra of graph-like thin manifolds*, J. Geom. Phys. **54** (2005), no. 1, 77–115, DOI 10.1016/j.geomphys.2004.08.003. MR2135966

[13] K. Fukaya, *Collapsing of Riemannian manifolds and eigenvalues of Laplace operator*, Invent. Math. **87** (1987), no. 3, 517–547, DOI 10.1007/BF01389241. MR874035

[14] A. Khrabustovskyi, *Periodic Riemannian manifold with preassigned gaps in spectrum of Laplace-Beltrami operator*, J. Differential Equations **252** (2012), no. 3, 2339–2369, DOI 10.1016/j.jde.2011.10.011. MR2860621

[15] J. Y. Li, *Spectrum of the Laplacian on a complete Riemannian manifold with nonnegative Ricci curvature which possess a pole*, J. Math. Soc. Japan **46** (1994), no. 2, 213–216, DOI 10.2969/jmsj/04620213. MR1264938

[16] F. Lledó and O. Post, *Existence of spectral gaps, covering manifolds and residually finite groups*, Rev. Math. Phys. **20** (2008), no. 2, 199–231, DOI 10.1142/S0129055X08003286. MR2400010

[17] J. Lott, *On the spectrum of a finite-volume negatively-curved manifold*, Amer. J. Math. **123** (2001), no. 2, 185–205. MR1828220

[18] P. Petersen, *Riemannian geometry*, 3rd ed., Graduate Texts in Mathematics, vol. 171, Springer, Cham, 2016, DOI 10.1007/978-3-319-26654-1. MR3469435

[19] O. Post, *Periodic manifolds with spectral gaps*, J. Differential Equations **187** (2003), no. 1, 23–45, DOI 10.1016/S0022-0396(02)00006-2. MR1946544

[20] O. Post, *Spectral convergence of quasi-one-dimensional spaces*, Ann. Henri Poincaré **7** (2006), no. 5, 933–973, DOI 10.1007/s00023-006-0272-x. MR2254756

[21] S. Rosenberg, *The Laplacian on a Riemannian manifold*, London Mathematical Society Student Texts, vol. 31, Cambridge University Press, Cambridge, 1997. An introduction to analysis on manifolds, DOI 10.1017/CBO9780511623783. MR1462892
[22] R. Schoen and H. Tran, *Complete manifolds with bounded curvature and spectral gaps*, J. Differential Equations **261** (2016), no. 4, 2584–2606, DOI 10.1016/j.jde.2016.05.002. MR3505201
[23] D. T. Zhou, *Essential spectrum of the Laplacian on manifolds of nonnegative curvature*, Internat. Math. Res. Notices **5** (1994), 209 ff., approx. 6 pp., DOI 10.1155/S1073792894000231. MR1270134

Department of Mathematics, Bellevue College, Bellevue, Washington 98007
Email address: h.leal@bellevuecollege.edu

Department of Mathematics, University of California, Irvine, Irvine, California 92697
Email address: zlu@uci.edu

Contemporary Mathematics
Volume **777**, 2022
https://doi.org/10.1090/conm/777/15627

Chen-Ricci inequalities for Riemannian maps and their applications

Jae Won Lee, Chul Woo Lee, Bayram Şahin, and Gabriel-Eduard Vîlcu

Abstract. Riemannian maps between Riemannian manifolds, originally introduced by A.E. Fischer in [Contemp. Math. 132 (1992), 331–366], provide an excellent tool for comparing the geometric structures of the source and target manifolds. Isometric immersions and Riemannian submersions are particular examples of such maps. In this work, we first prove a geometric inequality for Riemannian maps having a real space form as a target manifold. Applying it to the particular case of Riemannian submanifolds, we recover a classical result, obtained by B.-Y. Chen in [Glasgow Math. J. 41 (1999), 33–41], which nowadays is known as the Chen-Ricci inequality. Moreover, we extend this inequality in case of Riemannian maps with a complex space form as a target manifold. We also improve this inequality when the Riemannian map is Lagrangian. Applying it to Riemannian submanifolds, we recover the improved Chen-Ricci inequality for Lagrangian submanifolds in a complex space form, that is a basic inequality obtained by S. Deng in [Int. Electron. J. Geom. 2 (2009), 39-45] as an improvement of a geometric inequality stated by B.-Y. Chen in [Arch. Math. (Basel) 74 (2000), 154–160].

1. Introduction

The celebrated embedding theorem of Nash gave a strong motivation for the establishment of sharp relationships between the intrinsic and extrinsic invariants of Riemannian submanifolds, this being the main reason in the introduction of the Chen invariants in the early 90s (see [**C6**]). In 1993, Chen [**C1**] established an optimal inequality involving basic intrinsic and extrinsic invariants of a submanifold in a real space form, namely the scalar curvature, the sectional curvature and the squared mean curvature. Afterwards, he established a basic relationship between the Ricci curvature and the squared mean curvature for submanifolds in real

2020 *Mathematics Subject Classification.* Primary 53C05; Secondary 49K35, 62B10.

Key words and phrases. Riemannian map, isometric immersion, horizontal space, Chen-Ricci inequality, complex space form.

The first author was supported under the framework of international cooperation program managed by the National Research Foundation of Korea (2019K2A9A1A06097856). The third author was supported under the framework of international cooperation program managed by the Scientific and Technological Research Council of Turkey (TUBITAK) with project id: 119N087. The second author was supported by Basic Science Research Program through the National Research Foundation of Korea (NRF) funded by the Ministry of Education (2018R1D1A1B07040576). The fourth author was supported by a grant of the Ministry of Research, Innovation and Digitization, CNCS/CCCDI - UEFISCDI, project number PN-III-P4-ID-PCE-2020-0025, within PNCDI III.

space forms [**C2**], as well as for totally real submanifolds in complex space forms [**C3**], known as *Chen-Ricci inequality.* On the other hand, Mihai [**Mih**] proved a similar inequality for submanifolds of Sasakian space forms. For a Lagrangian submanifold in a complex space form, i.e. a totally real submanifold with smallest codimension, Oprea [**O**] improved the Chen-Ricci inequality by using an optimization technique on submanifolds. However, Oprea could not determine the equality condition. Later, Deng [**D**] provides a new proof for the improved Chen-Ricci inequality for Lagrangian submanifolds of complex space forms, based on some crucial algebraic inequalities and also discusses the equality case. After that, Chen-Ricci inequalities were obtained for several classes of submanifolds into various ambient spaces (see, e.g., [**LMSSJ, MK, MiA, MiR, MUS, T**]).

On the other hand, in 1992, Fischer introduced the concept of *Riemannian map* between Riemannian manifolds, as a generalization of the notions of isometric immersions and Riemannian submersions [**F**]. His notion plays an important role in comparing geometric structures of two Riemannian manifolds [**S3**]. Let (M^m, g_M) and (N^n, g_N) be two Riemannian manifolds and $F : (M^m, g_M) \longrightarrow (N^n, g_N)$ be a smooth map such that $0 \leq \mathrm{rank}F = r \leq \min\{\mathrm{m}, \mathrm{n}\}$. Then, the tangent bundle of M has the orthogonal complement decomposition

$$TM = \mathrm{ker}F_* \oplus (\mathrm{ker}F_*)^{\perp} = \mathrm{ker}F_* \oplus \mathcal{H},$$

where $\mathcal{H} = (\mathrm{ker}F_*)^{\perp}$. If $q = F(p) \in N$, $p \in M$, then we have an orthogonal complement decomposition for the tangent space T_qN, namely

$$T_qN = (\mathrm{Im}F_{*p}) \oplus (\mathrm{Im}F_{*p})^{\perp}.$$

A smooth map $F : (M^m, g_M) \longrightarrow (N^n, g_N)$ with $0 \leq \mathrm{rank}F = r \leq \min\{\mathrm{m}, \mathrm{n}\}$ is said to be a *Riemannian map at* $p \in M$ if the horizontal restriction F^h_{*p} : $(\mathrm{ker}F_*)^{\perp} \longrightarrow \mathrm{Im}F_{*p}$ is a linear isometry. Moreover, if F is a Riemannian map at each $p \in M$, then F is said to be *a Riemannian map.* In particular, if $\mathrm{ker}F_* = \{0\}$ $((\mathrm{Im}F_*)^{\perp} = \{0\})$, Riemannian maps are isometric immersions (Riemannian submersions), respectively. Fischer showed in [**F**] that a Riemannian map satisfies the eikonal equation, realizing a bridge between geometric and physical optics. He also presented a program of building a quantum model of the nature using Riemannian maps between Riemannian manifolds, therefore connecting harmonic maps and Lagrangian field theory on the mathematical side and Maxwell's and Shrödinger's equations on the physical side. On another hand, in [**S1**], Şahin obtained a basic Chen inequality for Riemannian maps from Riemannian manifolds to real space forms. This inequality is a generalization of both isometric immersion case [**C1**] and Riemannian submersion case [**C4, C5**]. Moreover, the authors of the present article obtained some optimal inequalities involving Casorati curvatures for Riemannian maps to real and complex space forms, recovering in particular cases some basic inequalities in the field (see [**LLSV**]).

In this paper, we first obtain a basic inequality for Riemannian maps in terms of the Ricci curvature of horizontal spaces. Applying it to Riemannian submanifolds (isometric immersions), we recover the classical Chen-Ricci inequality stated in [**C2**]. For Riemannian submersions, we obtain a new optimal inequality and give an example. We also improve this inequality for Riemannian maps when the target manifold is a complex space form, provided that the rank of the map F coincides with the dimension of $(\mathrm{Im}F_*)^{\perp}$. This improved inequality generalizes the results of [**C3**, **D**] obtained for isotropic and Lagrangian submanifolds in complex space forms.

2. Preliminaries

In this section, we review some basic concepts and results regarding as the second fundamental form, the shape operator and the Gauss equation of a Riemannian map.

Let (M, g_M) and (N, g_N) be Riemannian manifolds and suppose that $F : M \longrightarrow N$ is a smooth map. Then, the differential F_* can be considered as a section of the bundle $Hom\left(TM, F^{-1}TN\right)$ on M, where $F^{-1}TN$ is the pullback bundle, whose fibre $\left(F^{-1}TN\right)_p = T_{F(p)}N$, $p \in M$. Note that $Hom\left(TM, F^{-1}TN\right)$ has an induced connection ∇ from the Levi-Civita connection ∇^M on M. Then, the *second fundamental form* of F is given by

$$(\nabla F_*)(X,Y) = \nabla^F_X F_*(Y) - F_*\left(\nabla^M_X Y\right)$$

for $X, Y \in \Gamma(TM)$. Note that the second fundamental form is symmetric. Recall that F is said to be *a totally geodesic map* if $(\nabla F_*)(X,Y) = 0$ for all $X, Y \in \Gamma(TM)$.

If the smooth map F is Riemannian, Şahin [**S2**] showed that $(\nabla F_*)(X,Y)$ has no components in $\mathrm{Im}F_*$, provided that $X, Y \in \Gamma\left((\ker F_*)^\perp\right)$. More precisely,

$$(\nabla F_*)(X,Y) \in \Gamma\left((\mathrm{Im}F_*)^\perp\right), \forall X, Y \in \Gamma\left((\ker F_*)^\perp\right), \tag{2.1}$$

where $(\mathrm{Im}F_*)^\perp$ is the subbundle of $F^{-1}(TN)$ with a fiber $(F_*(T_pM))^\perp$, $p \in M$.

We now denote by ∇^N both the Levi-Civita connection on (N, g_N) and its pullback along F. Let $\nabla^{F^\perp}_X V$ be the orthogonal projection of $\nabla^N_X V$ onto $(\mathrm{Im}F_*)^\perp$, $X \in \Gamma(TM)$, $V \in \Gamma\left((\mathrm{Im}F_*)^\perp\right)$. In [**N**], Nore showed that $\nabla^{F^\perp}$ is a linear connection on $(F_*(TM))^\perp$ such that $\nabla^{F^\perp} g_N = 0$. Therefore, we can define the operator $\mathcal{S}_V$ on $F_*(TM)$ by

$$\nabla^N{}_X V = -\mathcal{S}_V F_* X + \nabla^{F^\perp}_X V, \tag{2.2}$$

where $-\mathcal{S}_V F_* X$ stands for the tangential component of $\nabla^N{}_X V$, hence it is a vector field along F. Remark that $\nabla^N{}_X V$ is obtained from the pullback connection of ∇^N. It is clear that $(\mathcal{S}_V F_* X)_p$ depends only on V_p and $F_{*p}(X_p)$, and $\mathcal{S}_V F_* X$ is bilinear in V and $F_* X$. By a direct computation, we have

$$g_N(\mathcal{S}_V F_* X, F_* Y) = g_N(V, (\nabla F_*)(X,Y)) \tag{2.3}$$

for all $X, Y \in \Gamma\left((\ker F_*)^\perp\right)$ and $V \in \Gamma\left((\mathrm{Im}F_*)^\perp\right)$. Since ∇F_* is symmetric, $\mathcal{S}_V$ is a symmetric linear transformation of $\mathrm{Im}F_*$, called the *shape operator* of the Riemannian map F [**S3**]. We say that F is *an umbilical Riemannian map at* $p \in M$ if

$$\mathcal{S}_V F_{*p}(X_p) = \lambda F_{*p}(X_p),$$

for $X \in \Gamma\left((\ker F_*)^\perp\right)$ and $V \in \Gamma\left((\mathrm{Im}F_*)^\perp\right)$, where λ is a smooth function on M. Also F is said to be *an umbilical Riemannian map* if F is umbilical at each $p \in M$. From [**S3**, Lemma 50], it is known that $F : (M, g_M) \longrightarrow (N, g_N)$ is an umbilical Riemannian map if and only if

$$(\nabla F_*)(X,Y) = g_M(X,Y)\,\Upsilon$$

for all $X, Y \in \Gamma\left((\ker F_*)^\perp\right)$, where Υ is a nowhere zero vector field on $(\mathrm{Im} F_*)^\perp$.

Let R^M and R^N be the curvature tensor fields of ∇^M and ∇^N, respectively. Then we have the Gauss formula given by [**S3**]

$$\begin{aligned} g_N\left(R^N(F_*X, F_*Y)F_*Z, F_*W\right) &= g_M\left(R^M(X,Y)Z,W\right) \\ &\quad + g_N\left((\nabla F_*)(X,Z),(\nabla F_*)(Y,W)\right) \\ &\quad - g_N\left((\nabla F_*)(X,W),(\nabla F_*)(Y,Z)\right) \end{aligned} \tag{2.4}$$

for all $X, Y, Z, W \in \Gamma\left((\ker F_*)^\perp\right)$.

Let$\{e_1, \ldots, e_r\}$ and $\{V_{r+1}, \ldots, V_n\}$ be orthonormal bases of $\mathcal{H}_p = \ker F_{*p}^\perp$ and $\left(\mathrm{Im} F_{*p}\right)^\perp$, $p \in M$, respectively. We define the Ricci curvature $\mathrm{Ric}^{\mathcal{H}}$ on the horizontal space $\mathcal{H}_p$ as

$$\mathrm{Ric}^{\mathcal{H}}(X) = \sum_{i=1}^{r} g_M\left(R^M(e_i, X)X, e_i\right), \quad X \in \mathcal{H}_p.$$

And we also define the scalar curvature $\tau^{\mathcal{H}}$ on the horizontal space $\mathcal{H}_p$ as

$$\tau^{\mathcal{H}} = \sum_{1 \le i < j \le r} g_M\left(R^M(e_i, e_j)e_j, e_i\right).$$

We put

$$\begin{aligned} \zeta^{\mathcal{H}\alpha}_{ij} &= g_N\left((\nabla F_*)(e_i, e_j), V_\alpha\right), i,j = 1, \ldots, r, \ \alpha = r+1, \ldots, n, \\ \|\zeta^{\mathcal{H}}\|^2 &= \sum_{i,j=1}^{r} g_N\left((\nabla F_*)(e_i,e_j),(\nabla F_*)(e_i,e_j)\right), \\ \mathrm{trace}\zeta^{\mathcal{H}} &= \sum_{i=1}^{r} (\nabla F_*)(e_i, e_i), \\ \|\mathrm{trace}\zeta^{\mathcal{H}}\|^2 &= g_N\left(\mathrm{trace}\zeta^{\mathcal{H}}, \mathrm{trace}\zeta^{\mathcal{H}}\right). \end{aligned} \tag{2.5}$$

3. Chen-Ricci inequality for Riemannian maps with a real space form

In this section, we state the Chen-Ricci inequality for a Riemannian map from a Riemannian manifold onto a real space form. In particular, if the Riemannian map is an isometric immersion, we recover the classical Chen-Ricci inequality originally established by B.-Y. Chen in [**C2**].

THEOREM 3.1. *Let F be a Riemannian map from a Riemannian manifold (M^m, g_M) onto a real space form $(N^n(c), g_N)$ with* $\mathrm{rank} F = r < n$ *and X be an unit horizontal vector field on M. Then we have*

$$\mathrm{Ric}^{\mathcal{H}}(X) \le (r-1)c + \frac{1}{4}\|\mathrm{trace}\zeta^{\mathcal{H}}\|^2. \tag{3.1}$$

Moreover:

(*i*) *The equality case of* (3.1) *holds for $X \in \mathcal{H}_p$ if and only if*

$$\begin{cases} (\nabla F_*)(X,Y) = 0 & \textit{for all } Y \in \mathcal{H}_p \textit{ orthogonal to } X, \\ (\nabla F_*)(X,X) = \frac{1}{2}\mathrm{trace}\zeta^{\mathcal{H}}, \end{cases} \tag{3.2}$$

(ii) The equality case of (3.1) *holds identically for all* $X \in \mathcal{H}_p$ *if and only if either* $\zeta^{\mathcal{H}} = 0$ *or* $r = 2$ *and* ${\zeta^{\mathcal{H}}}^{\alpha}_{11} = {\zeta^{\mathcal{H}}}^{\alpha}_{22}$, $\alpha = r+1, \ldots, n$.

PROOF. Since $N(c)$ is a real space form of constant sectional curvature c, the Riemannian curvature tensor R^N of $N(c)$ satisfies

$$g_N\left(R^N(X,Y)Z,W\right) = c\left\{g_N(X,W)\,g_N(Y,Z) - g_N(X,Z)\,g_N(Y,W)\right\}, \tag{3.3}$$

for any $X, Y, Z, W \in \Gamma(TN)$. From (2.4) and (3.3), taking into account that F is a Riemanian map, we derive

$$\begin{aligned} g_M\left(R^M(X,Y)Z,W\right) = {} & c\left\{g_M(X,W)\,g_M(Y,Z) - g_M(X,Z)\,g_M(Y,W)\right\} \\ & - g_N\left((\nabla F_*)(X,Z), (\nabla F_*)(Y,W)\right) \\ & + g_N\left((\nabla F_*)(X,W), (\nabla F_*)(Y,Z)\right) \end{aligned}$$

for all $X, Y, Z, W \in \Gamma\left((\mathrm{ker}F_*)^{\perp}\right)$.

Let $p \in M$ and the sets $\{e_1, e_2, \cdots, e_r\}$, with $e_1 = X$, and $\{V_{r+1}, V_{r+2}, \cdots, V_n\}$ be orthonormal bases of $\mathcal{H}_p$ and $\left(\mathrm{Im}F_{*p}\right)^{\perp}$, respectively. From the last equation and (2.5), we have

$$r(r-1)c = 2\tau^{\mathcal{H}}(p) - \|\mathrm{trace}\zeta^{\mathcal{H}}\|^2 + \|\zeta^{\mathcal{H}}\|^2, \tag{3.4}$$

where $\tau^{\mathcal{H}}(p)$ stands for the scalar curvature of $\mathcal{H}_p$.

On the other hand, we note that

$$\begin{aligned} \|\zeta^{\mathcal{H}}\|^2 = {} & \frac{1}{2}\|\mathrm{trace}\zeta^{\mathcal{H}}\|^2 + \frac{1}{2}\sum_{\alpha=r+1}^{n}\left({\zeta^{\mathcal{H}}}^{\alpha}_{11} - {\zeta^{\mathcal{H}}}^{\alpha}_{22} - \cdots - {\zeta^{\mathcal{H}}}^{\alpha}_{rr}\right)^2 \\ & + 2\sum_{\alpha=r+1}^{n}\sum_{i=2}^{r}\left({\zeta^{\mathcal{H}}}^{\alpha}_{1i}\right)^2 - 2\sum_{\alpha=r+1}^{n}\sum_{2\le i<j\le r}\left[{\zeta^{\mathcal{H}}}^{\alpha}_{ii}{\zeta^{\mathcal{H}}}^{\alpha}_{jj} - \left({\zeta^{\mathcal{H}}}^{\alpha}_{ij}\right)^2\right], \end{aligned} \tag{3.5}$$

and

$$\begin{aligned} \tau^{\mathcal{H}}(p) = {} & \mathrm{Ric}^{\mathcal{H}}(X) + \frac{(r-1)(r-2)}{2}c \\ & + \sum_{\alpha=r+1}^{n}\sum_{2\le i<j\le r}\left[{\zeta^{\mathcal{H}}}^{\alpha}_{ii}{\zeta^{\mathcal{H}}}^{\alpha}_{jj} - \left({\zeta^{\mathcal{H}}}^{\alpha}_{ij}\right)^2\right]. \end{aligned} \tag{3.6}$$

Combining (3.4), (3.5) and (3.6), we derive

$$\begin{aligned} \|\mathrm{trace}\zeta^{\mathcal{H}}\|^2 = {} & 4\mathrm{Ric}^{\mathcal{H}}(X) - 4(r-1)c \\ & + \sum_{\alpha=r+1}^{n}\left({\zeta^{\mathcal{H}}}^{\alpha}_{11} - {\zeta^{\mathcal{H}}}^{\alpha}_{22} - \cdots - {\zeta^{\mathcal{H}}}^{\alpha}_{rr}\right)^2 + \sum_{\alpha=r+1}^{n}\sum_{i=2}^{r}\left({\zeta^{\mathcal{H}}}^{\alpha}_{1i}\right)^2 \\ \ge {} & 4\mathrm{Ric}^{\mathcal{H}}(X) - 4(r-1)c \end{aligned} \tag{3.7}$$

and (3.1) is now clear from (3.7).

(i) The equality in (3.7) is valid if and only if

$$\begin{cases} {\zeta^{\mathcal{H}}}^{\alpha}_{11} = {\zeta^{\mathcal{H}}}^{\alpha}_{22} + \cdots + {\zeta^{\mathcal{H}}}^{\alpha}_{rr}, \\ {\zeta^{\mathcal{H}}}^{\alpha}_{12} = {\zeta^{\mathcal{H}}}^{\alpha}_{13} = \cdots = {\zeta^{\mathcal{H}}}^{\alpha}_{1r} = 0, \end{cases} \qquad \alpha \in \{r+1, r+2, \cdots, n\}, \tag{3.8}$$

which is equivalent to (3.2).

(ii) Now we assume the equality in (3.1) is true for all unit vectors $X \in \mathcal{H}_p$. Then from (3.8), for each $\alpha \in \{r+1, r+2, \cdots, n\}$, we obtain

$$\zeta^{\mathcal{H}\alpha}_{ij} = 0, \quad i \neq j, \tag{3.9}$$

$$2\zeta^{\mathcal{H}\alpha}_{ii} = \zeta^{\mathcal{H}\alpha}_{11} + \zeta^{\mathcal{H}\alpha}_{22} + \cdots + \zeta^{\mathcal{H}\alpha}_{rr}, \quad i \in \{1, 2, \cdots, r\}. \tag{3.10}$$

Now, it can be easily checked that (3.10) implies

$$(r-2)\left(\zeta^{\mathcal{H}\alpha}_{11} + \zeta^{\mathcal{H}\alpha}_{22} + \cdots + \zeta^{\mathcal{H}\alpha}_{rr}\right) = 0,$$

and hence, either $r = 2$ or

$$\zeta^{\mathcal{H}\alpha}_{11} + \zeta^{\mathcal{H}\alpha}_{22} + \cdots + \zeta^{\mathcal{H}\alpha}_{rr} = 0. \tag{3.11}$$

If (3.11) is satisfied, then we have $\zeta^{\mathcal{H}} = 0$ from (3.9) and (3.10). On the other hand, if $r = 2$, then from (3.10) we deduce

$$\zeta^{\mathcal{H}\alpha}_{11} = \zeta^{\mathcal{H}\alpha}_{22} = \frac{1}{2}\left(\zeta^{\mathcal{H}\alpha}_{11} + \zeta^{\mathcal{H}\alpha}_{22}\right).$$

As the converse is trivial, the proof is now complete. □

EXAMPLE 3.2. Let $\pi : \mathbb{S}^7 \to \mathbb{S}^4$ be the Hopf fibration and $f : \mathbb{S}^4 \to \mathbb{R}^5$ be the canonical embedding of the 4-sphere $\mathbb{S}^4$ as a hypersurface of $\mathbb{R}^5$. Then the composition $F = f \circ \pi : \mathbb{S}^7 \to \mathbb{R}^5$ is a totally umbilical Riemannian map which satisfies inequality (3.1).

REMARK 3.3. Any totally geodesic Riemannian map from a Riemannian manifold to a real space form satisfies identically the equality case of (3.1). For more details on the geometry of this very special class of Riemannian maps see [**GC**, Section 4.4 of Chapter 4] and [**S3**, Section 3 of Chapter 4].

Using Theorem 3.1 in the particular case when the Riemannian map F is an isometric immersion, we can recover the classical Chen-Ricci inequality between the Ricci curvature and the squared norm of mean curvature vector field for submanifolds in real space forms [**C2**].

COROLLARY 3.4. [**C2**, Theorem 4] *Let $F : M^n \longrightarrow R^m(c)$ be an isometric immersion of a Riemannian n-dimensional manifold M^n into a real space form $R^m(c)$. Then for each tangent vector $X \in T_pM$, we have*

$$\|H\|^2 \geq \frac{4}{n^2}\left\{Ric(X) - (n-1)c\right\} \tag{3.12}$$

where $\|H\|^2$ is the squared norm of mean curvature vector field of M^n.

Moreover, the equality case of (3.12) *holds identically for all unit tangent vectors at p if and only if either p is a totally geodesic point or $n = 2$ and p is a totally umbilical point.*

PROOF. Since $F : M^n \longrightarrow R^m(c)$ is an isometric immersion, we have

$$\|\mathrm{trace}\zeta^{\mathcal{H}}\| = n\|H\|$$

and thus we conclude the result via Theorem 3.1. □

On the other hand, Theorem 3.1 gives us immediately the following sharp estimate for the squared norm of horizontal mean curvature vector field in case of the target manifold is an Euclidean space of arbitrary codimension.

COROLLARY 3.5. *Let F be a Riemannian map from a Riemannian manifold (M^m, g_M) into an Euclidean space $\mathbb{E}^n$ such that* $\mathrm{rank}F = r < n$. *Then we have*

$$\|\mathrm{trace}\zeta^{\mathcal{H}}\|^2 \geq 4 \max_X \mathrm{Ric}^{\mathcal{H}}(X), \tag{3.13}$$

where X runs over all unit horizontal vectors at $p \in M$.

EXAMPLE 3.6. The isometric immersions of the spherical hypercylinder $S^2(r) \times \mathbb{E}^1$ and round hypercone in the Euclidean 4-space $\mathbb{E}^4$ (see [**C2**]) satisfy the equality cases of the inequalities (3.1), (3.12) and (3.13).

REMARK 3.7. For a Riemannian submersion $F : M^m \longrightarrow N^n$, we have the following basic notions. For more details, we may refer to [**FIP**].

Let $\chi^h(M)$ and $\chi^v(M)$ be the spaces of horizontal and vertical vector fields, respectively and the operator $A^{\mathcal{V}} : \chi^h(M) \times \chi^h(M) \longrightarrow \chi^v(M)$, defined by

$$A^{\mathcal{V}}(X, Y) = v\nabla_X Y$$

for all $X, Y \in \chi^h(M)$. Then, it is clear that $A^{\mathcal{V}}$ is an anti-symmetric operator on $\chi^h(M) \times \chi^h(M)$.

Denote by R^M the Riemannian curvature tensor of M and by R^* the (1,3) tensor field on $\chi^h(M)$ with values in $\chi^h(M)$ which associates to any $X, Y, Z \in \chi^h(M)$ and $p \in M$ the horizontal lift $(R^*(X,Y)Z)_p$ of $R^N_{F(p)}(F_{*p}(X_p), F_{*p}(Y_p))F_{*p}(Z_p)$, where R^N stands for the Riemannian curvature tensor of N.

Then we have the Gauss-Codazzi type equation [**FIP**]:

$$\begin{aligned} g_M(R^M(X,Y)W,Z) &= g_M(R^*(X,Y)W,Z) - 2g_M\left(A^{\mathcal{V}}(X,Y), A^{\mathcal{V}}(Z,W)\right) \\ &\quad + g_M\left(A^{\mathcal{V}}(Y,Z), A^{\mathcal{V}}(X,W)\right) - g_M\left(A^{\mathcal{V}}(X,Z), A^{\mathcal{V}}(Y,W)\right). \end{aligned} \tag{3.14}$$

Note that for a Riemannian submersion $F : M^m \longrightarrow N^n$ one can define a linear map $A_1^{\mathcal{V}} : \chi^h(M) \to \chi^v(M)^*$, where $\chi^v(M)^*$ is the dual vector space of the vertical vector space $\chi^v(M)$, by [**GMK**]

$$A_1^{\mathcal{V}}(X) = A^{\mathcal{V}}(X, \cdot),$$

for all $X \in \chi^h(M)$, where $A^{\mathcal{V}}(X, \cdot) \in \chi^v(M)^*$, $A^{\mathcal{V}}(X,Y) = v\nabla_X Y$, for $Y \in \chi^h(M)$. Then we can define the squared norm of the above linear map by

$$\|A_1^{\mathcal{V}}(X)\|^2 = \sum_{i=1}^n g_M\left(A^{\mathcal{V}}(e_i, X), A^{\mathcal{V}}(e_i, X)\right)$$

where $\{e_1, e_2, \cdots, e_n\}$ is orthonormal frame of $\chi^h(M)$.

Moreover, for any horizontal vector field X, we also put

$$Ric^*(X) = \sum_{i=1}^n g_M(R^*(e_i, X)X, e_i). \tag{3.15}$$

THEOREM 3.8. *Let $F : M^m \longrightarrow N^n(c)$ be a Riemannian submersion of a Riemannian manifold M^m onto a real space form $N^n(c)$. Then for any unit horizontal vector field X on M we have*

$$\mathrm{Ric}^{\mathcal{H}}(X) \leq (n-1)c. \tag{3.16}$$

Moreover, the equality holds in (3.16) *if and only if* $A_1^{\mathcal{V}}(X) = 0$.

PROOF. Using the anti-symmetry of $A^{\mathcal{V}}$, we derive from (3.14) that

$$\mathrm{Ric}^{\mathcal{H}}(X) = \mathrm{Ric}^*(X) - 3\|A_1^{\mathcal{V}}(X)\|^2. \tag{3.17}$$

On the other hand, choosing an orthonormal frame $\{e_1, e_2, \cdots, e_n\}$ of $\chi^h(M)$ with $e_1 = X$ and using (3.15), the projectability of the horizontal lifts and (3.3), we obtain by direct computation that

$$\mathrm{Ric}^*(X) = (n-1)c. \tag{3.18}$$

Using now (3.17) and (3.18), we deduce

$$\mathrm{Ric}^{\mathcal{H}}(X) = (n-1)c - 3\|A_1^{\mathcal{V}}(X)\|^2$$

and the conclusion is now clear. □

EXAMPLE 3.9. Let $M = M_1^{m_1}(c_1) \times M_2^{m_2}(c_2)$ be the Riemannian product of two real space forms $M_1^{m_1}(c_1)$ and $M_2^{m_2}(c_2)$. Then the canonical projections $\pi_1 : M \to M_1$ and $\pi_2 : M \to M_2$ are Riemannian submersions with totally geodesic fibres and integrable horizontal distributions (see [**FIP**]). Obviously, both submersions satisfy identically the equality case of the inequality (3.16).

4. Improved Chen-Ricci inequalities for Lagrangian Riemannian maps

A Riemannian map $F : (M^m, g_M) \to (N^n, J_n, g_N)$ from a Riemannian manifold (M^m, g_M) to an almost Hermitian manifold (N^n, J_N, g_N) is said to be *an anti-invariant Riemannian map at* $p \in M$ if $J_N(\mathrm{Im}F_{*p}) \subset (\mathrm{Im}F_{*p})^{\perp}$. Moreover, if the Riemannian map F is anti-invariant for every $p \in M$, then F is said to be *an anti-invariant Riemannian map* [**S2**]. For examples of such maps see [**S2**]. Obviously, the notion of anti-invariant Riemannian map generalizes the concept of totally real submanifold of an almost Hermitian manifold. Recall that an n-dimensional submanifold M^n of an almost Hermitian manifold (N^n, J_N, g_N) of complex dimension n is called *totally real* (or *isotropic*) if J_N maps each tangent space of M into the normal space. In particular, if $m = n$, then M^n is said to be a *Lagrangian submanifold.* We say that an anti-invariant Riemannian map $F : (M^m, g_M) \to (N^n, J_N, g_N)$ is *Lagrangian* if $\mathrm{Rank}F = \dim(\mathrm{Im}F_*)^{\perp}$. Clearly, a Lagrangian submanifold provides us a particular example of Lagrangian Riemannian map. For other examples of Lagrangian Riemannian maps, see [**S3**].

Recall that a non-totally geodesic Lagrangian submanifold M^n of a complex space form $(\bar{M}^n(c), J_N, g_N)$ is called *H-umbilical* if its second fundamental form h satisfies

$$\begin{aligned} h(e_i, e_i) &= \mu J_N e_1, \ h(e_1, e_i) = \mu J_N e_i, \ i = 2, \dots, n, \\ h(e_1, e_1) &= \lambda J_N e_1, \ h(e_i, e_j) = 0, \ 2 \le i \ne j \le n, \end{aligned} \tag{4.1}$$

for some functions μ, λ with respect to an orthonormal frame $\{e_1, \dots, e_n\}$. If the ratio of $\mu : \lambda$ is a constant r, then M^n is said to be an *H-umbilical submanifold of ratio r*. Note that within the family of Lagrangian submanifolds in complex space forms, the subfamily of H-umbilical submanifolds form the simplest class next to the class of Lagrangian totally geodesic submanifolds. This is due to the non-existence of (non-totally geodesic) totally umbilical Lagrangian submanifolds in complex space forms, as shown in [**CO**]. For examples of such submanifolds and recent results, see [**ALVY, CPW, CVW, V**].

REMARK 4.1. In the previous Section, we established a Chen-Ricci inequality for Riemannian maps to real space forms. In this Section we focuss on the case of anti-invariant Riemannian maps to complex space form. Therefore, if the target manifold of a Riemannian map F is a complex space form $(N^n(c), g_N)$ of constant holomorphic curvature c, then we can prove the following Chen-Ricci inequality whose proof will be omitted, because it is similar to that of Theorem 3.1.

THEOREM 4.2. *Let F be an anti-invariant Riemannian map from a Riemannian manifold (M^m, g_M) onto a complex space form $(N^n(c), g_N)$ with* rank$F = r$ *and X be an unit horizontal vector on M. Then we have*

$$\mathrm{Ric}^{\mathcal{H}}(X) \leq \frac{(r-1)c}{4} + \frac{1}{4}\|\mathrm{trace}\zeta^{\mathcal{H}}\|^2. \tag{4.2}$$

Moreover:

(*i*) *The equality case of* (4.2) *holds for $X \in \mathcal{H}_p$ if and only if*

$$\begin{cases} (\nabla F_*)(X,Y) = 0 & \text{for all } Y \in \mathcal{H}_p \text{ orthogonal to } X, \\ (\nabla F_*)(X,X) = \frac{1}{2}\mathrm{trace}\zeta^{\mathcal{H}}, \end{cases} \tag{4.3}$$

(*ii*) *The equality case of* (4.2) *holds identically for all $X \in \mathcal{H}_p$ if and only if either $\zeta^{\mathcal{H}} = 0$ or $r = 2$ and ${\zeta^{\mathcal{H}}}^{\alpha}_{11} = {\zeta^{\mathcal{H}}}^{\alpha}_{22}$, $\alpha = r+1, \ldots, n$.*

REMARK 4.3. Notice that Theorem 4.2 generalizes the main result of [**C3**]. Indeed, if F is an anti-invariant isometric immersion, then Theorem 4.2 reduces to the following result, which is [**C3**, Theorem 1].

COROLLARY 4.4. [**C3**] *Let (M^m, g_M) be an isotropic submanifold of a complex space form $(N^n(c), g_N)$. Then for each tangent vector $X \in T_pM$, we have*

$$\mathrm{Ric}(X) \leq \frac{(n-1)c}{4} + \frac{n^2}{4}\|H\|^2, \tag{4.4}$$

where $\|H\|^2$ is the squared mean curvature of M^m.

Moreover, the equality case of (4.4) *holds identically for all unit tangent vectors at p if and only if either p is a totally geodesic point or $n = 2$ and p is a totally umbilical point.*

EXAMPLE 4.5. As proved in [**C3**], there are large classes of totally real submanifolds in complex space forms satisfying (4.4) identically. For example, if N^{n-1} is a Lagrangian minimal submanifold of $\mathbb{C}^{n-1}$ and L^1 a (real) line in $\mathbb{C}^1$, then the Riemannian product $M^n = N^{n-1} \times L^1$ is an isotropic submanifold of $\mathbb{C}^n$ which satisfies (4.4) identically. For other examples, see [**C3**].

REMARK 4.6. If the anti-invariant Riemannian map F in Theorem 4.2 is Lagrangian, then we will prove that inequality (4.2) is not sharp. Next we are going to obtain an improved Chen-Ricci inequality for Lagrangian Riemannian maps.

The following lemmas play a key role in the proof of our next theorems.

LEMMA 4.7. [**D**, Lemma 2.2] *Let $f_1 : \mathbb{R}^n \to \mathbb{R}$ be a function given by*

$$f_1(x_1, \cdots, x_n) = x_1 \sum_{j=2}^{n} x_j - \sum_{j=2}^{n} x_j^2.$$

If $x_1 + x_2 + \cdots + x_n = 2(n+1)a$, *then we have*

$$f_1(x_1, \cdots, x_n) \leq \frac{n-1}{4n}\left(\sum_{j=1}^{n} x_j\right)^2.$$

Moreover, the equality case holds in the above inequality if and only if

$$\frac{1}{n+1}x_1 = x_2 = \cdots = x_n = a.$$

LEMMA 4.8. [**D**, Lemma 2.3] *Let* $f_2 : \mathbb{R}^n \to \mathbb{R}$ *be a function given by*

$$f_2(x_1, \cdots, x_n) = x_1 \sum_{j=2}^{n} x_j - x_1^2.$$

If $x_1 + x_2 + \cdots + x_n = 4a$, *then we have*

$$f_2(x_1, \cdots, x_n) \leq \frac{1}{8}\left(\sum_{j=1}^{n} x_j\right)^2.$$

Moreover, the equality case holds in the above inequality if and only if

$$x_1 = a, x_2 + \cdots + x_n = 3a.$$

THEOREM 4.9. *Let* $F : (M^m, g_M) \to (N^n(c), g_N)$ *be a Lagrangian Riemannian map from a Riemannian manifold* (M^m, g_M) *to a complex space form* $(N^n(c), J_N, g_N)$ *of complex dimension* $n > 1$. *Then, for any unit horizontal vector* X *at* $p \in M$, *we have*

$$\mathrm{Ric}^{\mathcal{H}}(X) \leq \frac{(n-1)c}{4} + \frac{n-1}{4n}\|\mathrm{trace}\zeta^{\mathcal{H}}\|^2. \tag{4.5}$$

Moreover, the equality holds in the above inequality for any unit horizontal vector at p *if and only if either*

(*i*) $\zeta^{\mathcal{H}} = 0$ *or*

(*ii*) $n = 2$ *and*

$$(\nabla F_*)(e_1, e_1) = 3\mu J e_1,\ (\nabla F_*)(e_2, e_2) = \mu J e_1,\ (\nabla F_*)(e_1, e_2) = \mu J e_2$$

for some function μ *with respect to some orthonormal local frame field* $\{e_1, e_2\}$.

PROOF. Since the target space $N^n(c)$ is a complex space form of constant holomorphic sectional curvature c, the Riemannian curvature tensor R^N of $N^n(c)$ satisfies

$$\begin{aligned} g_N\left(R^N(X,Y)Z, W\right) &= \frac{c}{4}\{g_N(X,W)g_N(Y,Z) - g_N(X,Z)g_N(Y,W) \\ &\quad + g_N(J_N X, W)g_N(J_N Y, Z) - g_N(J_N X, Z)g_N(J_N Y, W) \\ &\quad - 2g_N(X, J_N Y)g_N(Z, J_N W)\}, \end{aligned} \tag{4.6}$$

for any $X, Y, Z, W \in \Gamma(TN)$.

Let $p \in M$ and $\{e_1, \cdots, e_r\}$ be an orthonormal base of $\mathcal{H}_p$ with $e_1 = X$. Since F is a Lagrangian Riemannian map, it follows that $\{V_1 = J_N F_* e_1, \cdots, V_r = J_N F_* e_r\}$ is an orthonormal base of $(\mathrm{Im}F_*)^\perp$ and we derive easily that $r = n$ and we also have

$$\zeta^{\mathcal{H}\alpha}_{ij} = \zeta^{\mathcal{H}i}_{\alpha j}, \quad i, j, \alpha \in \{1, \cdots, n\}. \tag{4.7}$$

From (2.4) and (2.5), taking account of (4.6), we obtain

$$\begin{aligned}\frac{(n-1)c}{4} &= \mathrm{Ric}^{\mathcal{H}}(X) + \sum_{i=1}^{r} g_N\left((\nabla F_*)(e_i,e_1),(\nabla F_*)(e_i,e_1)\right)\\ &\quad - \sum_{i=1}^{r} g_N\left((\nabla F_*)(e_i,e_i),(\nabla F_*)(e_1,e_1)\right)\\ &= \mathrm{Ric}^{\mathcal{H}}(X) + \sum_{\alpha=r+1}^{n}\sum_{i=2}^{r}\left\{\left(\zeta^{\mathcal{H}}{}^{\alpha}_{1i}\right)^2 - \zeta^{\mathcal{H}}{}^{\alpha}_{11}\zeta^{\mathcal{H}}{}^{\alpha}_{ii}\right\}.\end{aligned}$$

Since

$$\sum_{\alpha=1}^{r}\sum_{i=2}^{r}\left(\zeta^{\mathcal{H}}{}^{\alpha}_{1i}\right)^2 \geq \sum_{i=2}^{r}\left(\zeta^{\mathcal{H}}{}^{1}_{1i}\right)^2 + \sum_{i=2}^{r}\left(\zeta^{\mathcal{H}}{}^{i}_{1i}\right)^2$$

it follows that

$$\begin{aligned}\mathrm{Ric}^{\mathcal{H}}(X) - \frac{(r-1)c}{4} &= \sum_{\alpha=1}^{r}\sum_{i=2}^{r}\left\{\zeta^{\mathcal{H}}{}^{\alpha}_{11}\zeta^{\mathcal{H}}{}^{\alpha}_{ii} - \left(\zeta^{\mathcal{H}}{}^{\alpha}_{1i}\right)^2\right\}\\ &\leq \sum_{\alpha=1}^{r}\sum_{i=2}^{r}\zeta^{\mathcal{H}}{}^{\alpha}_{11}\zeta^{\mathcal{H}}{}^{\alpha}_{ii} - \sum_{i=2}^{r}\left(\zeta^{\mathcal{H}}{}^{1}_{1i}\right)^2 - \sum_{i=2}^{r}\left(\zeta^{\mathcal{H}}{}^{i}_{1i}\right)^2.\end{aligned} \tag{4.8}$$

Using now (4.7) in (4.8), we derive

$$\mathrm{Ric}^{\mathcal{H}}(X) - \frac{(r-1)c}{4} \leq \sum_{\alpha=1}^{r}\sum_{i=2}^{r}\zeta^{\mathcal{H}}{}^{\alpha}_{11}\zeta^{\mathcal{H}}{}^{\alpha}_{ii} - \sum_{i=2}^{r}\left(\zeta^{\mathcal{H}}{}^{i}_{11}\right)^2 - \sum_{i=2}^{r}\left(\zeta^{\mathcal{H}}{}^{1}_{ii}\right)^2. \tag{4.9}$$

Next, we define r real functions $f_1, f_2, \ldots, f_r$ by

$$f_1(\zeta^{\mathcal{H}}{}^{1}_{11},\zeta^{\mathcal{H}}{}^{1}_{22},\cdots,\zeta^{\mathcal{H}}{}^{1}_{rr}) = \zeta^{\mathcal{H}}{}^{1}_{11}\sum_{i=2}^{r}\zeta^{\mathcal{H}}{}^{1}_{ii} - \sum_{i=2}^{r}\left(\zeta^{\mathcal{H}}{}^{1}_{ii}\right)^2,$$

$$f_\alpha(\zeta^{\mathcal{H}}{}^{\alpha}_{11},\zeta^{\mathcal{H}}{}^{\alpha}_{22},\cdots,\zeta^{\mathcal{H}}{}^{\alpha}_{rr}) = \zeta^{\mathcal{H}}{}^{\alpha}_{11}\sum_{j=2}^{r}\zeta^{\mathcal{H}}{}^{\alpha}_{jj} - \left(\zeta^{\mathcal{H}}{}^{\alpha}_{11}\right)^2, \quad \alpha\in\{2,\cdots,r\}.$$

Denoting

$$(\mathrm{trace}\zeta^{\mathcal{H}})^k = \zeta^{\mathcal{H}}{}^{k}_{11} + \zeta^{\mathcal{H}}{}^{k}_{22} + \cdots + \zeta^{\mathcal{H}}{}^{k}_{rr},$$

for $k = 1, \ldots, r$, then by Lemma 4.7 we have

$$f_1(\zeta^{\mathcal{H}}{}^{1}_{11},\zeta^{\mathcal{H}}{}^{1}_{22},\cdots,\zeta^{\mathcal{H}}{}^{1}_{rr}) \leq \frac{r-1}{4r}\left((\mathrm{trace}\zeta^{\mathcal{H}})^1\right)^2 \tag{4.10}$$

and by Lemma 4.8 we get for $2 \leq \alpha \leq r$ that

$$\begin{aligned}f_\alpha(\zeta^{\mathcal{H}}{}^{\alpha}_{11},\zeta^{\mathcal{H}}{}^{\alpha}_{22},\cdots,\zeta^{\mathcal{H}}{}^{\alpha}_{rr}) &\leq \frac{1}{8}\left((\mathrm{trace}\zeta^{\mathcal{H}})^\alpha\right)^2\\ &\leq \frac{r-1}{4r}\left((\mathrm{trace}\zeta^{\mathcal{H}})^\alpha\right)^2.\end{aligned} \tag{4.11}$$

Combining now (4.9) with (4.10) and (4.11), we obtain

$$\mathrm{Ric}^{\mathcal{H}}(X) - \frac{(r-1)c}{4} \leq \sum_{\alpha=1}^{r}\frac{r-1}{4r}\left((\mathrm{trace}\zeta^{\mathcal{H}})^\alpha\right)^2$$

or, equivalently,

$$\mathrm{Ric}^{\mathcal{H}}(X) \leq \frac{(r-1)c}{4} + \frac{r-1}{4r}\|\mathrm{trace}\zeta^{\mathcal{H}}\|^2,$$

due to the fact that

$$\sum_{\alpha=1}^{r}\left((\mathrm{trace}\zeta^{\mathcal{H}})^{\alpha}\right)^2=\|\mathrm{trace}\zeta^{\mathcal{H}}\|^2.$$

Hence, taking into account that $r=n$, the required inequality is proved.

Now, we assume the equality holds in inequality (4.5) for any unit horizontal vector X at $p\in M$. Since $r=n>1$, (4.11) implies $\left(\mathrm{trace}\zeta^{\mathcal{H}}\right)^{\alpha}=0$ for all $\alpha\geq 2$. Therefore, from Lemma 4.8 we obtain

$$\zeta^{\mathcal{H}}{}^{1}_{1i}=\zeta^{\mathcal{H}}{}^{i}_{11}=\frac{\left(\mathrm{trace}\zeta^{\mathcal{H}}\right)^{1}}{4}=0,\quad i\geq 2.$$

From (4.8), we get

$$\zeta^{\mathcal{H}}{}^{1}_{ij}=0\quad i,j\geq 2,\quad i\neq j.$$

On the other hand, Lemma 4.7 implies that $\left(\zeta^{\mathcal{H}}{}^{1}_{ij}\right)_{i,j=1\cdots r}$ must be a diagonal matrix with

$$\zeta^{\mathcal{H}}{}^{1}_{11}=\frac{(r+1)\left(\mathrm{trace}\zeta^{\mathcal{H}}\right)^{1}}{2r},\qquad \zeta^{\mathcal{H}}{}^{1}_{ii}=\frac{\left(\mathrm{trace}\zeta^{\mathcal{H}}\right)^{1}}{2r},\qquad i\geq 2.$$

By a direct calculation of $\mathrm{Ric}^{\mathcal{H}}(e_2)$ as it was done for $\mathrm{Ric}^{\mathcal{H}}(e_1)$ in (4.8), the corresponding equality implies

$$\zeta^{\mathcal{H}}{}^{j}_{2i}=\zeta^{\mathcal{H}}{}^{2}_{ij}=0,\quad i\neq 2,\quad j\neq 2,\quad i\neq j.$$

Therefore, applying again Lemma 4.7, we obtain

$$\frac{1}{r+1}\zeta^{\mathcal{H}}{}^{2}_{11}=\zeta^{\mathcal{H}}{}^{2}_{22}=\cdots=\zeta^{\mathcal{H}}{}^{2}_{rr}=\frac{\left(\mathrm{trace}\zeta^{\mathcal{H}}\right)^{2}}{2r}=0.$$

Now, with similar arguments for matrices $\left(\zeta^{\mathcal{H}}{}^{3}_{ij}\right)_{i,j=1\cdots r},\ldots,\left(\zeta^{\mathcal{H}}{}^{r}_{ij}\right)_{i,j=1\cdots r}$, we obtain

$$\zeta^{\mathcal{H}}{}^{2}_{2i}=\zeta^{\mathcal{H}}{}^{i}_{22}=\frac{\left(\mathrm{trace}\zeta^{\mathcal{H}}\right)^{i}}{2r}=0,\quad i\geq 3.$$

Thus, the matrices $\left(\zeta^{\mathcal{H}}{}^{\alpha}_{ij}\right)_{i,j=1\cdots r}$ can have a limited number of possible non-zero entries, the only eventual possibilities being

$$\zeta^{\mathcal{H}}{}^{\alpha}_{1\alpha}=\zeta^{\mathcal{H}}{}^{\alpha}_{\alpha 1}=\zeta^{\mathcal{H}}{}^{1}_{\alpha\alpha}=\frac{\left(\mathrm{trace}\zeta^{\mathcal{H}}\right)^{1}}{2r},\quad \alpha\geq 2.$$

Replacing now $X=W=e_2$ and $Y=Z=e_i$ in (2.4), we obtain

$$\begin{aligned}g_N\left(R^N(F_*e_2,F_*e_i)F_*e_i,F_*e_2\right)&=g_M\left(R^M(e_2,e_i)e_i,e_2\right)\\&+g_N\left((\nabla F_*)(e_2,e_i),(\nabla F_*)(e_2,e_i)\right)\\&-g_N\left((\nabla F_*)(e_2,e_2),(\nabla F_*)(e_i,e_i)\right),\end{aligned}\tag{4.12}$$

which implies

$$\begin{aligned}&g_N\left(R^N(F_*e_2,F_*e_i)F_*e_i,F_*e_2\right)\\&=g_M\left(R^M(e_2,e_i)e_i,e_2\right)-\left(\frac{\left(\mathrm{trace}\zeta^{\mathcal{H}}\right)^{1}}{2r}\right)^2,\quad i\geq 3.\end{aligned}\tag{4.13}$$

If $i = 1$ in (4.12), we obtain

$$\begin{aligned} & g_N\left(R^N\left(F_*e_2, F_*e_1\right)F_*e_1, F_*e_2\right) \\ & = g_M\left(R^M\left(e_2, e_1\right)e_1, e_2\right) - (r+1)\left(\frac{\left(\mathrm{trace}\zeta^{\mathcal{H}}\right)^1}{2r}\right)^2 + \left(\frac{\left(\mathrm{trace}\zeta^{\mathcal{H}}\right)^1}{2r}\right)^2. \end{aligned} \tag{4.14}$$

Combining (4.13) and (4.14), we get

$$\begin{aligned} & \mathrm{Ric}^{\mathcal{H}}\left(e_2\right) - \frac{(r-1)c}{4} \\ & = (r+1)\left(\frac{\left(\mathrm{trace}\zeta^{\mathcal{H}}\right)^1}{2r}\right)^2 - \left(\frac{\left(\mathrm{trace}\zeta^{\mathcal{H}}\right)^1}{2r}\right)^2 + (r-2)\left(\frac{\left(\mathrm{trace}\zeta^{\mathcal{H}}\right)^1}{2r}\right)^2 \\ & = 2\left(r-1\right)\left(\frac{\left(\mathrm{trace}\zeta^{\mathcal{H}}\right)^1}{2r}\right)^2. \end{aligned}$$

But, from the assumption of the equality case, we have

$$\mathrm{Ric}^{\mathcal{H}}\left(e_2\right) - \frac{(r-1)c}{4} = \frac{r-1}{4r}\|\mathrm{trace}\zeta^{\mathcal{H}}\|^2 = r\left(r-1\right)\left(\frac{\left(\mathrm{trace}\zeta^{\mathcal{H}}\right)^1}{2r}\right)^2.$$

Therefore, we obtain

$$r\left(r-1\right)\left(\frac{\left(\mathrm{trace}\zeta^{\mathcal{H}}\right)^1}{2r}\right)^2 = 2\left(r-1\right)\left(\frac{\left(\mathrm{trace}\zeta^{\mathcal{H}}\right)^1}{2r}\right)^2.$$

Taking into account that $r = n \neq 1$, we conclude that either $\left(\mathrm{trace}\zeta^{\mathcal{H}}\right)^1 = 0$ or $r = 2$.

If $\left(\mathrm{trace}\zeta^{\mathcal{H}}\right)^1 = 0$, then $\left(\zeta^{\mathcal{H}\alpha}_{ij}\right)_{i,j=1\cdots r}$ is a null matrix for $\alpha = 1, \ldots, r$, and hence $\zeta^{\mathcal{H}} = 0$.

If $r = n = 2$, then we derive that

$$\begin{aligned} \left(\nabla F_*\right)\left(e_1, e_1\right) &= \frac{3\left(\mathrm{trace}\zeta^{\mathcal{H}}\right)^1}{2r}Je_1, \\ \left(\nabla F_*\right)\left(e_2, e_2\right) &= \frac{\left(\mathrm{trace}\zeta^{\mathcal{H}}\right)^1}{2r}Je_1, \\ \left(\nabla F_*\right)\left(e_1, e_2\right) &= \frac{\left(\mathrm{trace}\zeta^{\mathcal{H}}\right)^1}{2r}Je_2 \end{aligned}$$

and the conclusion follows.

The converse is easy to prove by direct computation. □

Example 4.10. Let M_1 be the real projective space $\mathbb{R}P^n\left(\frac{c}{4}\right)$ of dimension $n > 2$, and M_2 be an arbitrary Riemannian manifold. Suppose $\iota : M_1 \to \mathbb{C}P^n(c)$ is the canonical imbedding of M_1 into the complex projective space $\mathbb{C}P^n(c)$ as a totally real and totally geodesic submanifold (see [**CO**]). Let $M = M_1 \times_f M_2$ be the warped product of M_1 and M_2 by a function $f > 0$ on M_1, and let $\pi_1 : M_1 \times_f M_2 \to M_1$ be the canonical projection defined by $\pi_1(p, q) = p$, for all $(p, q) \in M_1 \times M_2$. Then $\iota \circ \pi_1 : M_1 \times_f M_2 \to \mathbb{C}P^n(c)$ is an anti-invariant Riemannian map illustrating the main inequalities of this section.

REMARK 4.11. Using Theorem 4.9 in the particular case when the Lagrangian Riemannian map F is an isometric Lagrangian immersion, we recover the main result of [**D**], namely the improved Chen-Ricci inequality for Lagrangian submanifolds in complex space forms.

COROLLARY 4.12. [**D**, Theorem 3.1] *Let $F : M^n \longrightarrow N^n(c)$ be an isometric Lagrangian immersion of a Riemannian n-dimensional manifold M^n into a complex space form $(N^n(c), J_N, g_N)$. Then, for any unit tangent vector $X \in T_pM$, $p \in M$, the following inequality holds:*

$$\operatorname{Ric}(X) \leq \frac{n-1}{4}\left(c + n\|H\|^2\right). \tag{4.15}$$

Moreover, the equality case holds identically for all unit tangent vectors at p if and only if either p is a totally geodesic point, or $n = 2$ and p is an H-umbilical point with $\lambda = 3\mu$.

REMARK 4.13. Note that B.-Y. Chen, A. Prieto-Martín and X. Wang provided in [**CPW**] a method for constructing H-umbilical Lagrangian submanifolds with any given ratio r in $\mathbb{C}P^n(4)$ and $\mathbb{C}H^n(-4)$. In particular case when $n = 2$ and $r = 3$, we have examples of submanifolds attaining equality in the above inequality at all points. We also note that real projective spaces $\mathbb{R}P^n(1)$ and real hyperbolic spaces $H^n(-1)$ are totally geodesic Lagrangian submanifolds of the complex projective spaces $\mathbb{C}P^n(4)$ and the complex hyperbolic spaces $\mathbb{C}H^n(-4)$, respectively, and these provide natural examples of submanifolds attaining equality in (4.15) at all points. Finally, we remark that the Whitney 2-sphere in $\mathbb{C}^2$ also satisfies identically the equality case of (4.15) (see [**D**, Example 3.1]).

References

[ALVY] Mohd. Aquib, Jae Won Lee, Gabriel-Eduard Vîlcu, and Dae Won Yoon, *Classification of Casorati ideal Lagrangian submanifolds in complex space forms*, Differential Geom. Appl. **63** (2019), 30–49, DOI 10.1016/j.difgeo.2018.12.006. MR3896193

[C1] Bang-Yen Chen, *Some pinching and classification theorems for minimal submanifolds*, Arch. Math. (Basel) **60** (1993), no. 6, 568–578, DOI 10.1007/BF01236084. MR1216703

[C2] Bang-Yen Chen, *Relations between Ricci curvature and shape operator for submanifolds with arbitrary codimensions*, Glasg. Math. J. **41** (1999), no. 1, 33–41, DOI 10.1017/S0017089599970271. MR1689730

[C3] Bang-Yen Chen, *On Ricci curvature of isotropic and Lagrangian submanifolds in complex space forms*, Arch. Math. (Basel) **74** (2000), no. 2, 154–160, DOI 10.1007/PL00000420. MR1735232

[C4] Bang-Yen Chen, *Riemannian submersions, minimal immersions and cohomology class*, Proc. Japan Acad. Ser. A Math. Sci. **81** (2005), no. 10, 162–167 (2006). MR2196721

[C5] Bang-Yen Chen, *Examples and classification of Riemannian submersions satisfying a basic equality*, Bull. Austral. Math. Soc. **72** (2005), no. 3, 391–402, DOI 10.1017/S000497270003522X. MR2199641

[C6] Bang-Yen Chen, *Pseudo-Riemannian geometry, δ-invariants and applications*, World Scientific Publishing Co. Pte. Ltd., Hackensack, NJ, 2011. With a foreword by Leopold Verstraelen, DOI 10.1142/9789814329644. MR2799371

[CO] Bang-yen Chen and Koichi Ogiue, *Two theorems on Kaehler manifolds*, Michigan Math. J. **21** (1974), 225–229 (1975). MR367884

[CPW] Bang-Yen Chen, Alicia Prieto-Martín, and Xianfeng Wang, *Lagrangian submanifolds in complex space forms satisfying an improved equality involving* $\delta(2,2)$, Publ. Math. Debrecen **82** (2013), no. 1, 193–217, DOI 10.5486/PMD.2013.5405. MR3034376

[CVW] Bang-Yen Chen, Luc Vrancken, and Xianfeng Wang, *Lagrangian submanifolds in complex space forms satisfying equality in the optimal inequality involving* $\delta(2,\ldots,2)$, Beitr. Algebra Geom. **62** (2021), no. 1, 251–264, DOI 10.1007/s13366-020-00541-4. MR4249863

[D] Shangrong Deng, *An improved Chen-Ricci inequality*, Int. Electron. J. Geom. **2** (2009), no. 2, 39–45. MR2558711

[FIP] Maria Falcitelli, Stere Ianus, and Anna Maria Pastore, *Riemannian submersions and related topics*, World Scientific Publishing Co., Inc., River Edge, NJ, 2004, DOI 10.1142/9789812562333. MR2110043

[F] Arthur E. Fischer, *Riemannian maps between Riemannian manifolds*, Mathematical aspects of classical field theory (Seattle, WA, 1991), Contemp. Math., vol. 132, Amer. Math. Soc., Providence, RI, 1992, pp. 331–366, DOI 10.1090/conm/132/1188447. MR1188447

[GC] Eduardo García-Río and Demir N. Kupeli, *Semi-Riemannian maps and their applications*, Mathematics and its Applications, vol. 475, Kluwer Academic Publishers, Dordrecht, 1999, DOI 10.1007/978-94-017-2979-6. MR1700746

[GMK] M. Gülbahar, Ş. Eken Meriç, and E. Kiliç, *Sharp inequalities involving the Ricci curvature for Riemannian submersions*, Kragujevac J. Math. **41** (2017), no. 2, 279–293, DOI 10.5937/kgjmath1702279g. MR3741764

[LLSV] Chul Woo Lee, Jae Won Lee, Bayram Şahin, and Gabriel-Eduard Vîlcu, *Optimal inequalities for Riemannian maps and Riemannian submersions involving Casorati curvatures*, Ann. Mat. Pura Appl. (4) **200** (2021), no. 3, 1277–1295, DOI 10.1007/s10231-020-01037-7. MR4242128

[LMSSJ] Mehraj Ahmad Lone, Yoshio Matsuyama, Falleh R. Al-Solamy, Mohammad Hasan Shahid, and Mohammed Jamali, *Upper bounds for Ricci curvatures for submanifolds in Bochner-Kaehler manifolds*, Tamkang J. Math. **51** (2020), no. 1, 53–67, DOI 10.5556/j.tkjm.51.2020.2967. MR4075188

[MiA] Adela Mihai, *Inequalities on the Ricci curvature*, J. Math. Inequal. **9** (2015), no. 3, 811–822, DOI 10.7153/jmi-09-67. MR3345139

[MiR] Adela Mihai and Ioana N. Rădulescu, *An improved Chen-Ricci inequality for Kaehlerian slant submanifolds in complex space forms*, Taiwanese J. Math. **16** (2012), no. 2, 761–770, DOI 10.11650/twjm/1500406613. MR2892910

[MK] Fereshteh Malek and Mohammad Bagher Kazemi Balgeshir, *Slant submanifolds of almost contact metric 3-structure manifolds*, Mediterr. J. Math. **10** (2013), no. 2, 1023–1033, DOI 10.1007/s00009-012-0222-4. MR3045693

[Mih] Ion Mihai, *Ricci curvature of submanifolds in Sasakian space forms*, J. Aust. Math. Soc. **72** (2002), no. 2, 247–256, DOI 10.1017/S1446788700003888. MR1887135

[MUS] Abdulqader Mustafa, Siraj Uddin, and Falleh R. Al-Solamy, *Chen-Ricci inequality for warped products in Kenmotsu space forms and its applications*, Rev. R. Acad. Cienc. Exactas Fís. Nat. Ser. A Mat. RACSAM **113** (2019), no. 4, 3585–3602, DOI 10.1007/s13398-019-00718-0. MR3999036

[N] Thérèse Nore, *Second fundamental form of a map*, Ann. Mat. Pura Appl. (4) **146** (1987), 281–310, DOI 10.1007/BF01762368. MR916696

[O] Teodor Oprea, *Ricci curvature of Lagrangian submanifolds in complex space forms*, Math. Inequal. Appl. **13** (2010), no. 4, 851–858, DOI 10.7153/mia-13-61. MR2760505

[S1] Bayram Şahin, *Chen's first inequality for Riemannian maps*, Ann. Polon. Math. **117** (2016), no. 3, 249–258, DOI 10.4064/ap3958-7-2016. MR3557196

[S2] Bayram Şahin, *Invariant and anti-invariant Riemannian maps to Kähler manifolds*, Int. J. Geom. Methods Mod. Phys. **7** (2010), no. 3, 337–355, DOI 10.1142/S0219887810004324. MR2646767

[S3] Bayram Şahin, *Riemannian submersions, Riemannian maps in Hermitian geometry, and their applications*, Elsevier/Academic Press, London, 2017. MR3644540

[T] Mukut Mani Tripathi, *Improved Chen-Ricci inequality for curvature-like tensors and its applications*, Differential Geom. Appl. **29** (2011), no. 5, 685–698, DOI 10.1016/j.difgeo.2011.07.008. MR2831825

[V] Gabriel-Eduard Vîlcu, *An optimal inequality for Lagrangian submanifolds in complex space forms involving Casorati curvature*, J. Math. Anal. Appl. **465** (2018), no. 2, 1209–1222, DOI 10.1016/j.jmaa.2018.05.060. MR3809353

Department of Mathematics Education, Gyeongsang National University and RINS, Jinju 52828, South Korea

Email address: leejaew@gnu.ac.kr

Department of Mathematics, Kyungpook National University, Daegu 41566, South Korea

Email address: mathisu@knu.ac.kr

Department of Mathematics, Ege University, Izmir, Turkey

Email address: bayram.sahin@ymail.com

Department of Cybernetics, Economic Informatics, Finance and Accountancy, Petroleum-Gas University of Ploieşti, Bd. Bucureşti, Nr. 39, Ploieşti 100680, Romania

Current address: University of Bucharest, Faculty of Mathematics and Computer Science, Research Center in Geometry, Topology and Algebra, Str. Academiei, Nr. 14, Sector 1, Bucureşti 70109, Romania

Email address: gvilcu@upg-ploiesti.ro

Contemporary Mathematics
Volume **777**, 2022
https://doi.org/10.1090/conm/777/15633

Totally geodesic surfaces in the complex quadric

Marilena Moruz, Joeri Van der Veken, Luc Vrancken, and Anne Wijffels

ABSTRACT. We provide explicit descriptions of all totally geodesic surfaces of a complex quadric of arbitrary dimension. Totally geodesic submanifolds of complex quadrics were first studied by Chen and Nagano in 1977 and fully classified by Klein in 2008. In particular, we interpret some of these surfaces as Gaussian images of surfaces in a unit three-sphere and all others as elements of the Veronese sequence introduced by Bolton, Jensen, Rigoli and Woodward. We also briefly discuss how the classification can be translated to the non-compact dual of the complex quadric, namely the hyperbolic complex quadric.

1. Introduction

Totally geodesic submanifolds of the complex quadric $Q^n \subseteq \mathbb{C}P^{n+1}$ were studied by Chen and Nagano in [**4**]. The list that they obtained was later extended to a full classification by Klein in [**6**], by relating root space decompositions of symmetric spaces to root space decompositions of their symmetric subspaces. In particular, when restricting to two-dimensional submanifolds, he proved the existence of six types of totally geodesic surfaces in Q^n, adding one type to the list of Chen and Nagano, namely a totally geodesic round two-sphere of radius $\sqrt{10}/2$. As remarked in Chen's review of [**6**] on MathSciNet, no explicit description of this surface is given. In the present paper, we give explicit descriptions of all the totally geodesic surfaces in Q^n. In particular, we interpret some of them as Gaussian images of surfaces in $S^3(1)$ and we prove that all others, including Klein's new example, correspond to elements of the *Veronese sequence*, see [**1**]. In the last two sections of the paper, we interpret the results for $Q^2 \cong S^2(1/2) \times S^2(1/2)$ and we translate the classification to the non-compact dual symmetric space of the complex quadric, namely the hyperbolic complex quadric.

2020 *Mathematics Subject Classification.* Primary 53C40, 53C42.

Key words and phrases. Totally geodesic, surface, complex quadric, hyperbolic complex quadric.

All authors were supported by the KU Leuven Research Fund (BOF) under project no. 3E160361 at the time of writing the paper. The second author was supported by the Research Foundation Flanders (FWO) and the Fonds de la Recherche Scientifique (FNRS) under Excellence Of Science project no. G0H4518N. The first author was supported by a grant of the Romanian Ministry of Education and Research, CNCS-UEFISCDI, project number PN-III-P1-1.1-PD-2019-0253, within PNCDI III.

2. Preliminaries

Throughout the paper, we will always assume that manifolds are connected. Moreover, we use the following notation for spheres: $S^k(r)=\{x\in\mathbb{R}^{k+1} \mid \langle x,x\rangle=r^2\}$ for all positive integers k and for all positive real numbers r, where $\langle\cdot,\cdot\rangle$ is the Euclidean inner product on $\mathbb{R}^{k+1}$.

2.1. The complex quadric. Let $\mathbb{C}P^{n+1}(4)$ be the complex projective space of complex dimension $n+1$ equipped with the Fubini-Study metric g_{FS} of constant holomorphic sectional curvature 4. Then the Hopf fibration,

$$\pi : S^{2n+3}(1) \subseteq \mathbb{C}^{n+2} \to \mathbb{C}P^{n+1}(4) : z \mapsto [z],$$

where $[z]$ denotes the set of all non-zero complex multiples of z and $\mathbb{C}^{n+2}$ is identified with $\mathbb{R}^{2n+4}$ to contain the sphere $S^{2n+3}(1)$, is a Riemannian submersion. The complex structure J on $\mathbb{C}P^{n+1}(4)$ is induced from multiplication by i on $TS^{2n+3}(1)$ and it is well-known that $(\mathbb{C}P^{n+1}(4), g_{FS}, J)$ is a Kähler manifold.

We define the *complex quadric* of complex dimension n as the following complex hypersurface of $\mathbb{C}P^{n+1}(4)$:

$$Q^n = \{[(z_0,\ldots,z_{n+1})] \in \mathbb{C}P^{n+1}(4) \mid z_0^2+\ldots+z_{n+1}^2=0\}.$$

If Q^n is equipped with the induced metric $g_{FS}|_{Q^n}$, which we will denote by g, and the induced almost complex structure $J|_{Q^n}$, which we will again denote by J, then (Q^n, g, J) is of course a Kähler manifold itself. We mention that Q^n can be identified with the Grassmannian of oriented 2-planes in $\mathbb{R}^{n+2}$ and hence, as a homogeneous space, is

$$Q^n = \frac{\mathrm{SO}(n+2)}{\mathrm{SO}(n)\times\mathrm{SO}(2)}.$$

Denote by $\mathcal{A}$ the set of all shape operators of Q^n in $\mathbb{C}P^{n+1}(4)$ associated with unit normal vector fields. Since we need it in the next sections, we allow for elements of $\mathcal{A}$ to be defined only on a subset of Q^n. One can deduce the following (see for example [**10**] or [**11**]).

LEMMA 2.1. *Any $A\in\mathcal{A}$ is involutive, symmetric and anti-commutes with J.*

This implies in particular that $\mathcal{A}$ is a family of almost product structures. However, these almost product structures are not integrable.

The equation of Gauss for Q^n as a submanifold of $\mathbb{C}P^{n+1}(4)$ yields the following expression for the Riemann-Christoffel curvature tensor of Q^n:

$$\begin{aligned} R^{Q^n}(X,Y)Z &= g(Y,Z)X - g(X,Z)Y \\ &+ g(JY,Z)JX - g(JX,Z)JY - 2g(JX,Y)JZ \\ &+ g(AY,Z)AX - g(AX,Z)AY + g(JAY,Z)JAX - g(JAX,Z)JAY, \end{aligned} \tag{2.1}$$

where A is any element of $\mathcal{A}$. It follows directly from (2.1) that Q^n is Einstein and that Q^1 has constant Gaussian curvature $K=2$. In fact, Q^1 is isometric to $S^2(1/\sqrt{2})$.

2.2. Gauss maps of spherical hypersurfaces. Consider an orientable hypersurface $a : M^n \to S^{n+1}(1)$ of a unit sphere and fix a unit normal vector field $b : M^n \to TS^{n+1}(1)$ to this hypersurface. For every point $p\in M^n$, we can look at $a(p)$ and $b(p)$ as unit vectors in $\mathbb{R}^{n+2}$ and it is easy to check that the sum of the squares of the components of the vector $a(p) + ib(p) \in \mathbb{C}^{n+2}$ equals zero. Hence, the following map is well-defined:

$$G : M^n \to Q^n : p \mapsto [a(p) + ib(p)]. \tag{2.2}$$

This map is known as the *Gauss map* of the given oriented spherical hypersurface. As far as the auhtors know, this notion first appeared in [**9**], in a slighly different form, where also the following result was proven.

THEOREM 2.2. [**9**] *The Gauss map of an oriented hypersurface $M^n \to S^{n+1}(1)$ is a Lagrangian immersion $M^n \to Q^n$.*

Moreover, the following result was proven in [**7**].

THEOREM 2.3. [**7**] *A totally geodesic Lagrangian immersion $M^n \to Q^n$ is, up to an isometry of Q^n, the Gauss map of $a_1 : M^n \subseteq S^n(1) \to S^{n+1}(1) : x \mapsto (x, 0)$ or of $a_2 : M^n \subseteq S^k(1/\sqrt{2}) \times S^{n-k}(1/\sqrt{2}) \to S^{n+1}(1) : (x, y) \mapsto (x, y)$ for some $k \in \{1, \ldots, n-1\}$.*

Note that if we choose $b_1(x) = (0, 1)$ and $b_2(x, y) = (-x, y)$ as unit normal vector fields to the hypersurfaces a_1 and a_2 from Theorem 2.3 respectively, their Gauss maps are given by

$$G_1 : M^n \subseteq S^n(1) \to Q^n : x \mapsto [(x, i)] \tag{2.3}$$

and

$$G_2 : M^n \subseteq S^k(1/\sqrt{2}) \times S^{n-k}(1/\sqrt{2}) \to Q^n : (x, y) \mapsto [(x, iy)] \tag{2.4}$$

respectively. The Lagrangian immersions G_1 and G_2 are not isometric in general. In fact, G_1 induces a metric of constant sectional curvature 2 on the sphere, while G_2 only induces a metric of constant sectional curvature on the product of spheres if $n = 2$ and $k = 1$, in which case it induces a flat metric on $S^1(1/\sqrt{2}) \times S^1(1/\sqrt{2})$.

2.3. The Veronese sequence. We recall the following family of minimal surfaces in $\mathbb{C}P^k(4)$ from [**1**]. For $\ell \in \{0, \ldots, k\}$, let

$$\phi_\ell : \mathbb{C}P^1 \to \mathbb{C}P^k(4) : [(z_0, z_1)] \mapsto \left[\left(f_{\ell,0}\left(\frac{z_1}{z_0}\right), \ldots, f_{\ell,k}\left(\frac{z_1}{z_0}\right)\right)\right],$$

where

$$f_{\ell,j}(z) = \sqrt{\binom{k}{j}}\, z^{j-\ell} \sum_{m\in\mathbb{Z}} (-1)^m \binom{j}{\ell - m}\binom{k-j}{m} |z|^{2m}$$

for $j \in \{0, \ldots, k\}$. Then ϕ_ℓ is a conformal minimal immersion, which induces a metric of constant Gaussian curvature

$$K = \frac{4}{k + 2\ell(k - \ell)}$$

on $\mathbb{C}P^1 \approx S^2$ and has constant Kähler angle θ satisfying

$$\tan^2(\theta/2) = \frac{\ell(k - \ell + 1)}{(\ell + 1)(k - \ell)}.$$

Each of the immersions $\phi_0, \ldots, \phi_k : \mathbb{C}P^1 \to \mathbb{C}P^k(4)$ is an embedding, unless $k = 2\ell$, in which case ϕ_ℓ is a totally real immersion and is essentially the Veronese immersion of S^2 into $\mathbb{R}P^k$. The sequence of immersions $\phi_0, \ldots, \phi_k : \mathbb{C}P^1 \to \mathbb{C}P^k(4)$ is known as the *Veronese sequence.*

The following result is a local version of a theorem from [**8**], of which an elementary proof can also be found in [**2**].

THEOREM 2.4. [**8**] *Consider a simply connected oriented minimal surface in* $\mathbb{C}P^k(4)$ *with constant Gaussian curvature* $K \neq 0$ *and constant Kähler angle. If the immersion is linearly full in* $\mathbb{C}P^k(4)$*, then, up to holomorphic local isometrics of* $\mathbb{C}P^k(4)$*, it is an element of the Veronese sequence.*

3. Explicit descriptions of totally geodesic surfaces in Q^n

Interpreting the classification of [**6**] for two-dimensional submanifolds shows that there are six types of complete totally geodesic surfaces in Q^n, at least if n is sufficiently large. The following table shows the isometry classes of these surfaces and the minimal dimension n for which they appear and it specifies whether they are complex or totally real.

TABLE 1. Totally geodesic surfaces in Q^n

	isometry class	minimal value of n	complex or totally real
(1)	$Q^1 \cong S^2(1/\sqrt{2})$	1	complex
(2)	$\mathbb{C}P^1(4) \cong S^2(1/2)$	2	complex
(3)	$\mathbb{R}P^2(1)$	4	totally real
(4)	$S^1(1/\sqrt{2}) \times S^1(1/\sqrt{2})$	2	totally real
(5)	$S^2(1/\sqrt{2})$	2	totally real
(6)	$S^2(\sqrt{10}/2)$	3	neither

In order to give explicit descriptions of the embeddings of these surfaces, we first remark that for all integers k and n, with $1 \leq k \leq n$, the following isometric embedding is totally geodesic:

$$\iota_{k,n} : Q^k \to Q^n : [(z_0, \dots, z_{k+1})] \mapsto [(z_0, \dots, z_{k+1}, \underbrace{0, \dots, 0}_{n-k \text{ zeroes}})]. \tag{3.1}$$

One way of seeing this, inspired by [**4**], is noting that the image of $\iota_{k,n}$ is a connected component of a fixed point set of an isometry of Q^n. Indeed, the map

$$F : \mathbb{C}P^{n+1}(4) \to \mathbb{C}P^{n+1}(4) :$$
$$[(z_0, \dots, z_{k+1}, z_{k+2}, \dots, z_{n+1})] \mapsto [(z_0, \dots, z_{k+1}, -z_{k+2}, \dots, -z_{n+1})]$$

is an isometry of $\mathbb{C}P^{n+1}(4)$ which leaves Q^n globally invariant and hence the restriction $F|_{Q^n}$ is an isometry of Q^n. The image of $\iota_{k,n}$ is one of the connected components of the fixed point set of $F|_{Q^n}$.

In particular, this means that $\iota_{1,n} : Q^1 \to Q^n$ defines a totally geodesic surface of Q^n for all $n \geq 1$ and that, if $\iota : M^2 \to Q^k$ is a totally geodesic immersion of a surface into some Q^k, then $\iota_{k,n} \circ \iota : M^2 \to Q^n$ is a totally geodesic immersion of that surface into Q^n for all $n \geq k$. This explains why we gave a *minimal* value of de dimension n in Table 1.

We now give explicit descriptions of the types in Table 1.

3.1. Type (1). As explained above, the map

$$\iota_{(1)} = \iota_{1,n} : Q^1 \to Q^n : [(z_0, z_1, z_2)] \mapsto [(z_0, z_1, z_2, \underbrace{0, \dots, 0}_{n-1 \text{ zeroes}})]$$

is a totally geodesic isometric embedding. Note that the resulting embedded surface is linearly full in a totally geodesic $\mathbb{C}P^2(4) \subseteq \mathbb{C}P^{n+1}(4)$. In fact, it is minimal in this $\mathbb{C}P^2(4)$. Since it has constant Gaussian curvature $K = 2$ and constant Kähler angle $\theta = 0$, Theorem 2.4 implies that it must be locally congruent to the element of the Veronese sequence with $k = 2$ and $\ell = 0$.

3.2. Type (2). The map

$$\iota_{(2)} : \mathbb{C}P^1(4) \to Q^n : [(z_0, z_1)] \mapsto [(z_0, z_1, iz_0, iz_1, \underbrace{0, \dots, 0}_{n-2 \text{ zeroes}})]$$

is a totally geodesic isometric embedding of $\mathbb{C}P^1(4)$ into Q^n as a complex surface. Note that the resulting embedded surface is a totally geodesic $\mathbb{C}P^1(4) \subseteq \mathbb{C}P^{n+1}(4)$. Since it has constant Gaussian curvature $K = 4$ and constant Kähler angle $\theta = 0$, Theorem 2.4 implies that it must be locally congruent to the element of the Veronese sequence with $k = 1$ and $\ell = 0$ (i.e., the trivial embedding $\mathbb{C}P^1(4) \to \mathbb{C}P^1(4)$).

3.3. Type (3). The map

$$\iota_{(3)} : \mathbb{R}P^2(1) \to Q^n : [(x_0, x_1, x_2)] \mapsto [(x_0, x_1, x_2, ix_0, ix_1, ix_2, \underbrace{0, \dots, 0}_{n-4 \text{ zeroes}})]$$

is a totally geodesic isometric embedding of $\mathbb{R}P^2(1)$ into Q^n as a totally real surface. Note that it is actually the composition of the trivial totally geodesic embedding $\mathbb{R}P^2(1) \to \mathbb{C}P^2(4)$ with a totally geodesic embedding $\mathbb{C}P^2(4) \to Q^n$, very similar to $\iota_{(2)}$. The image is linearly full in a totally geodesic $\mathbb{C}P^2(4) \subseteq \mathbb{C}P^{n+1}(4)$. In fact, it is totally geodesic and hence minimal in this $\mathbb{C}P^2(4)$. Since the surface has constant Gaussian curvature $K = 1$ and constant Kähler angle $\theta = \pi/2$, Theorem 2.4 implies that it must be locally congruent to the element of the Veronese sequence with $k = 2$ and $\ell = 1$.

3.4. Type (4). The map

$$\iota_{(4)} : S^1(1/\sqrt{2}) \times S^1(1/\sqrt{2}) \to Q^n :$$
$$((x_1, x_2), (y_1, y_2)) \mapsto [(x_1, x_2, iy_1, iy_2, \underbrace{0, \dots, 0}_{n-2 \text{ zeroes}})]$$

is the composition of $G_2 : S^1(1/\sqrt{2}) \times S^1(1/\sqrt{2}) \to Q^2$ (see (2.4)) and $\iota_{2,n} : Q^2 \to Q^n$ (see (3.1)). Since both are totally geodesic, $\iota_{(4)}$ is also totally geodesic. Moreover, since both $\iota_{2,n}$ and G_2 are isometric when $S^1(1/\sqrt{2}) \times S^1(1/\sqrt{2})$ is equipped with the flat metric, $\iota_{(4)}$ is an isometric totally geodesic embedding of the flat torus into Q^n. Finally, since G_2 is Lagrangian and $\iota_{2,n}$ is complex, $\iota_{(4)}$ is a totally real embedding. The surface hence has constant Gaussian curvature $K = 0$ and constant Kähler angle $\theta = \pi/2$.

3.5. Type (5). Similarly to Type (4), we can compose $G_1 : S^2(1) \to Q^2$ for $n = 2$ (see (2.3)) with $\iota_{2,n}$ (see (3.1)). The resulting embedding will be totally geodesic. However, G_1 is only isometric if we change the metric on $S^2(1)$ to a metric of constant Gaussian curvature 2, i.e. if we shrink the radius of the sphere to $1/\sqrt{2}$. In that case, the map becomes $G_1 : S^2(1/\sqrt{2}) \to Q^2 : x \mapsto [(\sqrt{2}\,x, i)] = [(x, i/\sqrt{2})]$. The resulting isometric totally geodesic embedding is

$$\iota_{(5)} : S^2(1/\sqrt{2}) \to Q^n : (x_1, x_2, x_3) \mapsto [(x_1, x_2, x_3, i/\sqrt{2}, \underbrace{0, \dots, 0}_{n-2 \text{ zeroes}})].$$

The surface has constant Gaussian curvature $K = 2$ and constant Kähler angle $\theta = \pi/2$.

3.6. Type (6). In order to explicitly describe an immersion of Type (6), we start from the Veronese surface

$$f : S^2(1) \to S^4(1) :$$
$$(x_1, x_2, x_3) \mapsto \left(\sqrt{3}\,x_1x_2, \sqrt{3}\,x_1x_3, \sqrt{3}\,x_2x_3, \frac{\sqrt{3}}{2}(x_1^2 - x_2^2), \frac{1}{2}(x_1^2 + x_2^2) - x_3^2 \right).$$

We then consider isothermal coordinates on $S^2(1)\backslash\{(0,0,1)\}$ given by stereographic projection, i.e., we put

$$x_1 = \frac{2u}{u^2 + v^2 + 1}, \quad x_2 = \frac{2v}{u^2 + v^2 + 1}, \quad x_3 = \frac{u^2 + v^2 - 1}{u^2 + v^2 + 1}$$

and define

$$g : \mathbb{R}^2 \to Q^3 : (u, v) \mapsto [f_u(u, v) + i f_v(u, v)].$$

The fact that the image of g is contained in $Q^3 \subseteq \mathbb{C}P^4(4)$ follows immediately from the fact that u and v are isothermal coordinates for the original surface. Moreover, in terms of the complex coordinate $z = u + iv$ we have $f_u + i f_v = 2 f_{\bar z}$, which means that choosing different isothermal coordinates on $S^2(1)$ in a neighborhood of $(0, 0, 1)$ allows us to extend the map g to a map $\iota : S^2(1) \to Q^3$. We will check below that this map induces a metric of constant Gaussian curvature $K = 2/5$ on $S^2(1)$ and hence, in order to make it isometric, we have to enlarge the radius of the sphere to $\sqrt{10}/2$. We will also check below that the map defines a totally geodesic surface in Q^3 with constant Kähler angle. The resulting isometric totally geodesic embedding of $S^2(\sqrt{10}/2)$ into Q^n is

$$\iota_{(6)} = \iota_{3,n} \circ \iota : S^2(\sqrt{10}/2) \to Q^n.$$

Before continuing our computations, we note that the idea behind the above construction can be generalised to associate with any immersion of a Riemann surface into $\mathbb{R}^{n+2}$ a map from that Riemann surface to Q^n.

Again in the example at hand, consider the map

$$h : \mathbb{R}^3 \to S^9(1) \subseteq \mathbb{C}^5 : (u, v, w) \mapsto \frac{e^{iw}(u^2 + v^2 + 1)}{2\sqrt{6}}(f_u(u, v) + i f_v(u, v)).$$

If $\pi : S^9(1) \to \mathbb{C}P^4(4)$ is the Hopf fibration, then $\pi \circ h$ is independent of w. In fact, $(\pi \circ h)(u, v, w) = g(u, v)$ for all $u, v, w \in \mathbb{R}^3$. If one defines the following vector fields on $\mathbb{R}^3$:

$$U = \frac{\partial}{\partial u} + \frac{2v}{1 + u^2 + v^2}\frac{\partial}{\partial w}, \quad V = \frac{\partial}{\partial v} - \frac{2u}{1 + u^2 + v^2}\frac{\partial}{\partial w}, \quad W = \frac{\partial}{\partial w},$$

then a long but straightforward calculation yields

$$\begin{aligned} &\langle (dh)(U),(dh)(V)\rangle = \langle (dh)(U),(dh)(W)\rangle = \langle (dh)(V),(dh)(W)\rangle = 0,\\ &\langle (dh)(U),(dh)(U)\rangle = \langle (dh)(V),(dh)(V)\rangle = \frac{10}{(1+u^2+v^2)^2}. \end{aligned} \tag{3.2}$$

Since $(dh)(W)$ spans the space of vertical vector fields tangent to the image of h, the equations in the first line of (3.2) imply that $(dh)(U)$ and $(dh)(V)$ span the space of horizontal vector fields tangent to the image of h. Moreover, since π is a Riemannian submersion, formulas (3.2) imply that $(dh)(U)$ and $(dh)(V)$ give rise to isothermal coordinate vector fields on the image of g and that the Gaussian curvature induced by g, and hence by ι, is constant $K = 2/5$.

The fact that g, and hence ι, is totally geodesic in Q^3 follows from the fact that $D_X(dh)(Y)$, where D is the covariant derivative of $\mathbb{C}^5$ and $X, Y \in \{U, V\}$, is a linear combination of $(dh)(U)$, $(dh)(V)$, h, ih, $\bar{h}$ and $i\bar{h}$. Indeed, none of these vector fields corresponds under π to a vector field which is normal to the image of g but tangent to Q^3.

Next, we compute the Kähler angle θ of g. Using again that π is a Riemannian submersion and that the complex structure on $\mathbb{C}P^4(4)$ is induced from multiplication by i on $TS^9(1)$, we have

$$\cos\theta = \frac{\langle i(dh)(U),(dh)(V)\rangle}{\sqrt{\langle i(dh)(U), i(dh)(U)\rangle}\sqrt{\langle (dh)(V),(dh)(V)\rangle}} = \frac{\langle i(dh)(U),(dh)(V)\rangle}{\langle (dh)(U),(dh)(U)\rangle} = \frac{1}{5}.$$

We conclude that the Kähler angle of g, and hence of ι, is constant.

Finally, we note that $\iota_{(6)}$ is linearly full in a totally geodesic $\mathbb{C}P^4(4) \subseteq \mathbb{C}P^{n+1}(4)$ and that it is also minimal in this $\mathbb{C}P^4(4)$. Since the surface has constant Gaussian curvature $K = 2/5$ and constant Kähler angle satisfying $\cos\theta = 1/5$, which implies $\tan^2(\theta/2) = 2/3$, Theorem 2.4 implies that the surface is locally congruent the element of the Veronese sequence with $k = 4$ and $\ell = 1$.

REMARK 3.1. The Veronese surface $f : S^2(1) \to S^4(1)$ that we started from is a minimal surface with parallel second fundamental form. By a theorem of Vilms (see [**12**]), its Gauss map, taking values in the Grassmanian of oriented 2-planes in $\mathbb{R}^5$, which is nothing but Q^3, is totally geodesic. The totally geodesic surface of Type (6) can also be obtained in this way.

4. An alternative description in dimension two

As explained in [**3**], the complex quadric of complex dimension two is isometric to a product of spheres

$$Q^2 \cong S^2(1/2) \times S^2(1/2).$$

The complex structure on Q^2 corresponds to $(J_{S^2}, -J_{S^2})$ on the tangent space to the product. There are four obvious totally geodesic surfaces in $S^2(1/2)\times S^2(1/2)$:

- A slice
$$S^2(1/2) \to S^2(1/2)\times S^2(1/2) : x \mapsto (x, p_0),$$
where $p_0 \in S^2(1/2)$ is fixed. Since this surface has constant Gaussian curvature $K = 4$ and constant Kähler angle $\theta = 0$ (it is complex), it corresponds to Type (2) in Table 1.
- A product of geodesics
$$S^1(1/2)\times S^1(1/2) \to S^2(1/2)\times S^2(1/2).$$

Since this surface has constant Gaussian curvature $K = 0$ and constant Kähler angle $\theta = \pi/2$ (it is totally real), it corresponds to Type (4) in Table 1.

- The diagonal surface

$$S^2(1/\sqrt{2}) \to S^2(1/2) \times S^2(1/2) : x \mapsto (x, x)/\sqrt{2}.$$

Since this surface has constant Gaussian curvature $K = 2$ and constant Kähler angle $\theta = \pi/2$ (it is totally real), it corresponds to Type (5) in Table 1.

- The anti-diagonal surface

$$S^2(1/\sqrt{2}) \to S^2(1/2) \times S^2(1/2) : x \mapsto (x, -x)/\sqrt{2}.$$

Since this surface has constant Gaussian curvature $K = 2$ and constant Kähler angle $\theta = 0$ (it is complex), it corresponds to Type (1) in Table 1. In particular, the anti-diagonal corresponds to the image of $\iota_{1,2} : Q^1 \to Q^2$ under the identification $Q^2 \cong S^2(1/2) \times S^2(1/2)$.

5. Totally geodesic surfaces in the hyperbolic complex quadric

The hyperbolic complex quadric Q^{n*} is the non-compact dual of Q^n in the sense of symmetric spaces (see for example [**5**]). In particular,

$$Q^{n*} = \frac{\mathrm{SO}(2,n)^0}{\mathrm{SO}(2) \times \mathrm{SO}(n)},$$

where the superscript 0 denotes the connected component of the identity. One can also model it as a complex hypersurface, namely

$$Q^{n*} = \{[(z_0, \dots, z_{n+1})] \in \mathbb{C}H_1^{n+1}(-4) \mid -z_0^2 - z_1^2 + z_2^2 + \dots + z_{n+1}^2 = 0\}^0,$$

where $\mathbb{C}H_1^{n+1}(-4)$ is the complex anti-de Sitter space of constant holomorphic sectional curvature -4 and the superscript 0 again denotes a connected component, for example the one containing $[(1, i, 0 \dots, 0)]$.

Since one can use duality of symmetric spaces to translate any result on totally geodesic submanifolds of a symmetric space to a result on totally geodesic submanifolds of its dual symmetric space (see again [**5**]), there are six families of totally geodesic surfaces in Q^{n*}.

TABLE 2. Totally geodesic surfaces in Q^{n*}

	isometry class	minimal value of n	complex or totally real
(1)	$Q^{1*} \cong H^2(1/\sqrt{2})$	1	complex
(2)	$\mathbb{C}H^1(-4) \cong H^2(1/2)$	2	complex
(3)	$H^2(1)$	4	totally real
(4)	$H^1(1/\sqrt{2}) \times H^1(1/\sqrt{2})$	2	totally real
(5)	$H^2(1/\sqrt{2})$	2	totally real
(6)	$H^2(\sqrt{10}/2)$	3	neither

Note that $H^k(r)$ denotes the hyperbolic space of radius r,

$$H^k(r) = \{(x_0, x_1, \dots, x_k) \in \mathbb{R}_1^{k+1} \mid -x_0^2 + x_1^2 + \dots + x_k^2 = -r^2,\ x_0 > 0\},$$

which, for $k > 1$, has constant sectional curvature $-1/r^2$.

Obviously, the immersions of the surfaces in Table 2 can be explicitly described using the hypersurface model of Q^{n*} and these descriptions are completely analogous to those given in Section 3. Moreover, the immersions occurring for $n \leq 2$, can be alternatively described as immersions into $H^2(1/2) \times H^2(1/2) \cong Q^{2*}$. Again, they are completely analogous to those given in Section 4. We refer to the recent PhD thesis [**13**] for more details.

References

[1] John Bolton, Gary R. Jensen, Marco Rigoli, and Lyndon M. Woodward, *On conformal minimal immersions of S^2 into* $\mathbf{C}\mathrm{P}^n$, Math. Ann. **279** (1988), no. 4, 599–620, DOI 10.1007/BF01458531. MR926423

[2] J. Bolton and L. M. Woodward, *Minimal surfaces in* $\mathbf{C}\mathrm{P}^n$ *with constant curvature and Kähler angle*, Math. Proc. Cambridge Philos. Soc. **112** (1992), no. 2, 287–296, DOI 10.1017/S0305004100070973. MR1171165

[3] Ildefonso Castro and Francisco Urbano, *Minimal Lagrangian surfaces in* $\mathbb{S}^2 \times \mathbb{S}^2$, Comm. Anal. Geom. **15** (2007), no. 2, 217–248. MR2344322

[4] Bang-yen Chen and Tadashi Nagano, *Totally geodesic submanifolds of symmetric spaces. I*, Duke Math. J. **44** (1977), no. 4, 745–755. MR458340

[5] J. Carlos Díaz-Ramos, Miguel Domínguez-Vázquez, and Víctor Sanmartín-López, *Submanifold geometry in symmetric spaces of noncompact type*, São Paulo J. Math. Sci. **15** (2021), no. 1, 75–110, DOI 10.1007/s40863-019-00119-6. MR4258890

[6] Sebastian Klein, *Totally geodesic submanifolds of the complex quadric*, Differential Geom. Appl. **26** (2008), no. 1, 79–96, DOI 10.1016/j.difgeo.2007.11.004. MR2393975

[7] Haizhong Li, Hui Ma, Joeri Van der Veken, Luc Vrancken, and Xianfeng Wang, *Minimal Lagrangian submanifolds of the complex hyperquadric*, Sci. China Math. **63** (2020), no. 8, 1441–1462, DOI 10.1007/s11425-019-9551-2. MR4125728

[8] Yoshihiro Ohnita, *Minimal surfaces with constant curvature and Kähler angle in complex space forms*, Tsukuba J. Math. **13** (1989), no. 1, 191–207, DOI 10.21099/tkbjm/1496161017. MR1003602

[9] Bennett Palmer, *Hamiltonian minimality and Hamiltonian stability of Gauss maps*, Differential Geom. Appl. **7** (1997), no. 1, 51–58, DOI 10.1016/S0926-2245(96)00035-6. MR1441918

[10] Helmut Reckziegel, *On the geometry of the complex quadric*, Geometry and topology of submanifolds, VIII (Brussels, 1995/Nordfjordeid, 1995), World Sci. Publ., River Edge, NJ, 1996, pp. 302–315. MR1434581

[11] Brian Smyth, *Differential geometry of complex hypersurfaces*, Ann. of Math. (2) **85** (1967), 246–266, DOI 10.2307/1970441. MR206881

[12] J. Vilms, *Submanifolds of Euclidean space with parallel second fundamental form*, Proc. Amer. Math. Soc. **32** (1972), 263–267. MR0290298.

[13] A. Wijffels, *Submanifolds of complex quadrics and Gauss maps*, PhD thesis, KU Leuven, 2021.

Faculty of Mathematics, Al.I. Cuza University of Iaşi, Bd. Carol I, n. 11, 700506 Iaşi, Romania

Email address: marilena.moruz@uaic.ro

Department of Mathematics, KU Leuven, Celestijnenlaan 200B – Box 2400, 3001 Leuven, Belgium

Email address: joeri.vanderveken@kuleuven.be

LMI, Université Polytechnique Hauts-de-France, Campus du Mont Houy, 59313 Valenciennes Cedex 9, France and Department of Mathematics, KU Leuven, Celestijnenlaan 200B – Box 2400, 3001 Leuven, Belgium

Email address: luc.vrancken@uphf.fr, luc.vrancken@kuleuven.be

Department of Mathematics, KU Leuven, Celestijnenlaan 200B – Box 2400, 3001 Leuven, Belgium

Email address: anne.wijffels@kuleuven.be

Contemporary Mathematics
Volume **777**, 2022
https://doi.org/10.1090/conm/777/15630

Parallel Kähler submanifolds and R-spaces

Yoshihiro Ohnita

This article is dedicated to the memory of Professor Tadashi Nagano

ABSTRACT. *Parallel Kähler submanifolds* here mean complex submanifolds immersed in complex projective spaces with parallel second fundamental form. The classification of such submanifolds was done first by Hisao Nakagawa and Ryoichi Takagi (1976) and reproduced secondly and thirdly by Masaru Takeuchi (1978, 1984) by two different methods of unitary representation theory and Jordan triple systems. In this article we briefly survey such related submanifold theory and give the fourth proof for the classification theorem by a different approach based on the differential geometric characterization of R-spaces due to Carlos Olmos and Cristián U. Sánchez (1991).

1. Introduction

Let M^m be a complex m-dimensional complex submanifold immersed in a complex n-dimensional complex projective space $\mathbb{C}P^n$. Here $\mathbb{C}P^n$ is endowed with the Fubini-Study metric of constant holomorphic sectional curvature 4. Then M becomes intrinsically a Kähler manifold with respect to the metric g_M and complex structure J induced from $\mathbb{C}P^n$, thus M is also called a *Kähler submanifold* of $\mathbb{C}P^n$. When M is not contained in any proper totally geodesic complex submanifold $\mathbb{C}P^k$ $(0 \leq k \leq n-1)$ of $\mathbb{C}P^n$, we say that M is *fully* immersed in $\mathbb{C}P^n$. We denote by α^M the second fundamental form of M in $\mathbb{C}P^n$. The covariant derivative $\nabla^*\alpha^M$ of α^M in terms of the normal connection $\nabla^\perp$ and the Levi-Civita connection ∇^M is defined as

$$(\nabla^*\alpha^M)_X(Y,Z) := \nabla^\perp_X(\alpha^M(Y,Z)) - \alpha^M(\nabla^M_X Y, Z) - \alpha^M(Y, \nabla^M_X Z) \tag{1.1}$$

for any smooth vector fields X, Y, Z on N. If α^M satisfies the equation

$$\nabla^*\alpha^M = 0, \tag{1.2}$$

then we say that the submanifold M has *parallel second fundamental form.* A complex submanifold of $\mathbb{C}P^n$ with parallel second fundamental form is called simply a *parallel Kähler submanifold.* From the Gauss equation it is well-known that any Kähler submanifold M of $\mathbb{C}P^n$ with parallel second fundamental form is a locally

2020 *Mathematics Subject Classification.* Primary 53C40; Secondary 53C42, 53C55.

Key words and phrases. Kähler submanifolds, parallel second fundamental forms, R-spaces.

This work was partly supported by JSPS KAKENHI Grant Numbers JP17H06127, JP18K03307, JP18H03668 and by Osaka City University Advanced Mathematical Institute: MEXT Joint Usage/Research Center on Mathematics and Theoretical Physics JPMXP0619217849.

Hermitian symmetric space. In 1976 Nakagawa and Takagi [**13**] have classified parallel Kähler submanifolds of $\mathbb{C}P^n$ as follows:

THEOREM 1.1 (Nakagawa and Takagi [**13**]). *A complex submanifold M fully immersed in $\mathbf{C}P^n$ has parallel second fundamental form if and only if it is congruent to the open part of one of the following seven complex submanifolds:*

$$M_1 = \mathbb{C}P^m(4) \subset \mathbb{C}P^m(4) \textit{ totally geodesic}$$
$$M_2 = \mathbb{C}P^m(2) \subset \mathbb{C}P^{m+\frac{1}{2}m(m+1)}(4)$$
$$M_3 = \mathbb{C}P^{m-s}(4) \times \mathbb{C}P^s(4) \longrightarrow \mathbb{C}P^{m+s(m-s)}(4)$$
$$M_4 = Q_m(\mathbb{C}) \longrightarrow \mathbb{C}P^{m+1}(4)\ (m \geq 3)$$
$$M_5 = SU(s+2)/S(U(2)\times U(s)) \longrightarrow \mathbb{C}P^{s+\frac{1}{2}s(s+1)}(4)\ (s \geq 3)$$
$$M_6 = SO(10)/U(5) \longrightarrow \mathbb{C}P^{15}(4)$$
$$M_7 = E_6/((U(1)\times Spin(10))/\mathbb{Z}_4) \longrightarrow \mathbb{C}P^{26}(4)$$

Note that they are Hermitian symmetric spaces of compact type and rank at most 2. Several beautiful curvature characterizations related to those seven Kähler submanifolds are known as [**18**] etc., inspired by Ogiue's conjectures ([**14**]).

Theorem 1.1 was proved first by Nakagawa-Takagi ([**13**]) and was reproved secondly by Takeuchi ([**22**]) by means of the unitary representation theory for compact Lie groups. For a Kähler immersion $\varphi : M \to \mathbb{C}P^n$, the degree $d(\varphi)$ of φ (cf. [**13**]) is defined in terms of the higher order holomorphic osculating spaces along φ, and by definition $d(\varphi) = 1$ or 2 if and only if φ has parallel second fundamental form. By Calabi's rigidity and extension theorem ([**3**]), we may assume that M is a compact Hermitian symmetric space and φ is full, and then φ is an equivariant holomorphic map with respect to a unitary representation ρ of a maximal connected isometry group G on M into $SU(n+1)$. Moreover, if we decompose M into a direct product of irreducible compact Hermitian symmetric spaces M_i $(1 \leq i \leq \ell)$, then there is the p_i-th standard embedding $\varphi_i : M_i \to \mathbb{C}P^{n_i}$ for each i such that φ is expressed as a tensor product map of those equivariant holomorphic maps φ_i. If we denote by r_i the rank of each M_i, then they showed the degree formula $d(\varphi) = \sum_{i=1}^{\ell} p_i r_i$ ([**13**], [**21**]). By determining all Kähler immersions φ with $d(\varphi) = 1$ or 2, they obtained Theorem 1.1.

The third proof of Theorem 1.1 was given by Takeuchi ([**23**]) in 1984 by the algebraic method of Jordan triple systems. It is based on the correspondence between positive definite Hermitian Jordan triple systems and irreducible symmetric bounded domains, which is due to M. Koecher ([**11**]), see also I. Satake ([**19**]). A crucial point of [**23**] is to construct a positive definite Hermitian Jordan triple system with a Jordan product defined from the second fundamental form of a given parallel Kähler submanifold M^m of $\mathbb{C}P^n$. The corresponding symmetric bounded domain is an irreducible Hermitian symmetric space G^*/K of non-compact type. Let $\mathfrak{g}^* = \mathfrak{k} + \mathfrak{p}^*$ be the Cartan decomposition of $\mathfrak{g}^*$ and we have an identification $\mathfrak{p}^{1,0} \cong \mathbb{C}^{n+1}$, where $\mathfrak{p}^{\mathbb{C}} = \mathfrak{p}^{1,0} + \mathfrak{p}^{0,1}$ is the eigenspace decomposition of the complexification $\mathfrak{p}^{\mathbb{C}} = (\mathfrak{p}^*)^{\mathbb{C}}$ with respect to the complex structure tensor of the Hermitian symmetric space G^*/K. Take the highest weight vector $E^+(\neq 0) \in \mathfrak{p}^{1,0}$ relative to the maximal abelian subalgebra of $\mathfrak{k}$. Through the Hopf fibration $\pi : S^{2n+1}(1) \to \mathbb{C}P^n$, he showed that $\pi(\mathrm{Ad}(K)E^+) \subset \mathbb{C}P^n$ is a parallel Kähler submaifold and it is congruent to the original parallel Kähler submaifold M^m, which can be explicitly determined by the root system computation. Thus he obtained Theorem 1.1. In general a compact homogeneous space obtained as an orbit of the isotropy representation of a Riemannian symmetric space, i.e. an *s-representation*, is called an *R-space* (see Section 2), which was originally introduced by Jacques Tits [**25, 26**]. Since $\mathrm{Ad}(K)E^+$ is an R-space, he also obtained

THEOREM 1.2 (Takeuchi [**23**]). *Any parallel Kähler submanifold of $\mathbb{C}P^n$ can be obtained by the projection of an R-space obtained as an orbit of the isotropy representation of an irreducible Hermitian symmetric space.*

In 1991 Olmos and Sánchez ([**15**]) gave a differential geometric characterization of R-spaces standardly embedded in Euclidean spaces. They showed that a submanifold N immersed in a Euclidean space $\mathbb{R}^{m+k}$ is a standardly embedded R-space if and only if there exists a *canonical connection* ∇^c (see Section 2 for the definition) on the tangent vector bundle TN such that the second fundamental form α^N of N satisfies the equation

$$(\nabla^c\alpha^N)_X(Y,Z) := \nabla^\perp_X(\alpha^N(Y,Z)) - \alpha^N(\nabla^c_X Y, Z) - \alpha^N(Y, \nabla^c_X Z) = 0 \tag{1.3}$$

for each smooth vector field X, Y, Z on N. It successfully generalizes the classification theorem due to Ferus ([**8**],[**9**], see also [**5**]) for parallel submanifolds in Euclidean spaces by means of *symmetric R-spaces* in the case when ∇^c is the Levi-Civita connection ∇^N of N. Differential geometry of symmetric R-spaces has a long and fruitful history and it was first studied by a pioneering work of Tadashi Nagano [**12**] in 1965 from the viewpoint of transformation groups.

In this article we shall give the fourth proof of Theorems 1.1 and 1.2 by a different approach based on the differential geometric characterization of R-spaces due to Olmos-Sánchez. The main results of this article (Theorem 3.2) are the explicit construction of a *non-trivial* canonical connection (different from the Levi-Civita connection!) on the inverse image $\hat{M}$ of any complex submanifold M of $\mathbb{C}P^n$ under the Hopf fibration and that $\hat{M}$ satisfies the Olmos-Sánchez's condition (1.3) with respect to this canonical connection if and only if M is a parallel Kähler submanifold of $\mathbb{C}P^n$. Moreover by Olmos-Sánchez's theorem and some elementary arguments it will be shown that the inverse image of any parallel Kähler submanifold of $\mathbb{C}P^n$ is a standardly embedded (non-symmetric) R-space obtained as an orbit of the isotropy representation of an irreducible Hermitian symmetric space. By using Hyunjung Song's results ([**20**]) on differential geometric properties of R-spaces associated with irreducible Hermitian symmetric spaces, we can show the classification theorem of parallel Kähler submanifolds of $\mathbb{C}P^n$.

This article is organized as follows: In Section 2 we recall the definition of R-spaces and the standard embeddings constructed from an arbitrary given compact symmetric space G/K. And we explain the precise definition of a canonical connection on a Riemannian manifold and Olmos-Sánchez's theorem of differential geometric characterizations for R-spaces. In Section 3 we show our main results. Our main tool is the classical technique of Riemannian submersions ([**16**]) applied to the Hopf fibration restricted to a complex submanifold of $\mathbb{C}P^n$. Moreover we discuss relevant properties which imply Theorem 1.2. In Section 4 we explain how to classify parallel Kähler submanifolds of $\mathbb{C}P^n$ obtained as the projection of R-spaces associated with irreducible Hermitian symmetric pairs by means of results in [**20**]. It implies Theorem 1.1.

As other related topics, totally complex submanifolds in quaternionic projective spaces with parallel second fundamental form were classified by K. Tsukada in 1985 ([**27**]). More recently we have also obtained a similar result for such submanifolds. It will be described in detail in the forthcoming joint paper with Kaname Hashimoto (OCAMI) and Jong Taek Cho (Chonnam National University).

Throughout this article any manifold is smooth, connected and second countable.

2. R-spaces and Olmos-Sánchez's characterization

Let (G, K) be a compact symmetric pair associated with a compact symmetric space G/K. Here G is a connected compact Lie group with Lie algebra $\mathfrak{g}$ and K is a compact Lie subgroup of G with Lie algebra $\mathfrak{k}$. Let $\mathfrak{g} = \mathfrak{k} + \mathfrak{p}$ be the canonical decomposition of $\mathfrak{g}$ as a symmetric Lie algebra. The vector space $\mathfrak{p}$ can be regarded as a Euclidean space by the restriction of an $\mathrm{Ad}(G)$-invariant inner product $\langle\ ,\ \rangle$ of $\mathfrak{g}$ to $\mathfrak{p}$. The isotropy representation of a compact symmetric space G/K is given by an orthogonal representation of K on the vector space $\mathfrak{p}$

$$\mathrm{Ad}_{\mathfrak{p}} : K \ni a \longmapsto \mathrm{Ad}(a)|_{\mathfrak{p}} \in O(\mathfrak{p}),$$

which is also called an *s-representation.* It is well-known to be a polar representation. For any non-zero $H \in \mathfrak{p}$, we define a compact homogeneous space K/K_H diffeomorphic to an orbit of the isotropy representation of K through H by

$$\Phi_H : K/K_H \ni aK_H \longmapsto \mathrm{Ad}(a)H \in \mathrm{Ad}(K)H \subset \mathfrak{p}$$

where $K_H := \{a \in K \mid \mathrm{Ad}(a)H = H\}$. Then so obtained compact homogeneous space K/K_H is called an *R-space* and the embedding $\Phi_H : K/K_H \to \mathfrak{p}$ is called *the standard embedding* of R-space K/K_H. When (K, K_H) is a compact symmetric pair, K/K_H is called a *symmetric R-space.* Then the standard embedding $\Phi_H : K/K_H \to \mathfrak{p}$ has parallel second fundamental form and symmetric R-spaces give all submanifolds of Euclidean spaces with parallel second fundamental form except for affine subspaces (Ferus [**8**]). When $H \in \mathfrak{p}$ is a regular element (by definition $\mathrm{Ad}(K)H$ is of maximal dimension), K/K_H is called a *regular R-space.* Then $\Phi_H : K/K_H \to \mathfrak{p}$ is a homogeneous isoparametric submanifold of a Euclidean space and regular R-spaces give all homogeneous isoparametric submanifolds of Euclidean spaces ([**17**] and [**4**]).

Olmos and Sánchez ([**15**]) have showed that a general R-space standardly embedded into a Euclidean space can be characterized by the parallelism of the second fundamental form with respect to the normal connection and a *canonical connection* (not necessary the Levi-Civita connection!) on a given submanifold of a Euclidean space. Next we shall describe their results.

Let N be a connected submanifold immersed in the Euclidean space $\mathbb{R}^l$. Let g_N be a Riemannian metric on N induced from $\mathbb{R}^l$ and let ∇^N denote the Levi-Civita connection of a Riemannian manifold (N, g_N). An affine connection $\tilde{\nabla}$ on N is called a *metric connection* with respect to g if $\tilde{\nabla}$ satisfies the condition

$$\tilde{\nabla} g_N = 0. \tag{2.1}$$

Let D be a tensor field on N of type $(1, 2)$ defined by

$$D := \nabla^N - \tilde{\nabla}. \tag{2.2}$$

The metric condition (2.1) is equivalent to the condition that for each vector $X \in TN$ the linear endomorphism D_X is skew-symmetric with respect to g_N, that is,

$$g_N(D_X Y, Z) + g_N(Y, D_X Z) = 0 \quad (\forall Y, Z \in TN). \tag{2.3}$$

A metric connection ∇^c is called a *canonical connection* of a Riemannian manifold (N, g_N) if ∇^c satisfies the condition

$$\nabla^c D^c = 0, \tag{2.4}$$

where $D^c := \nabla^N - \nabla^c$ is a tensor field on N of type $(1,2)$. Note that the Levi-Civita connection itself is a *trivial* example of a canonical connection of (N, g_N) as $D^c = 0$ in this case.

The covariant derivative $\nabla^c \alpha^N$ of the second fundamental form α^N in terms of the normal connection $\nabla^\perp$ and a *canonical connection* ∇^c is defined by

$$(\nabla^c_X \alpha^N)(Y, Z) := \nabla^\perp_X(\alpha^N(Y, Z)) - \alpha^N(\nabla^c_X Y, Z) - \alpha^N(Y, \nabla^c_X Z) \tag{2.5}$$

for any smooth vector fields X, Y, Z on N. Then

THEOREM 2.1 (Olmos and Sánchez [**15**]). *Let N be a connected compact submanifold fully embedded in the Euclidean space $\mathbb{R}^l$. Then the following three conditions are equivalent each other:*

(1) *There is a canonical connection ∇^c on N such that*

$$\nabla^c \alpha^N = 0. \tag{2.6}$$

(2) *N is a homogeneous submanifold with constant principal curvatures (see* [**15**, p.127, Definition 1.2] *for the definition).*
(3) *N is an orbit of an s-representation, that is, an R-space standardly embedded in the Euclidean space.*

In this case we call a tensor field D^c on N of type $(1,2)$ defined by $D^c := \nabla^N - \nabla^c$ a *homogeneous structure tensor field* on a submanifold N. Notice that the argument of [**15**] also works to have the local version of this theorem. The proof of the implication (2) $\Rightarrow$ (3) uses the classification theorem of polar representations by Dadok [**4**], which clams that any orthogonal polar representation is orbit-equivalent to an s-representation. See also [**6**], [**7**], [**1**] for a conceptional proof of Dadok's result subject to some restriction and further works.

3. Homogeneous structure on the inverse images of parallel Kähler submanifolds under the Hopf fibration

Let $\mathbb{C}^{n+1}$ be an $n+1$-dimensional complex Euclidean space with the standard Hermitian inner product $\langle\langle \mathbf{x}, \mathbf{y} \rangle\rangle := \sum_{i=1}^{n+1} x_i \bar{y}_i$ for each $\mathbf{x} = (x_1, \cdots, x_{n+1}), \mathbf{y} = (y_1, \cdots, y_{n+1}) \in \mathbb{C}^{n+1}$. Let $\langle \mathbf{x}, \mathbf{y} \rangle := \mathrm{Re}\langle\langle \mathbf{x}, \mathbf{y} \rangle\rangle$ denote the standard real inner product of $\mathbb{C}^{n+1}$. Let $S^{2n+1}(1) := \{\mathbf{x} \in \mathbb{C}^{n+1} \mid \langle \mathbf{x}, \mathbf{x} \rangle = 1\}$ be the unit standard hypersphere of $\mathbb{C}^{n+1}$ and $\pi : S^{2n+1}(1) \longrightarrow \mathbb{C}P^n$ be the Hopf fibration over an n-dimensional complex projective space $\mathbb{C}P^n$. We endow $\mathbb{C}P^n$ with the Fubini-Study metric of constant holomorphic sectional curvature 4 so that $\pi : S^{2n+1}(1) \longrightarrow \mathbb{C}P^n$ is a Riemannian submersion. Then we have an orthogonal direct sum decomposition of the tangent vector bundle of $S^{2n+1}(1)$ into vertical and horizontal subbundles:

$$TS^{2n+1}(1) = \mathcal{V}S^{2n+1}(1) \oplus \mathcal{H}S^{2n+1}(1).$$

For a tangent vector or a vector field X on $\mathbb{C}P^n$, we denote by $\tilde{X}$ the horizontal lift of X to $S^{2n+1}(1)$. A horizontal vector field on $S^{2n+1}(1)$ is called *basic* if it is given as a horizontal lift of a vector field on $\mathbb{C}P^n$.

We denote also by $\mathbf{x}$ the position vector of a point in $\mathbb{C}^{n+1}$. The vertical subspace at each point $\mathbf{x} \in S^{2n+1}(1)$ is expressed as

$$\mathcal{V}_{\mathbf{x}} S^{2n+1}(1) = \mathbb{R}\sqrt{-1}\mathbf{x}.$$

At each point $\mathbf{x} \in S^{2n+1}(1)$, the restriction of the differential of π to the horizontal subspace $(d\pi)_{\mathbf{x}} : \mathcal{H}_{\mathbf{x}} S^{2n+1}(1) \to T_{\pi(\mathbf{x})}\mathbb{C}P^n$ is a linear isometry and we have $(d\pi)_{\mathbf{x}}(\tilde{X}) = X$ for each vector $X \in T_{\pi(\mathbf{x})}\mathbb{C}P^n$. Each horizontal subspace $\mathcal{H}_{\mathbf{x}} S^{2n+1}(1)$ is invariant under the scalar multiplication by $\sqrt{-1}$ on $\mathbb{C}^{n+1}$. The complex structure tensor J of $\mathbb{C}P^n$ is induced by this multiplication $\sqrt{-1}\times$ and it can be described as

$$JX = J(d\pi)_{\mathbf{x}}(\tilde{X}) = (d\pi)_{\mathbf{x}}(\sqrt{-1}\tilde{X}) \quad \text{or} \quad \widetilde{JX} = \sqrt{-1}\tilde{X} \in \mathcal{H}_{\mathbf{x}} S^{2n+1}(1).$$

Under the identification $T_{\pi(\mathbf{x})}\mathbb{C}P^n \cong T^{1,0}_{\pi(\mathbf{x})}\mathbb{C}P^n = \mathrm{Hom}_{\mathbb{C}}(\mathbb{C}\mathbf{x}, (\mathbb{C}\mathbf{x})^{\perp})$, for each $X \in T_{\pi(\mathbf{x})}\mathbb{C}P^n$ we have $\tilde{X} = X(\mathbf{x}) \in (\mathbb{C}\mathbf{x})^{\perp} = \mathcal{H}_{\mathbf{x}} S^{2n+1}(1) \subset \mathbb{C}^{n+1}$ and

$$(JX)(\mathbf{x}) = X(\sqrt{-1}\mathbf{x}) = \sqrt{-1}X(\mathbf{x}),$$

where $(\mathbb{C}\mathbf{x})^{\perp}$ denotes an n-dimensional complex vector subspace of $\mathbb{C}^{n+1}$ defined by

$$(\mathbb{C}\mathbf{x})^{\perp} := \{v \in \mathbb{C}^{n+1} \mid \langle\langle v, w \rangle\rangle = 0 \ (\forall w \in \mathbb{C}\mathbf{x})\}.$$

In this case the O'Neill tensors $\mathcal{T}$ and $\mathcal{A}$ for the Riemannian submersion ([**16**]) are given by $\mathcal{T} = 0$ and

$$\mathcal{A}_{\tilde{X}}\tilde{Y} = -\langle \sqrt{-1}\tilde{X}, \tilde{Y} \rangle \sqrt{-1}\mathbf{x} \tag{3.1}$$

for each horizontal vectors $\tilde{X}$,$\tilde{Y}$ at $\mathbf{x} \in S^{2n+1}(1)$.

Suppose that M^m is a complex m-dimensional complex submanifold immersed in $\mathbb{C}P^n$. The inverse image of the submanifold M under the Hopf fibration $\pi : S^{2n+1}(1) \longrightarrow \mathbb{C}P^n$ is defined as

$$\begin{aligned} \hat{M} =& \pi^{-1}(M) \\ =& \{(p, \mathbf{x}) \in M \times S^{2n+1}(1) \mid p \in M, \mathbf{x} \in \pi^{-1}(\varphi(p)) \subset S^{2n+1}(1)\}. \end{aligned}$$

$$\begin{array}{ccc} & & \mathbb{C}^{n+1} \\ & & \cup \\ \hat{M} = \pi^{-1}(M) & \xrightarrow{\ \hat{\varphi}\ } & S^{2n+1}(1) \\ \pi \downarrow S^1 & & \pi \downarrow S^1 \\ M & \xrightarrow{\ \varphi\ } & \mathbf{C}P^n \end{array}$$

Then $\hat{M}$ is a real $2m+1$-dimensional submanifold immersed in $S^{2n+1}(1) \subset \mathbb{C}^{n+1} \cong \mathbb{R}^{2n+2}$ and the projection $\pi : \hat{M} \to M$ is also a Riemannian submersion, so that we have an orthogonal direct sum decomposition of the tangent vector bundle of $\hat{M}$ into vertical and horizontal subbundles:

$$T\hat{M} = \mathcal{V}\hat{M} \oplus \mathcal{H}\hat{M}.$$

Here note that the vertical subspace at each point $\mathbf{x} \in \hat{M}$ is given by

$$\mathcal{V}_{\mathbf{x}}\hat{M} = \mathcal{V}_{\mathbf{x}}S^{2n+1}(1) = \mathbb{R}\sqrt{-1}\mathbf{x}.$$

We shall construct explicitly a homogeneous structure tensor field D in the sense of Olmos and Sánchez on the inverse image $\hat{M} = \pi^{-1}(M)$.

Now we define the tensor field D of type $(1,2)$ on $\hat{M} = \pi^{-1}(M) \subset S^{2n+1}(1)$ by

$$\begin{cases} D_{\tilde{X}}(\tilde{Y}) := -\langle\sqrt{-1}\tilde{X}, \tilde{Y}\rangle\sqrt{-1}\mathbf{x} \in \mathcal{V}_{\mathbf{x}}\hat{M}, \\ D_{\tilde{X}}(V) := \sqrt{-1}\tilde{X} = \widetilde{JX} \in \mathcal{H}_{\mathbf{x}}\hat{M}, \\ D_V(\tilde{X}) := \frac{1}{2}\sqrt{-1}\tilde{X} = \frac{1}{2}\widetilde{JX} \in \mathcal{H}_{\mathbf{x}}\hat{M}, \\ D_V(V) := 0 \end{cases} \tag{3.2}$$

for each horizontal vectors $\tilde{X}, \tilde{Y}$ and the vertical vector $V = \sqrt{-1}\mathbf{x}$ on $\hat{M}$.

REMARK 3.1. We can also define the tensor field $\mathbf{D}$ of type $(1,2)$ on $S^{2n+1}(1)$ by

$$\begin{cases} \mathbf{D}_{\tilde{X}}(\tilde{Y}) := -\langle\sqrt{-1}\tilde{X}, \tilde{Y}\rangle\sqrt{-1}\mathbf{x} \in \mathcal{V}_{\mathbf{x}}S^{2n+1}(1), \\ \mathbf{D}_{\tilde{X}}(V) := \sqrt{-1}\tilde{X} = \widetilde{JX} \in \mathcal{H}_{\mathbf{x}}S^{2n+1}(1), \\ \mathbf{D}_V(\tilde{X}) := \frac{1}{2}\sqrt{-1}\tilde{X} = \frac{1}{2}\widetilde{JX} \in \mathcal{H}_{\mathbf{x}}S^{2n+1}(1), \\ \mathbf{D}_V(V) := 0 \end{cases} \tag{3.3}$$

for each horizontal vectors $\tilde{X}, \tilde{Y}$ on $S^{2n+1}(1)$ and the vertical vector $V = \sqrt{-1}\mathbf{x}$ on $S^{2n+1}(1)$. Notice that if M is a complex submanifold of $\mathbb{C}P^n$, then the restriction of $\mathbf{D}$ to its inverse image $\hat{M} = \pi^{-1}(M)$ coincides with the tensor field D of type $(1,2)$ on $\hat{M}$ defined by (3.2). Also note that $\mathbf{D}_{\tilde{X}}(\tilde{Y}) = \mathcal{A}_{\tilde{X}}(\tilde{Y})$.

Then the main result of this article is described as follows:

THEOREM 3.2. *Suppose that M is a complex submanifold immersed in $\mathbb{C}P^n$. Let $\nabla^{\hat{M}}$ be the Levi-Civita connection of its inverse image $\hat{M} = \pi^{-1}(M) \subset S^{2n+1}(1)$ and D be a tensor field of type $(1,2)$ defined by (3.2) on $\hat{M} = \pi^{-1}(M)$. Then*

(1) *The affine connection $\nabla^c := \nabla^{\hat{M}} - D$ of $\hat{M}$ is a (non-trivial) canonical connection on $\hat{M}$.*
(2) *M has parallel second fundamental form if and only if $\hat{M}$ satisfies*

$$\nabla^c \alpha^{\hat{M}} = 0. \tag{3.4}$$

We prove this theorem by showing the following Lemmas 3.3, 3.4 and 3.5.

LEMMA 3.3. *The affine connection ∇^c is a metric connection on $\hat{M}$, that is.*

$$g_{\hat{M}}(D_u v, w) + g_{\hat{M}}(v, D_u w) = 0 \tag{3.5}$$

for all vectors $u, v, w \in T\hat{M}$.

PROOF. Following the definition of D, for each horizontal vectors $\tilde{X}$, $\tilde{Y}$, $\tilde{Z}$ and a vertical vector $V = \sqrt{-1}\mathbf{x}$ we compute

$$\begin{aligned}
g_{\hat{M}}(D_{\tilde{X}}\tilde{Y},\tilde{Z}) + g_{\hat{M}}(\tilde{Y},D_{\tilde{X}}\tilde{Z}) =&0+0=0,\\
g_{\hat{M}}(D_{\tilde{X}}V,\tilde{Z}) + g_{\hat{M}}(V,D_{\tilde{X}}\tilde{Z}) =&\langle\sqrt{-1}\tilde{X},\tilde{Z}\rangle - \langle\sqrt{-1}\tilde{X},\tilde{Z}\rangle = 0,\\
g_{\hat{M}}(D_{\tilde{X}}V,V) + g_{\hat{M}}(V,D_{\tilde{X}}V) =&0+0=0,\\
g_{\hat{M}}(D_{V}\tilde{Y},\tilde{Z}) + g_{\hat{M}}(\tilde{Y},D_{V}\tilde{Z}) =&\frac{1}{2}\langle\sqrt{-1}\tilde{Y},\tilde{Z}\rangle + \frac{1}{2}\langle\tilde{Y},\sqrt{-1}\tilde{Z}\rangle\\
=&\frac{1}{2}\langle\sqrt{-1}\tilde{Y},\tilde{Z}\rangle - \frac{1}{2}\langle\sqrt{-1}\tilde{Y},\tilde{Z}\rangle\\
=&0,\\
g_{\hat{M}}(D_{V}V,\tilde{Z}) + g_{\hat{M}}(V,D_{V}\tilde{Z}) =&0+0=0,\\
g_{\hat{M}}(D_{V}V,V) + g_{\hat{M}}(V,D_{V}V) =&0.
\end{aligned}$$

□

LEMMA 3.4. *The tensor field D on $\hat{M}$ satisfies the equation*

$$\nabla^{\hat{M}} D = D \cdot D. \tag{3.6}$$

PROOF. Let $\tilde{X}$, $\tilde{Y}$, $\tilde{Z}$ be any basic horizontal vector fields on $\hat{M}$ and $V = \sqrt{-1}\mathbf{x}$ be a vertical vector field on $\hat{M}$. First we show the equation

$$(\nabla^{\hat{M}}_{\tilde{Z}} D)_{\tilde{X}}(\tilde{Y}) = (D_{\tilde{Z}} \cdot D)_{\tilde{X}}(\tilde{Y}). \tag{3.7}$$

By the definition of D we compute

$$\begin{aligned}
&(\nabla^{\hat{M}}_{\tilde{Z}} D)_{\tilde{X}}(\tilde{Y})\\
=&\nabla^{\hat{M}}_{\tilde{Z}}(D_{\tilde{X}}\tilde{Y}) - D_{\nabla^{\hat{M}}_{\tilde{Z}}\tilde{X}}\tilde{Y} - D_{\tilde{X}}(\nabla^{\hat{M}}_{\tilde{Z}}\tilde{Y})\\
=&\nabla^{\hat{M}}_{\tilde{Z}}(D_{\tilde{X}}\tilde{Y}) - D_{-\langle\sqrt{-1}\tilde{Z},\tilde{X}\rangle V}\tilde{Y} - D_{\tilde{X}}(-\langle\sqrt{-1}\tilde{Z},\tilde{Y}\rangle V)\\
=&-\langle\sqrt{-1}\tilde{X},\tilde{Y}\rangle\sqrt{-1}\tilde{Z} + \langle\sqrt{-1}\tilde{Z},\tilde{X}\rangle\frac{\sqrt{-1}}{2}\tilde{Y} + \langle\sqrt{-1}\tilde{Z},\tilde{Y}\rangle\sqrt{-1}\tilde{X}.
\end{aligned}$$

On the other hand, we compute

$$\begin{aligned}
&(D_{\tilde{Z}} \cdot D)_{\tilde{X}}(\tilde{Y})\\
=&D_{\tilde{Z}}(D_{\tilde{X}}\tilde{Y}) - D_{D_{\tilde{Z}}\tilde{X}}\tilde{Y} - D_{\tilde{X}}(D_{\tilde{Z}}\tilde{Y})\\
=&-\langle\sqrt{-1}\tilde{X},\tilde{Y}\rangle\sqrt{-1}\tilde{Z} + \langle\sqrt{-1}\tilde{Z},\tilde{X}\rangle\frac{\sqrt{-1}}{2}\tilde{Y} + \langle\sqrt{-1}\tilde{Z},\tilde{Y}\rangle\sqrt{-1}\tilde{X}
\end{aligned}$$

Hence we obtain the equation (3.7). Following the definition of D, we can check all of the following other equations:

$$(\nabla^{\hat{M}}_{V} D)_{\tilde{X}}(\tilde{Y}) = (D_{V} \cdot D)_{\tilde{X}}(\tilde{Y}) \tag{3.8}$$

$$(\nabla^{\hat{M}}_{\tilde{Z}} D)_{\tilde{X}}(V) = (D_{\tilde{Z}} \cdot D)_{\tilde{X}}(V) \tag{3.9}$$

$$(\nabla^{\hat{M}}_{\tilde{Z}} D)_{V}(\tilde{Y}) = (D_{\tilde{Z}} \cdot D)_{V}(\tilde{Y}) \tag{3.10}$$

$$(\nabla^{\hat{M}}_{V} D)_{\tilde{X}}(V) = (D_{V} \cdot D)_{\tilde{X}}(V) \tag{3.11}$$

$$(\nabla^{\hat{M}}_{\tilde{Z}} D)_V(V) = (D_{\tilde{Z}} \cdot D)_V(V) \tag{3.12}$$

$$(\nabla^{\hat{M}}_{V} D)_V(\tilde{Y}) = (D_{V} \cdot D)_V(\tilde{Y}) \tag{3.13}$$

$$(\nabla^{\hat{M}}_{V} D)_V(V) = (D_{V} \cdot D)_V(V) \tag{3.14}$$

However they are also quite elementary computations. □

Since (3.6) is equivalent to (2.4), it follows from Lemmas 3.3 and 3.4 that ∇^c is a canonical connection on $\hat{M}$.

LEMMA 3.5. *$\nabla^*\alpha^M = 0$ if and only if $\nabla^c\alpha^{\hat{M}} = 0$, that is,*

$$(\nabla^*_u \alpha^{\hat{M}})(v,w) + \alpha^{\hat{M}}(D_u v, w) + \alpha^{\hat{M}}(v, D_u w) = 0 \tag{3.15}$$

for all vectors $u, v, w \in T\hat{M}$.

PROOF. We use some results from the fundamental equations of Riemannian submersions (cf. [**16**]). The second fundamental forms of $\hat{M}$ and M are related as

$$\alpha^{\hat{M}}(\tilde{X}, \tilde{Y}) = (\alpha^M(X,Y))^{\sim} \tag{3.16}$$

for any tangent vectors X, Y on M. By tanking the covariant derivative of the equation (3.16) in the horizontal direction $\tilde{Z}$ we have

$$(\nabla^*_{\tilde{Z}} \alpha^{\hat{M}})(\tilde{X}, \tilde{Y}) + \alpha^{\hat{M}}(\mathcal{A}_{\tilde{Z}}\tilde{X}, \tilde{Y}) + \alpha^{\hat{M}}(\tilde{X}, \mathcal{A}_{\tilde{Z}}\tilde{Y}) = ((\nabla^*_Z \alpha^M)(X,Y))^{\sim} \tag{3.17}$$

for any tangent vectors X, Y, Z on M. Since the fibers of $\pi : S^{2n+1}(1) \to \mathbb{C}P^n$ are totally geodesic, we have

$$\alpha^{\hat{M}}(V,W) = 0 \tag{3.18}$$

for any vertical vectors V, W on $\hat{M}$. It follows from (3.18) that

$$(\nabla^*_U \alpha^{\hat{M}})(V,W) = 0 \tag{3.19}$$

for any vertical vectors U, V, W on $\hat{M}$. Moreover, since M is a complex submanifold of $\mathbb{C}P^n$, for each normal vector field ξ to M we compute

$$\begin{aligned}
g_{S^{2n+1}}(\alpha^{\hat{M}}(V,\tilde{Y}), \tilde{\xi}) =& g_{S^{2n+1}}(\nabla^{S^{2n+1}}_{\tilde{Y}} V, \tilde{\xi}) \\
=& - g_{S^{2n+1}}(V, \nabla^{S^{2n+1}}_{\tilde{Y}} \tilde{\xi}) \\
=& - g_{S^{2n+1}}(V, \mathcal{A}_{\tilde{Y}}\tilde{\xi}) \\
=& - g_{S^{2n+1}}(V, -\langle\sqrt{-1}\tilde{Y}, \tilde{\xi}\rangle\sqrt{-1}\mathbf{x}) \\
=& g_{S^{2n+1}}(V, \langle\widetilde{JY}, \tilde{\xi}\rangle\sqrt{-1}\mathbf{x}) \\
=& \langle\widetilde{JY}, \tilde{\xi}\rangle g_{S^{2n+1}}(V, \sqrt{-1}\mathbf{x}) \\
=& g_{\mathbb{C}P^n}(JY, \xi) g_{S^{2n+1}}(V, \sqrt{-1}\mathbf{x}) = 0.
\end{aligned}$$

Hence we have

$$\alpha^{\hat{M}}(V, \tilde{Y}) = 0 \tag{3.20}$$

for any vertical vector V and any horizontal vector $\tilde{Y}$ on M. It follows from (3.20) that

$$(\nabla^*_U \alpha^{\hat{M}})(V, \tilde{Y}) = 0 \tag{3.21}$$

for any vertical vectors U, V and any horizontal vector $\tilde{Y}$ on $\hat{M}$. It follows from (3.18) and (3.20) that

$$(\nabla^*_{\tilde{X}}\alpha^{\hat{M}})(V,W)=0 \tag{3.22}$$

for any vertical vectors V, W and any horizontal vector $\tilde{X}$ on $\hat{M}$.

By (3.17), (3.18), (3.20) and the definition of D, we have

$$\begin{aligned}&(\nabla^*_{\tilde{Z}}\alpha^{\hat{M}})(\tilde{X},\tilde{Y})+\alpha^{\hat{M}}(D_{\tilde{Z}}\tilde{X},\tilde{Y})+\alpha^{\hat{M}}(\tilde{X},D_{\tilde{Z}}\tilde{Y})\\ =&(\nabla^*_{\tilde{Z}}\alpha^{\hat{M}})(\tilde{X},\tilde{Y})=((\nabla^*_{Z}\alpha^{M})(X,Y))\tilde{}.\end{aligned} \tag{3.23}$$

By differentiating (3.16) in the vertical direction $V=\sqrt{-1}\mathbf{x}$, we have

$$\begin{aligned}&g_{S^{2n+1}}((\nabla^*_V\alpha^{\hat{M}})(\tilde{X},\tilde{Y}),\tilde{\xi})\\ &+g_{S^{2n+1}}(\alpha^{\hat{M}}(\widetilde{JX},\tilde{Y}),\tilde{\xi})+g_{S^{2n+1}}(\alpha^{\hat{M}}(\tilde{X},\widetilde{JY}),\tilde{\xi})+g_{S^{2n+1}}(\alpha^{\hat{M}}(\tilde{X},\tilde{Y}),\widetilde{J\xi})=0\end{aligned}$$

for any tangent vectors X, Y on M and any normal vector ξ to M. By (3.16), the identity for a Kähler submanifold M and the definition of D we compute

$$\begin{aligned}&g_{S^{2n+1}}(\alpha^{\hat{M}}(\widetilde{JX},\tilde{Y}),\tilde{\xi})+g_{S^{2n+1}}(\alpha^{\hat{M}}(\tilde{X},\widetilde{JY}),\tilde{\xi})+g_{S^{2n+1}}(\alpha^{\hat{M}}(\tilde{X},\tilde{Y}),\widetilde{J\xi})\\ =&g_{\mathbb{C}P^n}(\alpha^M(JX,Y),\xi)\circ\pi+g_{\mathbb{C}P^n}(\alpha^M(X,JY),\xi)\circ\pi+g_{\mathbb{C}P^n}(\alpha^M(X,Y),J\xi)\circ\pi\\ =&g_{\mathbb{C}P^n}(\alpha^M(JX,Y),\xi)\circ\pi\\ =&\frac{1}{2}g_{\mathbb{C}P^n}(\alpha^M(JX,Y),\xi)\circ\pi+\frac{1}{2}g_{\mathbb{C}P^n}(\alpha^M(X,JY),\xi)\circ\pi\\ =&\frac{1}{2}g_{S^{2n+1}}(\alpha^{\hat{M}}(\widetilde{JX},\tilde{Y}),\tilde{\xi})+\frac{1}{2}g_{S^{2n+1}}(\alpha^{\hat{M}}(\tilde{X},\widetilde{JY}),\tilde{\xi})\\ =&g_{S^{2n+1}}(\alpha^{\hat{M}}(\frac{1}{2}\sqrt{-1}\tilde{X},\tilde{Y}),\tilde{\xi})+g_{S^{2n+1}}(\alpha^{\hat{M}}(\tilde{X},\frac{1}{2}\sqrt{-1}\tilde{Y}),\tilde{\xi})\\ =&g_{S^{2n+1}}(\alpha^{\hat{M}}(D_V\tilde{X},\tilde{Y}),\tilde{\xi})+g_{S^{2n+1}}(\alpha^{\hat{M}}(\tilde{X},D_V\tilde{Y}),\tilde{\xi}).\end{aligned}$$

Hence we have

$$(\nabla^*_V\alpha^{\hat{M}})(\tilde{X},\tilde{Y})+\alpha^{\hat{M}}(D_V\tilde{X},\tilde{Y})+\alpha^{\hat{M}}(\tilde{X},D_V\tilde{Y})=0. \tag{3.24}$$

It follows from (3.19) and the definition of D that

$$\begin{aligned}(\nabla^*_U\alpha^{\hat{M}})(V,W)+\alpha^{\hat{M}}(D_UV,W)+\alpha^{\hat{M}}(V,D_UW)=&(\nabla^*_U\alpha^{\hat{M}})(V,W)\\ =&0.\end{aligned} \tag{3.25}$$

By using (3.20), (3.22) and the definition of D, we compute

$$\begin{aligned}(\nabla^*_{\tilde{X}}\alpha^{\hat{M}})(V,W)+\alpha^{\hat{M}}(D_{\tilde{X}}V,W)+\alpha^{\hat{M}}(V,D_{\tilde{X}}W)=&(\nabla^*_{\tilde{X}}\alpha^{\hat{M}})(V,W)\\ =&0.\end{aligned} \tag{3.26}$$

By differentiating (3.20) in the horizontal direction $\tilde{X}$, we have

$$(\nabla^*_{\tilde{X}}\alpha^{\hat{M}})(V,\tilde{Y})+\alpha^{\hat{M}}((\mathcal{A}_{\tilde{X}}V)^{T\hat{M}},\tilde{Y})=0.$$

By (3.18) and the definition of D it becomes

$$\begin{aligned}&(\nabla^*_{\tilde{X}}\alpha^{\hat{M}})(V,\tilde{Y})+\alpha^{\hat{M}}(D_{\tilde{X}}V,\tilde{Y})+\alpha^{\hat{M}}(V,D_{\tilde{X}}\tilde{Y})\\ =&(\nabla^*_{\tilde{X}}\alpha^{\hat{M}})(V,\tilde{Y})+\alpha^{\hat{M}}(D_{\tilde{X}}V,\tilde{Y})\\ =&(\nabla^*_{\tilde{X}}\alpha^{\hat{M}})(V,\tilde{Y})+\alpha^{\hat{M}}((\mathcal{A}_{\tilde{X}}V)^{T\hat{M}},\tilde{Y})=0.\end{aligned} \tag{3.27}$$

By using (3.20) and (3.21) we compute

$$(\nabla^*_W \alpha^{\hat{M}})(V,\tilde{Y}) + \alpha^{\hat{M}}(D_W V,\tilde{Y}) + \alpha^{\hat{M}}(V, D_W\tilde{Y}) = (\nabla^*_W \alpha^{\hat{M}})(V,\tilde{Y}) = 0. \tag{3.28}$$

By those six equations (3.23), (3.24), (3.25), (3.26), (3.27) and (3.28), we obtain Lemma 3.5. □

Since (3.15) is equivalent to (3.4), we obtain Theorem 3.2. Therefore by Olmos and Sánchez's theorem 2.1 $\hat{M}$ is obtained as a standardly embedded (non-symmetric) R-space, that is, an orbit of the isotropy representation of a Riemannian symmetric pair (G,K).

In order to construct extrinsic symmetries of $\hat{M} \subset S^{2n+1}(1) \subset \mathbb{C}^{n+1}$, we recall the argument of [**15**]. For any $p,q \in \hat{M}$, let $\tau = \tau(t)$ $(t \in [a,b])$ be a piecewise smooth curve on $\hat{M}$ joining from $p = \tau(a)$ to $q = \tau(b)$. Let

$$\tau_a^b : T_p\hat{M} \longrightarrow T_q\hat{M}$$

denote the parallel displacement along τ with respect to the canonical connection ∇^c. Since the curvature tensor field R^c and the torsion tensor field T^c of ∇^c satisfy $\nabla^c R^c = 0$ and $\nabla^c T^c = 0$, there is a local isometry $s : U_p \to U_q$ of $\hat{M}$ such that

$$\begin{cases} s(p) = q, \\ (ds)_p = \tau_a^b. \end{cases}$$

Let

$$(\tau^\perp)_a^b : T_p^\perp\hat{M} \longrightarrow T_q^\perp\hat{M}$$

denote the parallel displacement along τ with respect to the normal connection $\nabla^\perp$. By using the orthogonal direct sum decompositions as real vector subspaces

$$\begin{aligned} \mathbb{C}^{n+1} \cong \mathbb{R}^{2n+2} &= \mathbb{R}\mathbf{x}(p) \oplus T_p\hat{M} \oplus T_p^\perp\hat{M} \\ &= \mathbb{R}\mathbf{x}(q) \oplus T_q\hat{M} \oplus T_q^\perp\hat{M}, \end{aligned}$$

we define an isometry $\tilde{s}$ of $S^{2n+1}(1)$, that is, $\tilde{s} \in SO(2n+2)$ by

$$\begin{cases} \tilde{s}(\mathbf{x}(p)) := \mathbf{x}(q), \\ \tilde{s}|_{T_p\hat{M}} := (ds)_p = \tau_a^b, \\ \tilde{s}|_{T_p^\perp\hat{M}} := (\tau^\perp)_a^b. \end{cases}$$

Then it follows from the condition $\nabla^c\alpha^{\hat{M}} = 0$ and the linearity of $\tilde{s}$ that

$$\begin{cases} s(p) = \tilde{s}(p) = q, \\ s(p') = \tilde{s}(p') \quad (\forall p' \in U_p), \\ (ds)_p = (d\tilde{s})_p|_{T_p\hat{M}} = \tilde{s}|_{T_p\hat{M}} = \tau_a^b, \\ (d\tilde{s})_p|_{T_p^\perp\hat{M}} = \tilde{s}|_{T_p^\perp\hat{M}} = (\tau^\perp)_a^b. \end{cases}$$

Moreover we can show

LEMMA 3.6.

$$\tilde{s}(\sqrt{-1}\mathbf{x}) = \sqrt{-1}\tilde{s}(\mathbf{x}) \quad (\forall \mathbf{x} \in \mathbb{C}^{n+1})$$

and hence $\tilde{s} \in U(n+1)$.

PROOF. We consider the trivial vector bundle

$$\underline{\mathbb{C}}^{n+1} := \hat{M} \times \mathbb{C}^{n+1}$$

over $\hat{M}$ with fiber $\mathbb{C}^{n+1}$ as a *real* vector bundle. For each point $\mathbf{x} \in \hat{M}$ we have orthogonal direct sum decompositions

$$\begin{aligned}\mathbb{C}^{n+1} =& \mathbb{R}\mathbf{x} \oplus T_{\mathbf{x}}\hat{M} \oplus T^{\perp}_{\mathbf{x}}\hat{M} \\ =& \mathbb{R}\mathbf{x} \oplus (\mathcal{V}_{\mathbf{x}}\hat{M} \oplus \mathcal{H}_{\mathbf{x}}\hat{M}) \oplus T^{\perp}_{\mathbf{x}}\hat{M}.\end{aligned}$$

Let

$$\mathcal{R}\hat{M} := \coprod_{\mathbf{x}\in\hat{M}} \mathbb{R}\mathbf{x}$$

be the real trivial line bundle over $\hat{M}$ equipped with the subbundle connection $\partial^{\mathcal{R}\hat{M}}$ on $\mathcal{R}\hat{M} \subset \underline{\mathbb{C}}^{n+1}$ induced from the trivial connection ∂, so that the position vector $\mathbf{x}$ of points of $\hat{M}$ gives a parallel global section of $\mathcal{R}\hat{M}$ with respect to $\partial^{\mathcal{R}\hat{M}}$. Then we have the following orthogonal direct sum decompositions of $\underline{\mathbb{C}}^{n+1}$ as real vector subbundles:

$$\begin{aligned}\underline{\mathbb{C}}^{n+1} =& \mathcal{R}\hat{M} \oplus T\hat{M} \oplus T\hat{M} \\ =& \mathcal{R}\hat{M} \oplus (\mathcal{V}\hat{M} \oplus \mathcal{H}\hat{M}) \oplus T^{\perp}\hat{M}.\end{aligned}$$

Along the first orthogonal decomposition we endow the vector bundle $\underline{\mathbb{C}}^{n+1}$ with the direct sum connection

$$\partial^{\mathcal{R}\hat{M}} \oplus \nabla^{c} \oplus \nabla^{\perp}.$$

Let $\mathrm{End}_{\mathbb{R}}(\underline{\mathbb{C}}^{n+1})$ be the vector bundles of $\mathbb{R}$-linear endomorphisms at each fiber of $\underline{\mathbb{C}}^{n+1}$. The multiplication $\sqrt{-1}\times : \mathbb{C}^{n+1} \to \mathbb{C}^{n+1}$ by $\sqrt{-1}$ on $\mathbb{C}^{n+1}$ can be regarded as $\mathbb{R}$-linear endomorphisms at each fiber of $\underline{\mathbb{C}}^{n+1}$, and thus it defines a smooth section $\sqrt{-1}\times$ of the real vector bundle $\mathrm{End}_{\mathbb{R}}(\underline{\mathbb{C}}^{n+1})$. Then we have only to show that *this section* $\sqrt{-1}\times$ *of* $\mathrm{End}_{\mathbb{R}}(\underline{\mathbb{C}}^{n+1})$ *is parallel with respect to the connection* $\partial^{\mathcal{R}\hat{M}} \oplus \nabla^{c} \oplus \nabla^{\perp}$.

Since M is a complex submanifold of $\mathbb{C}P^n$, at each point $\mathbf{x} \in \hat{M}$ the real vector subspaces $\mathbb{R}\mathbf{x} \oplus \mathbb{R}\sqrt{-1}\mathbf{x}$, $\mathcal{H}_{\mathbf{x}}\hat{M}$ and $T^{\perp}_{\mathbf{x}}\hat{M}$ are invariant under the multiplication by $\sqrt{-1}$, respectively. Thus three real vector subbundles $\mathcal{R}\hat{M} \oplus \mathcal{V}\hat{M}$, $\mathcal{H}\hat{M}$ and $T^{\perp}\hat{M}$ are invariant under the action of the section $\sqrt{-1}\times$ of $\mathrm{End}_{\mathbb{R}}(\underline{\mathbb{C}}^{n+1})$, respectively.

Set $(V)_{\mathbf{x}} := \sqrt{-1}\mathbf{x}$ $(\forall \mathbf{x} \in \hat{M})$, which is a vertical vector field on $\hat{M}$. First we observe that $V = \sqrt{-1}\mathbf{x}$ is a parallel vector field on $\hat{M}$ with respect to the canonical connection ∇^{c}. Indeed, since $\nabla^{\hat{M}}_{V} V = 0$ and $D_V V = 0$, we have $\nabla^{c}_{V} V = \nabla^{\hat{M}}_{V} V - D_V V = 0$. Since $\nabla^{\hat{M}}_{\tilde{X}} V = (\partial_{\tilde{X}} V)^{T\hat{M}} = \sqrt{-1}\tilde{X}$ and $D_{\tilde{X}} V = \sqrt{-1}\tilde{X}$ we have $\nabla^{c}_{\tilde{X}} V = \nabla^{\hat{M}}_{\tilde{X}} V - D_{\tilde{X}} V = \sqrt{-1}\tilde{X} - \sqrt{-1}\tilde{X} = 0$. It implies that

$$\nabla^{c}(\sqrt{-1}\mathbf{x}) = 0 = \sqrt{-1}\partial^{\mathcal{R}\hat{M}}\mathbf{x}. \tag{3.29}$$

Particularly the vertical subbundle $\mathcal{V}\hat{M}$ and horizontal subbundle $\mathcal{H}\hat{M}$ are invariant under parallel translations relative to ∇^{c} respectively. Moreover, by $\nabla^{c} D_V = 0$ and elementary computations we have

$$\nabla^{c}(\sqrt{-1}\tilde{X}) = \nabla^{c}(2D_V\tilde{X}) = 2D_V(\nabla^{c}\tilde{X}) = \sqrt{-1}\nabla^{c}\tilde{X}, \tag{3.30}$$

$$\nabla^{\perp}(\sqrt{-1}\tilde{\xi}) = \sqrt{-1}\nabla^{\perp}\tilde{\xi} \tag{3.31}$$

for any tangent vector field X and normal vector field ξ on M. Those equations (3.29), (3.30) and (3.31) mean the parallelism of $\sqrt{-1}\times \in \mathrm{End}_{\mathbb{R}}(\underline{\mathbb{C}}^{n+1})$ with respect to $\partial^{\mathcal{R}\hat{M}} \oplus \nabla^c \oplus \nabla^{\perp}$. Since a linear isomorphism $\tilde{s} : (\underline{\mathbb{C}}^{n+1})_{\mathbf{x}(p)} \to (\underline{\mathbb{C}}^{n+1})_{\mathbf{x}(q)}$ is a parallel displacement along τ with respect to $\partial^{\mathcal{R}\hat{M}} \oplus \nabla^c \oplus \nabla^{\perp}$, it preserves the parallel section $\sqrt{-1}\times$, namely, we obtain

$$\tilde{s}(\sqrt{-1}\mathbf{x}) = \sqrt{-1}\tilde{s}(\mathbf{x}) \quad (\forall \mathbf{x} \in \mathbb{C}^{n+1}).$$

□

Hence (G, K) can be taken as a Hermitian symmetric pair. And by the following lemma we also see that (G, K) is irreducible.

LEMMA 3.7. *Let N be a complex submanifold of $\mathbb{C}P^n$ and $\hat{N} = \pi^{-1}(N) \subset S^{2n+1}(1)$ be the inverse image of N under the Hopf fibration $\pi : S^{2n+1}(1) \to \mathbb{C}P^n$. Suppose that*

$$\mathbb{C}^{n+1} = \mathbb{C}^{\ell_1+1} \oplus \mathbb{C}^{\ell_2+1}$$

and

$$\hat{N} = \pi^{-1}(N) = \hat{N}_1 \times \hat{N}_2 \subset S^{2\ell_1+1}(r_1) \times S^{2\ell_2+1}(r_2) \subset \mathbb{C}^{\ell_1+1} \oplus \mathbb{C}^{\ell_2+1},$$

where $\hat{N}_1 \times \hat{N}_2$ is a Riemannian direct product of submanifolds $\hat{N}_i \subset S^{2\ell_i+1}(r_i) \subset \mathbb{C}^{\ell_i+1}$ with $r_i \geq 0$ for $i = 1, 2$ and $(r_1)^2 + (r_2)^2 = 1$. Then $r_1 = 0$ or $r_2 = 0$.

PROOF. Let $\mathbf{x} \in \hat{N}$. By the Riemannian product $\hat{N} = \hat{N}_1 \times \hat{N}_2$, the position vector $\mathbf{x} \in \hat{N}$ can be orthogonally decomposed as

$$\mathbf{x} = \mathbf{x}_1 + \mathbf{x}_2, \quad \mathbf{x}_i \in \hat{N}_i \subset S^{2\ell_i+1}(r_i) \subset \mathbb{C}^{\ell_i+1} \ (i = 1, 2)$$

and the tangent space of $\hat{N}$ at $\mathbf{x}$ can be orthogonally decomposed as

$$T_{\mathbf{x}}\hat{N} = T_{\mathbf{x}_1}\hat{N}_1 \oplus T_{\mathbf{x}_2}\hat{N}_2.$$

By using the orthogonal decomposition of the tangent vector space into vertical and horizontal subspaces

$$T_{\mathbf{x}}\hat{N} = \mathcal{V}_{\mathbf{x}}\hat{N} \oplus \mathcal{H}_{\mathbf{x}}\hat{N},$$

we have

$$\sqrt{-1}\mathbf{x}_1 + \sqrt{-1}\mathbf{x}_2 = \sqrt{-1}\mathbf{x} \in \mathcal{V}_{\mathbf{x}}\hat{N} \subset T_{\mathbf{x}}\hat{N} = T_{\mathbf{x}_1}\hat{N}_1 \oplus T_{\mathbf{x}_2}\hat{N}_2 \subset \mathbb{C}^{\ell_1+1} \oplus \mathbb{C}^{\ell_2+1}.$$

From this equation we see that

$$\sqrt{-1}\mathbf{x}_1 \in T_{\mathbf{x}_1}\hat{N}_1, \ \sqrt{-1}\mathbf{x}_2 \in T_{\mathbf{x}_2}\hat{N}_2.$$

If we set

$$\mathcal{H}_{\mathbf{x}_i}\hat{N}_i := \{v \in T_{\mathbf{x}_i}\hat{N}_i \mid \langle v, \sqrt{-1}\mathbf{x}_i\rangle = 0\} \quad (i = 1, 2),$$

then each tangent vector spaces can be orthogonally decomposed as

$$T_{\mathbf{x}_1}\hat{N}_1 = \mathbb{R}\sqrt{-1}\mathbf{x}_1 \oplus \mathcal{H}_{\mathbf{x}_1}\hat{N}_1, \quad T_{\mathbf{x}_2}\hat{N}_2 = \mathbb{R}\sqrt{-1}\mathbf{x}_2 \oplus \mathcal{H}_{\mathbf{x}_2}\hat{N}_2,$$

On the other hand, since $\hat{N}$ is the inverse image of N we have

$$T_{\mathbf{x}}S^{2n+1}(1) = \mathcal{V}_{\mathbf{x}}S^{2n+1}(1) \oplus \mathcal{H}_{\mathbf{x}}S^{2n+1}(1),$$

and

$$\mathcal{V}_{\mathbf{x}}\hat{N} = \mathcal{V}_{\mathbf{x}}S^{2n+1}(1) = \mathbb{R}\sqrt{-1}\mathbf{x} = \mathbb{R}\sqrt{-1}(\mathbf{x}_1 + \mathbf{x}_2).$$

Assume that $r_1 > 0$ and $r_2 > 0$. Then we can take a non-zero vector

$$\sqrt{-1}\left(\frac{1}{r_1^2}\mathbf{x}_1 - \frac{1}{r_2^2}\mathbf{x}_2\right) \in T_{\mathbf{x}_1}\hat{N}_1 \oplus T_{\mathbf{x}_1}\hat{N}_2 = T_{\mathbf{x}}\hat{N},$$

so that we have

$$\sqrt{-1}\mathbf{x} = \sqrt{-1}(\mathbf{x}_1 + \mathbf{x}_2) \perp \sqrt{-1}\left(\frac{1}{r_1^2}\mathbf{x}_1 - \frac{1}{r_2^2}\mathbf{x}_2\right).$$

Because

$$\begin{aligned}&\left\langle \sqrt{-1}\mathbf{x}, \sqrt{-1}\left(\frac{1}{r_1^2}\mathbf{x}_1 - \frac{1}{r_2^2}\mathbf{x}_2\right)\right\rangle \\ =&\left\langle \sqrt{-1}(\mathbf{x}_1 + \mathbf{x}_2), \sqrt{-1}\left(\frac{1}{r_1^2}\mathbf{x}_1 - \frac{1}{r_2^2}\mathbf{x}_2\right)\right\rangle \\ =&\frac{1}{r_1^2}\langle \mathbf{x}_1, \mathbf{x}_1\rangle - \frac{1}{r_2^2}\langle \mathbf{x}_2, \mathbf{x}_2\rangle = \frac{1}{r_1^2}r_1^2 - \frac{1}{r_2^2}r_2^2 = 1 - 1 = 0.\end{aligned}$$

Hence we obtain

$$\sqrt{-1}\left(\frac{1}{r_1^2}\mathbf{x}_1 - \frac{1}{r_2^2}\mathbf{x}_2\right) \in \mathcal{H}_{\mathbf{x}}\hat{N}.$$

However, since N is a complex submanifold of $\mathbb{C}P^n$, we have $\sqrt{-1}\mathcal{H}_{\mathbf{x}}\hat{N} = \mathcal{H}_{\mathbf{x}}\hat{N}$ and thus

$$\begin{aligned}-\frac{1}{r_1^2}\mathbf{x}_1 + \frac{1}{r_2^2}\mathbf{x}_2 =&\sqrt{-1}\sqrt{-1}\left(\frac{1}{r_1^2}\mathbf{x}_1 - \frac{1}{r_2^2}\mathbf{x}_2\right) \\ \in&\sqrt{-1}\mathcal{H}_{\mathbf{x}}\hat{N} = \mathcal{H}_{\mathbf{x}}\hat{N} \subset T_{\mathbf{x}}\hat{N} = T_{\mathbf{x}_1}\hat{N}_1 \oplus T_{\mathbf{x}_2}\hat{N}_2 \subset \mathbb{C}^{\ell_1+1} \oplus \mathbb{C}^{\ell_2+1}.\end{aligned}$$

Hence $\mathbf{x}_1 \in T_{\mathbf{x}_1}\hat{N}_1$ and $\mathbf{x}_2 \in T_{\mathbf{x}_2}\hat{N}_2$, a contradiction. Therefore we obtain that $r_1 = 0$ or $r_2 = 0$. □

Therefore we obtain a result of Takeuchi (Theorem 1.2) as a corollary:

COROLLARY 3.8. *Assume that M is a parallel Kähler submanifold of $\mathbb{C}P^n$. Then its inverse image $\hat{M} = \pi^{-1}(M) \subset S^{2n+1}(1) \subset \mathbb{C}^{n+1}$ is a standardly embedded R-space which is obtained as an orbit of the isotropy representation of an irreducible Hermitian symmetric pair (G, K).*

4. Classification of parallel Kähler submanifolds

Here we shall explain how we can determine explicitly all parallel Kähler submanifolds of $\mathbb{C}P^n$ obtained as the projection of R-spaces associated with irreducible Hermitian symmetric pairs by using results of [**20**]. it implies Theorem 1.1.

We refer [**10**] and its references for fundamental results on Hermitian symmetric spaces. Any irreducible Hermitian symmetric pair (G, K) of compact type is obtained in the following way ([**28**]). Let G be a compact connected simple Lie group with Lie algebra $\mathfrak{g}$. Let $\mathfrak{h}$ be a maximal abelian subalgebra of $\mathfrak{g}$ and

$$\mathfrak{g}^{\mathbb{C}} = \mathfrak{h}^{\mathbb{C}} + \sum_{\alpha \in \Delta_{\mathfrak{g}}} \mathfrak{g}^{\alpha}$$

be the root decomposition of $\mathfrak{g}^{\mathbb{C}}$ relative to $\mathfrak{h}^{\mathbb{C}}$, where $\Delta_{\mathfrak{g}}$ denotes the set of all roots $\alpha \in (\mathfrak{h}^{\mathbb{C}})^*$ of $\mathfrak{g}^{\mathbb{C}}$ relative to $\mathfrak{h}^{\mathbb{C}}$ and each root space is defined as

$$\mathfrak{g}^{\alpha} := \{X \in \mathfrak{g}^{\mathbb{C}} \mid (\mathrm{ad}H)X = \alpha(H)X \ (\forall H \in \mathfrak{h}^{\mathbb{C}})\}.$$

We denote by $(\ , \ )$ the inner product of $\mathfrak{h}^{\mathbb{C}}$ restricted from the Killing-Cartan form of $\mathfrak{g}^{\mathbb{C}}$. For $\alpha \in (\mathfrak{h}^{\mathbb{C}})^*$ we define the dual element $H_\alpha \in \mathfrak{h}^{\mathbb{C}}$ by $\alpha(H) = (H_\alpha, H)$ $(\forall H \in \mathfrak{h}^{\mathbb{C}})$. Note that if $\alpha \in \Delta_{\mathfrak{g}}$, then $H_\alpha \in \sqrt{-1}\mathfrak{h}$. Fix a lexicographic linear order $<$ on $\mathfrak{h}^*$. Let $\Delta_{\mathfrak{g}}^+$ denote the set of all positive roots $\alpha \in \Delta_{\mathfrak{g}}$. Choose a non-zero vector $X_\alpha \in \mathfrak{g}^\alpha$ for each $\alpha \in \Delta_{\mathfrak{g}}$ such that

$$X_\alpha - X_{-\alpha}, \sqrt{-1}(X_\alpha + X_{-\alpha}) \in \mathfrak{g} \text{ and } [X_\alpha, X_{-\alpha}] = \frac{2}{\alpha(H_\alpha)} H_\alpha.$$

Then we have

$$\mathfrak{g} = \mathfrak{h} + \sum_{\alpha \in \Delta_{\mathfrak{g}}^+} \left(\mathbb{R}(X_\alpha - X_{-\alpha}) + \mathbb{R}\sqrt{-1}(X_\alpha + X_{-\alpha})\right).$$

Let

$$\Pi(G) := \{\alpha_1, \cdots, \alpha_l\}$$

be the fundamental root system (Dynkin diagram) of $\Delta_{\mathfrak{g}}$ relative to $<$ and

$$\{\Lambda_1, \cdots, \Lambda_l\} \subset (\sqrt{-1}\mathfrak{h})^*$$

be the corresponding fundamental weight system defined by

$$\frac{2(\Lambda_i, \alpha_j)}{(\alpha_j, \alpha_j)} = \delta_{ij} \quad (\forall i, j = 1, \cdots, l).$$

Express the highest root $\tilde{\alpha} \in \Delta^+$ as

$$\tilde{\alpha} = m_1\alpha_1 + \cdots + m_i\alpha_l, \quad m_i \in \mathbb{Z},\ m_i \geq 1.$$

Now we choose a vertex α_{i_0} of the Dynkin diagram $\Pi(G)$ with $m_{i_0} = 1$. Set

$$\Delta_{\mathfrak{k}} := \{\alpha \in \Delta_{\mathfrak{g}} \mid (\alpha, \Lambda_{i_0}) = 0\} = \Delta_{\mathfrak{g}} \cap \left(\bigoplus_{i \neq i_0} \mathbb{Z}\alpha_i\right), \ \Delta_{\mathfrak{k}}^+ := \Delta_{\mathfrak{k}} \cap \Delta_{\mathfrak{g}}^+$$

and $\Delta_{\mathfrak{p}} := \Delta \setminus \Delta_{\mathfrak{k}}$, $\Delta_{\mathfrak{p}}^+ := \Delta^+ \setminus \Delta_{\mathfrak{k}}^+$. Define a Lie subalgebra $\mathfrak{k}$, a vector subspace $\mathfrak{p}$ of $\mathfrak{g}$ and their complexifications as

$$\begin{aligned}
\mathfrak{k} &:= \mathfrak{h} + \sum_{\alpha \in \Delta_{\mathfrak{k}}^+} \left(\mathbb{R}(X_\alpha - X_{-\alpha}) + \mathbb{R}\sqrt{-1}(X_\alpha + X_{-\alpha})\right) \\
\mathfrak{p} &:= \sum_{\alpha \in \Delta_{\mathfrak{p}}^+} \left(\mathbb{R}(X_\alpha - X_{-\alpha}) + \mathbb{R}\sqrt{-1}(X_\alpha + X_{-\alpha})\right), \\
\mathfrak{k}^{\mathbb{C}} &= \mathfrak{h}^{\mathbb{C}} + \sum_{\alpha \in \Delta_{\mathfrak{k}}} \mathfrak{g}^\alpha, \quad \mathfrak{p}^{\mathbb{C}} = \sum_{\alpha \in \Delta_{\mathfrak{p}}} \mathfrak{g}^\alpha.
\end{aligned}$$

Then the Lie algebra $\mathfrak{k}$ is the centralizer

$$\mathfrak{c}_{\mathfrak{g}}(\sqrt{-1}H_{\Lambda_{i_0}}) = \{X \in \mathfrak{g} \mid [X, \sqrt{-1}H_{\Lambda_{i_0}}] = 0\}$$

of $\mathfrak{g}$ to $\sqrt{-1}H_{\Lambda_{i_0}} \in \mathfrak{h} \subset \mathfrak{k}$ and has the center $\mathfrak{z}(\mathfrak{k}) = \mathbb{R}\sqrt{-1}H_{\Lambda_{i_0}}$. The centralizer $K := C_G(\sqrt{-1}H_{\Lambda_{i_0}})$ of G to $\sqrt{-1}H_{\Lambda_{i_0}}$ is a connected compact Lie subgroup of G with Lie algebra $\mathfrak{k}$. The Dynkin diagram of $\mathfrak{k}$ is

$$\Pi(K) := \{\alpha_j \in \Pi(G) \mid j \neq i_0\}.$$

Then (G, K) is an irreducible Hermitian symmetric pair of compact type with an involutive automorphism $\sigma := \mathrm{Ad}(\exp\left(\frac{2\pi}{(\alpha_{i_0}, \alpha_{i_0})}\sqrt{-1}H_{\Lambda_{i_0}}\right))$. If we set $Z :=$

$\frac{2}{(\alpha_{i_0},\alpha_{i_0})}\sqrt{-1}H_{\Lambda_{i_0}}$, then $(\mathrm{ad}_{\mathfrak{p}}Z)^2 = -\mathrm{Id}_{\mathfrak{p}}$ and thus $\mathrm{ad}_{\mathfrak{p}}Z$ defines an invariant orthogonal complex structure $J_{G/K}$ on $\mathfrak{p}$. Hence we have an identification $\mathfrak{p} \cong \mathbb{C}^{n+1}$ and for each $\alpha \in \Delta_{\mathfrak{p}}^+$ we have

$$J_{G/K}(X_\alpha - X_{-\alpha}) = \sqrt{-1}(X_\alpha + X_{-\alpha}),$$
$$J_{G/K}(\sqrt{-1}(X_\alpha + X_{-\alpha})) = (-1)(X_\alpha - X_{-\alpha}).$$

Let $\langle\ ,\ \rangle$ be an invariant inner product on $\mathfrak{p}$ defined by the (-1)-times Killing-Cartan form of $\mathfrak{g}$. Denote by $\pi : S^{2n+1}(1) \to \mathbb{C}P^n$ the Hopf fibration of the unit hypersphere $S^{2n+1}(1) \subset \mathfrak{p}$ over $\mathbb{C}P^n$.

Note that $\tilde{\alpha} \in \Delta_{\mathfrak{p}}^+$. Then we know that there exists a maximal system

$$\Gamma = \{\gamma_1 = \tilde{\alpha}, \gamma_2, \cdots, \gamma_r\} \subset \Delta_{\mathfrak{p}}^+$$

of strongly orthogonal roots in $\Delta_{\mathfrak{p}}^+$ in the sense that $\gamma_i \pm \gamma_j \notin \Delta$ $(1 \leq i, j \leq r)$, and thus a vector subspace

$$\mathfrak{a} = \sum_{i=1}^{r} \mathbb{R}\sqrt{-1}(X_{\gamma_i} + X_{-\gamma_i})$$

is a maximal abelian subspace of $\mathfrak{p}$, where set $r := \mathrm{rank}(G, K)$.

The *Cayley transform* is an inner automorphism of Lie algebra $\mathfrak{g}$ defined by

$$\nu = \mathrm{Ad}\left(\exp\left(\frac{\pi}{4}\sum_{i=1}^{r}(X_{\gamma_i} - X_{-\gamma_i})\right)\right).$$

Then we have

$$\nu\left(\sqrt{-1}(X_{\gamma_i} + X_{-\gamma_i})\right) = \frac{2\sqrt{-1}}{\gamma_i(H_{\gamma_i})}H_{\gamma_i} \quad (1 \leq i \leq r)$$

and thus

$$\nu(\mathfrak{a}) = \sum_{i=1}^{r} \mathbb{R}\sqrt{-1}H_{\gamma_i} \subset \mathfrak{h}.$$

We denote by $\bar{(\ )}$ the restriction of a linear form on $\mathfrak{h}^{\mathbb{C}}$ to $\nu(\mathfrak{a})^{\mathbb{C}}$. Then $R := \{\bar{\alpha} \mid \alpha \in \Delta_{\mathfrak{g}}\}$ corresponds to the restricted root system of the Hermitian symmetric pair (G, K) via ν^{-1}. Set $R^+ := \{\bar{\alpha} \mid \alpha \in \Delta_{\mathfrak{g}}^+\}$. Then we know that

$$R^+ = \left\{\bar{\gamma}_i\ (1 \leq i \leq r),\ \frac{1}{2}(\bar{\gamma}_i \pm \bar{\gamma}_j)\ (1 \leq i < j \leq r)\right\} \text{ (type } C)$$

or

$$R^+ = \left\{\bar{\gamma}_i\ (1 \leq i \leq r),\ \frac{1}{2}(\bar{\gamma}_i \pm \bar{\gamma}_j)\ (1 \leq i \leq j \leq r),\ \frac{1}{2}\bar{\gamma}_i\ (1 \leq i \leq r)\right\} \text{ (type } BC).$$

For each $E \in \mathfrak{a} \cap S^{2n+1}(1)$ consider an *R-space*

$$\hat{M}_E := \mathrm{Ad}_{\mathfrak{p}}(K)E \subset S^{2n+1}(1) \subset \mathfrak{p} \cong \mathbb{C}^{n+1}.$$

and its projection under π

$$M_E := \pi(\mathrm{Ad}_{\mathfrak{p}}(K)E) = K \cdot \pi(E) \subset \mathbb{C}P^n.$$

After a suitable change of E under the Weyl group action of (G, K) on $\mathfrak{a}$ we may assume that E satisfies $\gamma_1(\nu(E)) \geq \gamma_2(\nu(E)) \geq \cdots \geq \gamma_r(\nu(E)) \geq 0$. Set

$$R_E^+ := \{\bar{\gamma} \in R^+ \mid \bar{\gamma}(\nu(E)) = 0\}.$$

In these setting we can describe H. Song's results characterizing R-spaces whose projections under π are parallel Kähler submanifolds as follows:

THEOREM 4.1 ([20], p.287 Theorem 3.2 (iii), p.294 Theorem 4.4). *M_E is a complex submanifold of $\mathbb{C}P^n$ if and only if*

$$R_E^+ = \left\{\frac{1}{2}(\bar\gamma_i + \bar\gamma_j) \quad (2 \le i \le j \le r)\right\} \quad \text{(type } C\text{)}$$

or

$$R_E^+ = \left\{\frac{1}{2}(\bar\gamma_i + \bar\gamma_j) \quad (2 \le i \le j \le r),\ \frac{1}{2}\bar\gamma_i \quad (2 \le i \le r)\ \right\} \quad \text{(type } BC\text{)}\ .$$

In this case M_E always has parallel second fundamental form.

Suppose that M_E is a complex submanifold of $\mathbb{C}P^n$. Then by Theorem 4.1 we have $\nu(E) \in \mathbb{R}\sqrt{-1}H_{\gamma_1}$. Since $\gamma_1 = \tilde\alpha$, we obtain

$$\begin{aligned} E =&\nu^{-1}\left(\frac{\sqrt{-1}}{(\tilde\alpha,\tilde\alpha)^{1/2}} H_{\tilde\alpha}\right) \\ =&\frac{(\tilde\alpha,\tilde\alpha)^{1/2}}{2}\sqrt{-1}(X_{\tilde\alpha} + X_{-\tilde\alpha}) \in \mathfrak{a} \cap S^{2n+1}(1) \subset \mathfrak{p}. \end{aligned}$$

Under the identification $\mathfrak{p} \cong \mathbb{C}^{n+1}$, we denote a 1-dimensional complex vector subspace of $\mathbb{C}^{n+1}$ spanned by the vector E by

$$\mathbb{C}E := \pi(E) = \mathbb{R}\sqrt{-1}(X_{\tilde\alpha} + X_{-\tilde\alpha}) + \mathbb{R}(X_{\tilde\alpha} - X_{-\tilde\alpha}) \in M_E \subset \mathbb{C}P^n.$$

We denote by $\mathfrak{k}^{ss}$ the semisimple part of the Lie algebra $\mathfrak{k}$ and take an orthogonal direct sum decomposition $\mathfrak{h} = \mathfrak{z}(\mathfrak{k}) \oplus \mathfrak{h}'$, where $\mathfrak{h}' := \mathfrak{h} \cap \mathfrak{k}^{ss}$. Then we have

$$\begin{aligned} \mathfrak{k} =&\mathfrak{z}(\mathfrak{k}) \oplus \mathfrak{k}^{ss} \\ =&\mathbb{R}\sqrt{-1}H_{\Lambda_{i_0}} \oplus \left(\mathfrak{h}' + \sum_{\alpha\in\Delta_{\mathfrak{k}}^+} (\mathbb{R}\sqrt{-1}(X_\alpha + X_{-\alpha}) + \mathbb{R}(X_\alpha - X_{-\alpha}))\right). \end{aligned}$$

Let K^{ss} denote a compact connected Lie subgroup of K with Lie algebra $\mathfrak{k}^{ss}$. Notice that

$$\hat M_E = \mathrm{Ad}_{\mathfrak{p}}(K)(E) = \mathrm{Ad}_{\mathfrak{p}}(K^{ss})(E)$$

(see [24]) and thus

$$M_E = \pi(\mathrm{Ad}_{\mathfrak{p}}(K)H) = \pi(\mathrm{Ad}_{\mathfrak{p}}(K^{ss})H).$$

The isotropy subgroup $K_{\pi(E)} := \{a \in K \mid (\mathrm{Ad}_{\mathfrak{p}}a)\mathbb{C}E = \mathbb{C}E\}$ of K at $\pi(E) \in M_E \subset \mathbb{C}P^{n+1}$ has Lie algebra $\mathfrak{k}_{\pi(E)} := \{S \in \mathfrak{k} \mid (\mathrm{ad}_{\mathfrak{p}}S)\mathbb{C}E \subset \mathbb{C}E\}$ and the isotropy subgroup $K^{ss}_{\pi(E)} := \{a \in K^{ss} \mid (\mathrm{Ad}_{\mathfrak{p}}a)\mathbb{C}E = \mathbb{C}E\}$ of K^{ss} at $\pi(E) \in M_E \subset \mathbb{C}P^{n+1}$ has Lie algebra $\mathfrak{k}^{ss}_{\pi(E)} := \{S \in \mathfrak{k}^{ss} \mid (\mathrm{ad}_{\mathfrak{p}}S)\mathbb{C}E \subset \mathbb{C}E\}$. Then

$$\begin{aligned} \mathfrak{k}_{\pi(E)} =&\mathfrak{z}(\mathfrak{k}) \oplus \mathfrak{k}^{ss}_{\pi(E)} \\ =&(\mathbb{R}\sqrt{-1}H_{\Lambda_{i_0}}) \oplus \left(\mathfrak{h}' + \sum_{\alpha\in\Delta_{\mathfrak{k}}^+,(\alpha,\tilde\alpha)=0} (\mathbb{R}\sqrt{-1}(X_\alpha + X_{-\alpha}) + \mathbb{R}(X_\alpha - X_{-\alpha}))\right) \\ =&\mathfrak{z}(\mathfrak{k}) \oplus \mathfrak{c}_{\mathfrak{k}^{ss}}(\sqrt{-1}H^{ss}_{\tilde\alpha}) = \mathfrak{c}_{\mathfrak{k}}(\sqrt{-1}H_{\tilde\alpha}). \end{aligned}$$

Since any $a \in K_{\pi(E)}$ commutes with $J_{G/K}$, for any $\alpha \in \Delta_{\mathfrak{p}}^{+}$ we compute

$$\begin{aligned}-\frac{2}{\alpha(H_\alpha)}\mathrm{Ad}(a)\,(\sqrt{-1}H_\alpha) =&\mathrm{Ad}(a)[\sqrt{-1}(X_\alpha + X_{-\alpha}), X_\alpha - X_{-\alpha}]\\ =&[\mathrm{Ad}(a)\sqrt{-1}(X_\alpha + X_{-\alpha}), \mathrm{Ad}(a)(X_\alpha - X_{-\alpha})]\\ =&[\sqrt{-1}(X_\alpha + X_{-\alpha}), X_\alpha - X_{-\alpha}]\\ =&-\frac{2}{\alpha(H_\alpha)}\sqrt{-1}H_\alpha.\end{aligned}$$

Hence we obtain

$$K_{\pi(E)} = C_K(\sqrt{-1}H_{\tilde{\alpha}}) \quad \text{and} \quad K^{ss}_{\pi(E)} = C_{K^{ss}}(\sqrt{-1}H^{ss}_{\tilde{\alpha}})$$

and they have the Dynkin diagram

$$\begin{aligned}&\Pi(K_{\pi(E)}) = \Pi(K^{ss}_{\pi(E)})\\ =&\{\alpha_j \in \Pi(K) \mid (\alpha_j, \tilde{\alpha}) = 0\} = \{\alpha_j \in \Pi(G) \mid j \neq i_0,\, (\alpha_j, \tilde{\alpha}) = 0\}.\end{aligned}$$

Therefore $(K, K_{\pi(E)})$ and $(K^{ss}, K^{ss}_{\pi(E)})$ are compact Hermitian symmetric pairs with an involutive automorphism $\tau = \mathrm{Ad}(\exp\left(\frac{2\pi}{(\tilde{\alpha},\tilde{\alpha})}\sqrt{-1}H_{\tilde{\alpha}}\right))$, and $M_E = K/K_{\pi(E)} \cong K^{ss}/K^{ss}_{\pi(E)}$ is embedded in a complex projective space $\mathbb{C}P^n$ as a parallel Kähler submanifold.

Now they can be determined explicitly by the linear expression of the highest root in terms of the fundamental root system and the extended Dynkin diagram of $\mathfrak{g}$ as follows: Here we use the notations of Bourbaki's table ([**2**]).

(1) In the case when $G = SU(m+2)$,

$$\begin{aligned}&\Pi(G) = \{\alpha_i = \varepsilon_i - \varepsilon_{i+1} \mid 1 \leq i \leq m+1\} \cong \mathrm{A}_{m+1},\\ &\tilde{\alpha} = \alpha_1 + \alpha_2 + \cdots + \alpha_{m+1} = \varepsilon_1 - \varepsilon_{m+2}.\end{aligned}$$

Take $\{\alpha_{s+1}\} \subset \Pi(G)$. Here $0 \leq s \leq m$.

$$\begin{aligned}\Pi(K) =&\{\alpha_i \mid 1 \leq i \leq s\} \sqcup \{\alpha_i \mid s+2 \leq i \leq m+1\}\\ \cong&\mathrm{A}_s \oplus \mathrm{A}_{m-s},\\ \Pi(K_{\pi(E)}) =&\{\alpha_i \mid 2 \leq i \leq s\} \sqcup \{\alpha_i \mid s+2 \leq i \leq m\}\\ \cong&\mathrm{A}_{s-1} \oplus \mathrm{A}_{m-s-1}.\end{aligned}$$

Hence $(G, K) = (SU(m+2), S(U(s+1) \times U(m-s+1)))$, $K^{ss} = SU(s+1) \times SU(m-s+1)$ and $K^{ss}_{\pi(E)} = S(U(1) \times U(s)) \times S(U(1) \times U(m-s))$. Hence we obtain

$$M_E = \frac{SU(s+1)}{S(U(1) \times U(s))} \times \frac{SU(m-s+1)}{S(U(1) \times U(m-s))} = \mathbb{C}P^s \times \mathbb{C}P^{m-s}.$$

(2) In the case when $G = Sp(m+1)$,

$$\begin{aligned}&\Pi(G) = \{\alpha_i = \varepsilon_i - \varepsilon_{i+1}\ (i = 1, \cdots, m),\, \alpha_{m+1} = 2\varepsilon_{m+1}\} \cong \mathrm{C}_{m+1},\\ &\tilde{\alpha} = 2\alpha_1 + 2\alpha_2 + \cdots + 2\alpha_m + \alpha_{m+1} = 2\varepsilon_1.\end{aligned}$$

Take $\{\alpha_{m+1}\} \subset \Pi(G)$. Then

$$\begin{aligned}&\Pi(K) = \{\alpha_i\ (1 \leq i \leq m)\} \cong \mathrm{A}_m,\\ &\Pi(K_{\pi(E)}) = \{\alpha_i\ (2 \leq i \leq m)\} \cong \mathrm{A}_{m-1}.\end{aligned}$$

Hence $(G,K) = (Sp(m+1), U(m+1))$, $K^{ss} = SU(m+1)$ and $K^{ss}_{\pi(E)} = SU(m)$. Therefore we obtain

$$M_E = \frac{SU(m+1)}{S(U(1)\times U(m))} = \mathbb{C}P^m.$$

(3) In the case when $G = SO(2(s+2)+1)\,(s\geq 1)$,

$$\Pi(G) = \{\alpha_i = \varepsilon_i - \varepsilon_{i+1}\ (i=1,\cdots,s+1),\ \alpha_{s+2} = \varepsilon_{s+2}\} \cong \mathrm{B}_{s+2},$$
$$\tilde{\alpha} = \alpha_1 + 2\alpha_2 + \cdots + 2\alpha_s + 2\alpha_{s+1} + 2\alpha_{s+2} = \varepsilon_1 + \varepsilon_2.$$

Take $\{\alpha_1\} \subset \Pi(G)$. Then

$$\Pi(K) = \{\alpha_i\ (i=2,\cdots,s+1),\ \alpha_{s+2}\} \cong \mathrm{B}_{s+1},$$
$$\Pi(K_{\pi(E)}) = \{\alpha_i\ (i=3,\cdots,s+1),\ \alpha_{s+2}\} \cong \mathrm{B}_s.$$

Hence $(G,K) = (SO(2s+5), SO(2)\times SO(2s+3))$, $K^{ss} = SO(2s+3)$ and $K^{ss}_{\pi(E)} = SO(2s+1)$. Therefore we obtain

$$M_E = \frac{SO(2s+3))}{SO(2)\times SO(2s+1)} = Q_{2s+1}(\mathbb{C}).$$

(4) In the case when $G = SO(2(s+2))\ (s\geq 2)$,

$$\Pi(G) = \{\alpha_i = \varepsilon_i - \varepsilon_{i+1}\ (i=1,\cdots,s+1),\ \alpha_{s+2} = \varepsilon_{s+1} + \varepsilon_{s+2}\} \cong \mathrm{D}_{s+2},$$
$$\tilde{\alpha} = \alpha_1 + 2\alpha_2 + \cdots + 2\alpha_s + \alpha_{s+1} + \alpha_{s+2} = \varepsilon_1 + \varepsilon_2.$$

Take $\{\alpha_1\} \subset \Pi(G)$. Then

$$\Pi(K) = \{\alpha_i\ (i=2,\cdots,s+1),\ \alpha_{s+2}\} \cong \mathrm{D}_{s+1},$$
$$\Pi(K_{\pi(E)}) = \{\alpha_i\ (i=3,\cdots,s+1),\ \alpha_{s+2}\} \cong \mathrm{D}_s.$$

Hence $(G,K) = (SO(2(s+2)), SO(2)\times SO(2(s+1)))$, $K^{ss} = SO(2(s+1))$ and $K^{ss}_{\pi(E)} = SO(2s)$. Therefore we obtain

$$M_E = \frac{SO(2(s+1))}{SO(2)\times SO(2s)} = Q_{2s}(\mathbb{C}).$$

Take $\{\alpha_{s+1}\} \subset \Pi(G)$. Then

$$\Pi(K) = \{\alpha_i\ (i=1,\cdots,s),\ \alpha_{s+2}\} \cong \mathrm{A}_{s+1},$$
$$\Pi(K_{\pi(E)}) = \{\alpha_1\} \sqcup \{\alpha_i\ (3\leq i\leq s),\ \alpha_{s+2}\} \cong \mathrm{A}_1 \oplus \mathrm{A}_{s-1}.$$

Hence $(G,K) = (SO(2(s+2)), U(s+2))$, $K^{ss} = SU(s+2)$ and $K^{ss}_{\pi(E)} = SU(2)\times SU(s)$. Therefore we obtain

$$M_E = \frac{SU(s+2)}{S(U(2)\times U(s))}.$$

Take $\{\alpha_{s+2}\} \subset \Pi(G)$. Similarly we obtain

$$M_E = \frac{SU(s+2)}{S(U(2)\times U(s))}.$$

(5) In the case when $G = E_6$,

$$\begin{aligned}\Pi(G) = \{\alpha_1 &= \frac{1}{2}(\varepsilon_1 + \varepsilon_8) - \frac{1}{2}(\varepsilon_2 + \varepsilon_3 + \varepsilon_4 + \varepsilon_5 + \varepsilon_6 + \varepsilon_7)\\ &\alpha_2 = \varepsilon_1 + \varepsilon_2,\ \alpha_3 = \varepsilon_2 - \varepsilon_1,\ \alpha_4 = \varepsilon_3 - \varepsilon_2,\\ &\alpha_5 = \varepsilon_4 - \varepsilon_3,\ \alpha_6 = \varepsilon_5 - \varepsilon_4\} \cong \mathrm{E}_6,\\ \tilde{\alpha} &= \alpha_1 + 2\alpha_2 + 2\alpha_3 + 3\alpha_4 + 2\alpha_5 + \alpha_6\\ &= \frac{1}{2}(\varepsilon_1 + \varepsilon_2 + \varepsilon_3 + \varepsilon_4 + \varepsilon_5 - \varepsilon_6 - \varepsilon_7 + \varepsilon_8).\end{aligned}$$

Take $\{\alpha_1\} \subset \Pi(G)$. Then

$$\begin{aligned}\Pi(K) &= \{\alpha_2,\ \alpha_3,\ \alpha_4,\ \alpha_5,\ \alpha_6\} \cong \mathrm{D}_5,\\ \Pi(K_{\pi(E)}) &= \{\alpha_3,\ \alpha_4,\ \alpha_5,\ \alpha_6\} \cong \mathrm{A}_4.\end{aligned}$$

Hence $(G, K) = (E_6, Spin(10) \cdot U(1))$, $K^{ss} = SO(10)$ and $K^{ss}_{\pi(E)} = SU(5)$. Therefore we obtain

$$M_E = \frac{SO(10)}{U(5)}.$$

(6) In the case when $G = E_7$,

$$\begin{aligned}\Pi(G) = \{\alpha_1 &= \frac{1}{2}(\varepsilon_1 + \varepsilon_8) - \frac{1}{2}(\varepsilon_2 + \varepsilon_3 + \varepsilon_4 + \varepsilon_5 + \varepsilon_6 + \varepsilon_7)\\ &\alpha_2 = \varepsilon_1 + \varepsilon_2,\ \alpha_3 = \varepsilon_2 - \varepsilon_1,\ \alpha_4 = \varepsilon_3 - \varepsilon_2,\\ &\alpha_5 = \varepsilon_4 - \varepsilon_3,\ \alpha_6 = \varepsilon_5 - \varepsilon_4,\ \alpha_7 = \varepsilon_6 - \varepsilon_5\},\\ \tilde{\alpha} &= 2\alpha_1 + 2\alpha_2 + 3\alpha_3 + 4\alpha_4 + 3\alpha_5 + 2\alpha_6 + \alpha_7 = \varepsilon_8 - \varepsilon_7.\end{aligned}$$

Take $\{\alpha_7\} \subset \Pi(G)$. Then

$$\begin{aligned}\Pi(K) &= \{\alpha_1,\ \alpha_2,\ \alpha_3,\ \alpha_4,\ \alpha_5,\ \alpha_6\} \cong \mathrm{E}_6,\\ \Pi(K_{\pi(E)}) &= \{\alpha_2,\ \alpha_3,\ \alpha_4,\ \alpha_5,\ \alpha_6\} \cong \mathrm{D}_5.\end{aligned}$$

Hence $(G, K) = (E_7, E_6 \cdot U(1))$, $K^{ss} = E_6$ and $K^{ss}_{\pi(E)} = Spin(10)$. Therefore we obtain

$$M_E = \frac{E_6}{Spin(10) \cdot U(1)}.$$

We use Ichiro Yokota [**29**] for the precise homogeneous space expression of exceptional Hermitian symmetric spaces.

(G, K)	n	M_E	m
$(SU(m+2), S(U(m+1) \times U(1)))$	m	$\mathbb{C}P^m(4)$	m
$(Sp(m+1), U(m+1))$	$m + \frac{m(m+1)}{2}$	$\mathbb{C}P^m(2)$	m
$(SU(m+2), S(U(s+1) \times U(m-s+1)))$	$m + s(m-s)$	$\mathbb{C}P^s(1) \times \mathbb{C}P^{m-s}(1)$	m
$(SO(m+4), SO(m+2) \times SO(2))$	$m+1$	$Q_m(\mathbb{C})$	m
$(SO(2(s+2)), U(s+2))$	$s + \frac{s(s+1)}{2}$	$SU(s+2)/S(U(2) \times U(s))$	$2s$
$(E_6, (Spin(10) \times U(1))/\mathbb{Z}_4)$	15	$SO(10)/U(5)$	10
$(E_7, (E_6 \times U(1))/\mathbb{Z}_3)$	26	$E_6/(Spin(10) \times U(1))/\mathbb{Z}_4)$	16

Acknowledgments

This article was contributed to the Proceedings of the AMS Special Session on Geometry of Submanifolds in January, 2020, in Memoriam Professor Tadashi Nagano. The author sincerely would like to thank the organizers, especially Professors B.-Y. Chen, B. D. Suceavă, M. S. Tanaka and H. Tamaru for their excellent organization and kind invitation to the contribution. He also would like to thank Professors T. Sakai, J. T. Cho and Dr. K. Hashimoto for valuable discussion and interests in this work.

References

[1] Isabel Bergmann, *Reducible polar representations*, Manuscripta Math. **104** (2001), no. 3, 309–324, DOI 10.1007/s002290170029. MR1828877

[2] N. Bourbaki, *Éléments de mathématique. Fasc. XXXIV. Groupes et algèbres de Lie. Chapitre IV: Groupes de Coxeter et systèmes de Tits. Chapitre V: Groupes engendrés par des réflexions. Chapitre VI: systèmes de racines* (French), Actualités Scientifiques et Industrielles [Current Scientific and Industrial Topics], No. 1337, Hermann, Paris, 1968. MR0240238

[3] Eugenio Calabi, *Isometric imbedding of complex manifolds*, Ann. of Math. (2) **58** (1953), 1–23, DOI 10.2307/1969817. MR57000

[4] Jiri Dadok, *Polar coordinates induced by actions of compact Lie groups*, Trans. Amer. Math. Soc. **288** (1985), no. 1, 125–137, DOI 10.2307/2000430. MR773051

[5] J.-H. Eschenburg and E. Heintze, *Extrinsic symmetric spaces and orbits of s-representations*, Manuscripta Math. **88** (1995), no. 4, 517–524, DOI 10.1007/BF02567838. MR1362935

[6] J.-H. Eschenburg and E. Heintze, *Polar representations and symmetric spaces*, J. Reine Angew. Math. **507** (1999), 93–106, DOI 10.1515/crll.1999.507.93. MR1670274

[7] J.-H. Eschenburg and E. Heintze, *On the classification of polar representations*, Math. Z. **232** (1999), no. 3, 391–398, DOI 10.1007/PL00004763. MR1719714

[8] Dirk Ferus, *Immersions with parallel second fundamental form*, Math. Z. **140** (1974), 87–93, DOI 10.1007/BF01218650. MR370437

[9] Dirk Ferus, *Symmetric submanifolds of Euclidean space*, Math. Ann. **247** (1980), no. 1, 81–93, DOI 10.1007/BF01359868. MR565140
New York-London, 1978.

[10] Sigurdur Helgason, *Differential geometry, Lie groups, and symmetric spaces*, Pure and Applied Mathematics, vol. 80, Academic Press, Inc. [Harcourt Brace Jovanovich, Publishers], New York-London, 1978. MR514561

[11] Max Koecher, *An elementary approach to bounded symmetric domains*, Rice University, Houston, Tex., 1969. MR0261032

[12] Tadashi Nagano, *Transformation groups on compact symmetric spaces*, Trans. Amer. Math. Soc. **118** (1965), 428–453, DOI 10.2307/1993971. MR182937

[13] Hisao Nakagawa and Ryoichi Takagi, *On locally symmetric Kaehler submanifolds in a complex projective space*, J. Math. Soc. Japan **28** (1976), no. 4, 638–667, DOI 10.2969/jmsj/02840638. MR417463

[14] Koichi Ogiue, *Differential geometry of Kaehler submanifolds*, Advances in Math. **13** (1974), 73–114, DOI 10.1016/0001-8708(74)90066-8. MR346719

[15] Carlos Olmos and Cristián Sánchez, *A geometric characterization of the orbits of s-representations*, J. Reine Angew. Math. **420** (1991), 195–202. MR1124572

[16] Barrett O'Neill, *The fundamental equations of a submersion*, Michigan Math. J. **13** (1966), 459–469. MR200865

[17] Richard S. Palais and Chuu-Lian Terng, *A general theory of canonical forms*, Trans. Amer. Math. Soc. **300** (1987), no. 2, 771–789, DOI 10.2307/2000369. MR876478

[18] Antonio Ros, *A characterization of seven compact Kaehler submanifolds by holomorphic pinching*, Ann. of Math. (2) **121** (1985), no. 2, 377–382, DOI 10.2307/1971178. MR786353

[19] Ichirô Satake, *Algebraic structures of symmetric domains*, Kanô Memorial Lectures, vol. 4, Iwanami Shoten, Tokyo; Princeton University Press, Princeton, N.J., 1980. MR591460

[20] Hyunjung Song, *Some differential-geometric properties of R-spaces*, Tsukuba J. Math. **25** (2001), no. 2, 279–298, DOI 10.21099/tkbjm/1496164288. MR1869763

[21] Ryoichi Takagi and Masaru Takeuchi, *Degree of symmetric Kählerian submanifolds of a complex projective space*, Osaka Math. J. **14** (1977), no. 3, 501–518. MR467632

[22] Masaru Takeuchi, *Homogeneous Kähler submanifolds in complex projective spaces*, Japan. J. Math. (N.S.) **4** (1978), no. 1, 171–219, DOI 10.4099/math1924.4.171. MR528871

[23] Masaru Takeuchi, *Parallel projective manifolds and symmetric bounded domains*, Osaka J. Math. **21** (1984), no. 3, 507–544. MR759479

[24] Hiroyuki Tasaki and Osami Yasukura, *R-spaces associated with a Hermitian symmetric pair*, Tsukuba J. Math. **10** (1986), no. 1, 165–170, DOI 10.21099/tkbjm/1496160400. MR846427
Étude géométrique d'une classe d'espaces homogènes.

[25] Jacques Tits, *Étude géométrique d'une classe d'espaces homogènes* (French), C. R. Acad. Sci. Paris **239** (1954), 466–468. MR66394

[26] Jacques Tits, *Sur les R-espaces* (French), C. R. Acad. Sci. Paris **22** (1954), 850–852. MR66395

[27] Kazumi Tsukada, *Parallel submanifolds in a quaternion projective space*, Osaka J. Math. **22** (1985), no. 2, 187–241. MR800968

[28] Joseph A. Wolf, *On the classification of hermitian symmetric spaces*, J. Math. Mech. **13** (1964), 489–495. MR0160850

[29] I. Yokota, *Exceptional Lie Groups*, `arXiv:0902.0431v1[mathDG]`,2009.

Osaka City University Advanced Mathematical Institute (OCAMI) & Department of Mathematics, Osaka City University, 3-3-138 Sugimoto, Sumiyoshi-ku, Osaka, 558-8585, Japan

Email address: `ohnita@sci.osaka-cu.ac.jp`

Contemporary Mathematics
Volume **777**, 2022
https://doi.org/10.1090/conm/777/15636

A survey on natural Γ-symmetric structures on R-spaces

Peter Quast and Takashi Sakai

Honoring the Memory of Professor Tadashi Nagano

ABSTRACT. This is a report upon the authors' 2020 paper on Γ-symmetric structures on R-spaces, which continues various aspects of Nagano's work. We first briefly recall some generalizations of symmetric spaces. Next we show that certain R-spaces admit Γ-symmetric structures appearing in a natural way as a generalization of symmetric R-spaces. Moreover we describe the maximal antipodal sets of these natural Γ-symmetric structures.

1. Introduction

Since their discovery by Elie Cartan in the 1920s symmetric spaces, a distinguished class of Riemannian manifolds, attracted the attention of many mathematicians from various fields such as differential geometry, algebraic topology, representation theory or harmonic analysis. The well-established theory of symmetric spaces is exposed in [**11**] and [**18**]. Generalizations of symmetric spaces have been proposed in various directions: Ledger [**16**] and Ledger–Obata [**17**] initiated the study of (regular) s-manifolds. These are (Riemannian) manifolds M which admit at each point $x \in M$ a symmetry s_x having x as an isolated fixed point. If s_x is of finite order k, a regular s-manifold is called a k-symmetric space. As a further generalization, Lutz [**19**] introduced Γ-symmetric space in 1981, where Γ is a finite abelian group. These are manifolds M with the following structure: To each point $x \in M$ one assigns in a suitable way a group Γ_x isomorphic to Γ which acts effectively on M and has x as an isolated fixed point. If Γ is isomorphic to $\mathbb{Z}_2$, then, in a Riemannian setting, a Γ-symmetric space is just a symmetric space. E. Cartan already observed the correspondence between symmetric spaces and symmetric pairs. In analogy, Lutz [**19**] introduced Γ-symmetric triples (see also [**10**] and [**1**]). Many examples of Γ-symmetric structures on homogeneous spaces have been constructed using this method.

An R-space is a quotient of a noncompact semisimple real Lie group by a parabolic subgroup. Every R-space can be realized as an orbit of the linear isotropy representation of a Riemannian symmetric space G/K of compact type and is therefore canonically a submanifold of a sphere in a Euclidean space. In 1964, 1965,

2020 *Mathematics Subject Classification.* Primary 53C30; Secondary 53C35, 17B40.

Key words and phrases. Γ-symmetric spaces, R-spaces, antipodal sets, extrinsic symmetric spaces and generalizations.

The second author was partly supported by the Grant-in-Aid for Science Research (C) No. 17K05223 and No. 21K03250, JSPS.

Nagano [**20**], Takeuchi [**25**] and Kobayashi and Nagano [**13**] initiated the study of compact symmetric spaces which are also R-spaces. These spaces are nowadays known as symmetric R-spaces. Symmetric R-spaces have various remarkable geometric properties, intrinsically and extrinsically. Ferus [**9**] has shown that compact extrinsically symmetric spaces in a Euclidean space (these are submanifolds of a Euclidean space which are invariant under the reflections across their affine normal spaces) are essentially canonically embedded symmetric R-spaces and vice-versa.

In this survey article we report on certain natural Γ-symmetric structures on R-spaces recently described by the authors in [**22**], where Γ is a power of $\mathbb{Z}_2$. We give a necessary and sufficient condition for an R-space to admit a certain natural Γ-symmetric structure in terms of the restricted root system of G/K. Our criterion allows us to classify all R-spaces admitting such a Γ-symmetric structure. For $\Gamma = \mathbb{Z}_2$ we recover the symmetric R-spaces. Our natural Γ-symmetric structures also act canonically on the ambient Euclidean space in such the way, that the normal space at a point x is the fixed space of Γ_x. Therefore these spaces can be considered as extrinsically Γ-symmetric spaces, a natural generalization of extrinsically symmetric spaces.

In 1988, Chen and Nagano [**8**] studied antipodal sets of compact symmetric spaces. These are finite sets and their maximal cardinality is a geometric invariant, called the 2-number, of the compact symmetric space at hand. The notion of an antipodal set naturally extends to Γ-symmetric spaces. We showed in [**22**] that any maximal antipodal set of our natural Γ-structure on an R-space can be described as an orbit of the Weyl group of the restricted root system of G/K. Hence all maximal antipodal sets are conjugate to each other. Consequently the maximal cardinality of an antipodal set of an R-space equipped with our natural Γ-symmetric structure equals the sum of its $\mathbb{Z}_2$-Betti numbers. This generalizes similar results by Tanaka and Tasaki [**28**] and Takeuchi [**26**] for symmetric R-spaces.

2. Γ-symmetric spaces

We recall that a Riemannian symmetric space is a (connected) Riemannian manifold M which has a geodesic symmetry s_x at each point $x \in M$. Here s_x is the (unique) isometry of M which reverses all geodesics through x, hence x is an isolated fixed point of s_x in M. The group generated by all geodesic symmetries acts transitively on M. Hence a Riemannian symmetric space is a complete, homogeneous Riemannian manifold. The set of geodesic symmetries $\{s_x\}_{x\in M}$ induces a multiplication on M. From this algebraic structure, we have a definition of symmetric spaces without a Riemannian metric.

DEFINITION 2.1 ([**18**] Vol. I). A *symmetric space* is a C^∞-manifold with a differentiable multiplication $\mu : M\times M \to M$, written as $\mu(x,y) = x\cdot y$ for $x,y\in M$, which satisfies the following conditions:

(1) $x\cdot x = x$,
(2) $x\cdot(x\cdot y) = y$,
(3) $x\cdot(y\cdot z) = (x\cdot y)\cdot(x\cdot z)$, and
(4) every x has a neighborhood U such that for all $y \in U$ we have $x\cdot y = y$ implies $y = x$.

For $x \in M$, define the map $s_x : M \to M$ by $s_x(y) := \mu(x,y) = x\cdot y$ for $y\in M$. Then the four conditions in Definition 2.1 are:

(1) $s_x(x) = x$,
(2) s_x is an involutive diffeomorphism, i.e., $s_x^2 = \mathrm{id}_M$,
(3) $s_x \circ s_y = s_{s_x(y)} \circ s_x$, and
(4) x is isolated in the fixed point set $F(s_x, M) := \{y \in M \mid s_x(y) = y\}$ of s_x.

Hence s_x is called the *symmetry* at x.

REMARK 2.2. In Definition 2.1, the conditions (1), (2) and (3) imply that a symmetric space is a *quandle* due to [**12**]. Those conditions also appear in [**24**] as the notion of (Kei).

We get the definition of a *regular s-manifold* by replacing the conditions (2) and (4) in Definition 2.1 with

(2)′ s_x is a diffeomorphism, and
(4)′ for each $x \in M$, the tangent map $(s_x)_{*x} : T_xM \to T_xM$ has no fixed vectors except the null vector.

If each symmetry s_x is of finite order k, then M is called a regular s-manifold of order k, or simply a k-symmetric space. We refer to [**15**] for the theory of regular s-manifolds. In 1981, Lutz [**19**] introduced the notion of Γ-symmetric spaces generalizing k-symmetric spaces.

DEFINITION 2.3 ([**19**]). Let M be a connected C^∞-manifold and Γ a finite abelian group. A *Γ-symmetric structure* on M is a family $\mu = \{\mu^\gamma\}_{\gamma\in\Gamma}$ of smooth maps $\mu^\gamma : M \times M \to M$ which satisfy the following conditions:

(1) for each $x \in M$, the map

$$\iota_x : \Gamma \to \mathrm{Diff}(M);\ \gamma \mapsto \gamma_x := \mu^\gamma(x, \cdot)$$

is an injective homomorphism,
(2) x is isolated in the fixed point set $\{y \in M \mid \gamma_x(y) = y \text{ for all } \gamma \in \Gamma\}$, and
(3) for any $x \in M$ and $\gamma \in \Gamma$, γ_x is an automorphism of μ, i.e.

$$\mu^\delta(\gamma_x(y), \gamma_x(z)) = \gamma_x(\mu^\delta(y, z)) \quad \text{for all } y, z \in M \text{ and for all } \delta \in \Gamma.$$

We call $\gamma_x \in \mathrm{Diff}(M)$ the *symmetry* at $x \in M$ corresponding to $\gamma \in \Gamma$. For each point $x \in M$ the subgroup $\Gamma_x := \iota_x(\Gamma) = \{\gamma_x : \gamma \in \Gamma\}$ of $\mathrm{Diff}(M)$ is isomorphic to Γ by condition (1). We call Γ_x the *symmetric transformation group* at $x \in M$. We denote the fixed point set of Γ_x (as in condition (2)) by

$$F(\Gamma_x, M) := \{y \in M \mid \gamma_x(y) = y \text{ for all } \gamma \in \Gamma\}.$$

Condition (3) can be rewritten as

$$\gamma_x \circ \delta_y = \delta_{\gamma_x(y)} \circ \gamma_x \quad \text{for all } x, y \in M \text{ and for all } \gamma, \delta \in \Gamma. \tag{2.1}$$

Hence it is a generalization of condition (3) in Definition 2.1. When $\Gamma = \mathbb{Z}_k$, a Γ-symmetric space is a k-symmetric space. $\mathbb{Z}_2 \times \mathbb{Z}_2$-symmetric spaces have been classified by Bahturin and Goze [**1**] and Kollross [**14**].

Lutz [**19**] introduced Γ-symmetric triples to construct Γ-symmetric spaces (see also [**10**] and [**1**]). Let K be a connected Lie group and Γ a finite abelian subgroup of the automorphism group $\mathrm{Aut}(K)$ of K. Let H be a closed subgroup of K satisfying

$$F(\Gamma, K)_0 \subset H \subset F(\Gamma, K). \tag{2.2}$$

Then the triple (K, H, Γ) is called a *Γ-symmetric triple*. Here we denote the fixed point set of Γ in K by $F(\Gamma, K) := \{k \in K \mid \gamma(k) = k \text{ for all } \gamma \in \Gamma\}$, and its identity component by $F(\Gamma, K)_0$. From a Γ-symmetric triple (K, H, Γ), one gets a K-equivariant Γ-symmetric structure on the homogeneous space K/H as follows: For any point $kH \in K/H$ and for each $\gamma \in \Gamma$ consider the (well defined) diffeomorphism

$$\gamma_{kH} : K/H \to K/H;\ \tilde{k}H \mapsto k\gamma(k^{-1}\tilde{k})H.$$

Note that $\gamma_{eH}(\tilde{k}H) = \gamma(\tilde{k})H$ for any $\tilde{k}H \in K/H$, where e is the neutral element of K. We have the following K-equivariance law:

$$\gamma_{kH} = L_k \circ \gamma_{eH} \circ L_{k^{-1}} \quad \text{for any } k \in K,$$

where L_k denotes the left action of k on K/H. Given $\gamma \in \Gamma$, the multiplication map

$$\mu^\gamma : K/H \times K/H \to K/H;\ (kH, \tilde{k}H) \mapsto \gamma_{kH}(\tilde{k}H) = k\gamma(k^{-1}\tilde{k})H$$

is obviously smooth.

Proposition 2.4 (Lutz). *The family $\mu = \{\mu^\gamma\}_{\gamma\in\Gamma}$ is a Γ-symmetric structure on K/H.*

Proof. We have to check the three conditions of Definition 2.3. For a any $kH \in K/H$, the surjective map $\iota_{kH} : \Gamma \to \Gamma_{kH};\ \gamma \mapsto \gamma_{kH}$ is obviously a group homeomorphism. By K-equivariance it is sufficient to verify that ι_{eH} is injective, that is that ι_{eH} has trivial kernel. The differential of any element $\gamma \in \Gamma$ at $e \in K$ is an automorphism γ_* of the Lie algebra $\mathfrak{k}$ of K, that is Γ also acts faithfully on $\mathfrak{k}$ by automorphisms. Condition (2.2) implies

$$\mathfrak{h} = \{X \in \mathfrak{k} : \gamma_*(X) = X \text{ for all } \gamma \in \Gamma\}, \tag{2.3}$$

where $\mathfrak{h}$ is the Lie algebra of H. Since Γ is finite, we can choose a Γ-invariant scalar product on $\mathfrak{k}$ by averaging. The orthogonal complement $\mathfrak{m}$ of $\mathfrak{h}$ in $\mathfrak{k}$ is therefore invariant under any automorphism γ_* with $\gamma \in \Gamma$. The differential at e of the canonical projection $\pi : K \to K/H; k \mapsto kH$ restricted to $\mathfrak{m}$, denoted by $\pi_* : \mathfrak{m} \to T_{eH}K/H$, is bijective and satisfies

$$\gamma_{eH*} = \pi_* \circ \gamma_*|_{\mathfrak{m}} \circ \pi_*^{-1}$$

for all $\gamma \in \Gamma$, where γ_{eH*} is the differential of γ_{eH} at eH. Now, if $\iota_{eH}(\gamma) = \gamma_{eH}$ is the identity on K/H, then γ_{eH*} is the identity on $T_{eH}K/H$ and γ_* the identity on $\mathfrak{m}$. By (2.3) γ_* is the identity on $\mathfrak{k}$. Since K is connected, γ is the identity automorphism of K. This shows that ι_{eH} has trivial kernel.

We now prove (2). Our arguments here are similar to [**15**, p. 46], and another proof is in [**32**]. The previous considerations show that for all $X \in T_{eH}K/H$ with $X \neq 0$ there exists $\gamma \in \Gamma$ such that $\gamma_{eH*}(X) \neq X$. Since Γ is finite, we can choose a Γ_{eH}-invariant Riemannian metric on K/H (by averaging). Take an open ball B in $T_{eH}K/H$ around the origin of sufficiently small radius such that the Riemannian exponential map Exp_{eH} of K/H at eH is a diffeomorphism from B onto its image $\mathrm{Exp}_{eH}(B)$. Since every map γ_{eH}, $\gamma \in \Gamma$, is an isometry of K/H fixing eH, we have

$$\gamma_{eH} \circ \mathrm{Exp}_{eH} = \mathrm{Exp}_{eH} \circ \gamma_{eH*}$$

for all $\gamma \in \Gamma$. Thus $\mathrm{Exp}_{eH}(B) \cap F(\Gamma_{eH}, K/H) = \{eH\}$. This shows (2) by K-homogeneity.

The verification of condition (3) amounts to a straight forward check of (2.1) using the commutativity of Γ. □

3. R-spaces and symmetric R-spaces

We refer to [**11**] and [**18**] for further details on the theory of symmetric spaces. Let P be a simply connected Riemannian symmetric space of compact type and G the identity component of the isometry group of P. We take a base point $o \in P$ and denote the isotropy subgroup at o by $K = \{g \in G \mid g(o) = o\}$. Then P can be identified with the coset space G/K by the fibration $\pi : G \to G/K \cong P; g \mapsto gK = g \cdot o$. Define $\sigma : G \to G; g \mapsto s_o g s_o^{-1}$, where s_o is the geodesic symmetry of P at o. Then σ is an involutive automorphism of G and K is the identity component $F(\sigma, G)_0$ of the fixed point set $F(\sigma, G)$ of σ in G, hence (G, K, σ) is a symmetric pair. The differential σ_* of σ at the identity is an involutive automorphism of the Lie algebra $\mathfrak{g}$ of G. The Lie algebra $\mathfrak{k}$ of K is the fixed point set of σ_*, and we denote by $\mathfrak{p}$ the (-1)-eigenspace of σ_*. Then we have the canonical direct sum decomposition $\mathfrak{g} = \mathfrak{k} \oplus \mathfrak{p}$, and the tangent space T_oP of P at o can be identified with $\mathfrak{p}$ by $\pi_*|_{\mathfrak{p}}$.

Using the identification by $\pi_*|_{\mathfrak{p}}$, the adjoint action of K on $\mathfrak{p}$ is equivalent to the linear isotropy representation of K on T_oP. We take a maximal abelian subspace $\mathfrak{a}$ of $\mathfrak{p}$. We denote the rank of P by $r := \dim \mathfrak{a}$.

Let $\mathfrak{a}^*$ be the dual vector space of $\mathfrak{a}$. For each $\alpha \in \mathfrak{a}^*$, we set

$$\mathfrak{g}_\alpha := \{Y \in \mathfrak{g} \otimes \mathbb{C} \mid [H, Y] = \sqrt{-1}\alpha(H)Y \text{ for all } H \in \mathfrak{a}\}.$$

A nonzero vector $\alpha \in \mathfrak{a}^* \setminus \{0\}$ is called a root of P, if $\mathfrak{g}_\alpha \neq \{0\}$. Then

$$\mathcal{R} := \{\alpha \in \mathfrak{a}^* \setminus \{0\} \mid \mathfrak{g}_\alpha \neq \{0\}\}$$

forms the root system of P. Take a fundamental system $\Sigma := \{\alpha_1, \dots, \alpha_r\}$ of $\mathcal{R}$ and let $\mathcal{R}^+ \subset \mathcal{R}$ be the set of positive roots with respect to Σ. Then Σ is a basis of $\mathfrak{a}^*$ and every root $\alpha \in \mathcal{R}^+$ can be expressed as

$$\alpha = \sum_{j=1}^{r} c_j^\alpha \alpha_j \qquad \text{with } c_j^\alpha \in \mathbb{N} \cup \{0\} \text{ for all } j \in \{1, \dots, r\}. \tag{3.1}$$

We have the following root space decompositions:

$$\begin{aligned} \mathfrak{p} &= \mathfrak{a} \oplus \sum_{\alpha \in \mathcal{R}^+} \mathfrak{p}_\alpha \quad \text{with} \quad \mathfrak{p}_\alpha = (\mathfrak{g}_\alpha \oplus \mathfrak{g}_{-\alpha}) \cap \mathfrak{p}, \\ \mathfrak{k} &= \mathfrak{m} \oplus \sum_{\alpha \in \mathcal{R}^+} \mathfrak{k}_\alpha \quad \text{with} \quad \mathfrak{k}_\alpha = (\mathfrak{g}_\alpha \oplus \mathfrak{g}_{-\alpha}) \cap \mathfrak{k}, \end{aligned}$$

where $\mathfrak{m} := \{Y \in \mathfrak{k} \mid [H, Y] = 0 \text{ for all } H \in \mathfrak{a}\}$. Let $\{\xi_1, \dots, \xi_r\}$ be the basis of $\mathfrak{a}$ dual to Σ, i.e. $\alpha_i(\xi_j) = \delta_{ij}$. Hence, by (3.1), we have $\alpha(\xi_j) = c_j^\alpha$ for each $\alpha \in \mathcal{R}^+$ and each $j \in \{1, \dots, r\}$.

We define

$$g_j := \exp(\pi \xi_j), \qquad j \in \{1, \dots, r\}. \tag{3.2}$$

Then the inner automorphism $\mathrm{Int}_G(g_j)$ of G is involutive. The differential of $\mathrm{Int}_G(g_j)$ is the involutive automorphisms $\mathrm{Ad}_G(g_j)$ of $\mathfrak{g}$ which commutes with σ_*. Thus $\mathfrak{k}$ is invariant under $\mathrm{Ad}_G(g_j)$ and $\mathrm{Int}_G(g_j)$ induces an involutive automorphism of K:

$$\gamma^j := \mathrm{Int}_G(g_j)|_K \in \mathrm{Aut}(K), \quad j \in \{1, \dots, r\}. \tag{3.3}$$

Every R-space can be realized as an orbit of an s-representation, that is an orbit of the linear isotropy representation of a Riemannian symmetric space $P \cong G/K$ of compact type (see [**25**]). For a non-empty subset I of $I_{\text{reg}} := \{1, \dots, r\}$, we set

$$\xi_I := \sum_{i \in I} \xi_i \in \mathfrak{a},$$

We denote the orbit of the s-representation through ξ_I by

$$X_I := \mathrm{Ad}_G(K)\xi_I \subset \mathfrak{p}.$$

Then X_I is identified with the coset space K/H_I where H_I is the isotropy subgroup of ξ_I in K. It is known that every R-space in $\mathfrak{p}$ is equivariantly isomorphic to one of these X_I (see [**5**] for detail). If I and $\tilde{I}$ are non-empty subsets of I_{reg} satisfying $\tilde{I} \subset I$, then $H_I \subset H_{\tilde{I}}$ holds (see [**22**, Lemma 1]). Hence the choice of a subset $I \subset I_{\text{reg}}$ describes the stratification of orbit types of the s-representation.

If $\eta \in \mathfrak{a} \setminus \{0\}$ satisfies $\alpha(\eta) \in \{-1, 0, 1\}$ for every root α, then the corresponding R-space $X_\eta = \mathrm{Ad}_G(K)\eta$ is called a *symmetric R-space.* These spaces were first considered by Nagano [**20**] and Takeuchi [**25**] and classified and described by Kobayashi and Nagano [**13**]. The involutive automorphism $\rho := \mathrm{Ad}_G(\exp(\pi\eta))$ of $\mathfrak{g}$ preserves $\mathfrak{p}$ and we have

$$F(\rho, \mathfrak{p}) := \{Y \in \mathfrak{p} : \rho(Y) = Y\} = \{Y \in \mathfrak{p} : [Y, \eta] = 0\}.$$

Thus $F(\rho, \mathfrak{p})$ is the normal space of X_η at η and ρ is the reflection across the normal space of X_η at η (see e.g. [**3**, eq. (3.2), p. 50]). An easy calculation (see [**9**, p. 82]) shows that ρ preserves M. By homogeneity X_η is an extrinsically symmetric space (see also [**3**, Example 3.7]). Ferus [**9**] proved that every compact extrinsically symmetric space can be realized as a symmetric R-space and that every non-compact extrinsically symmetric space is just a cylinder over a compact extrinsically symmetric one.

Coming back to our previous notation we call an element ξ_j with $j \in I_{\text{reg}}$ *extrinsically symmetric*, if α_j has coefficient 0 or 1 in any positive root.

4. Natural Γ-symmetric structures on R-spaces

In this section, we explain natural Γ-symmetric structures on R-spaces as a generalization of symmetric R-spaces obtained in [**22**]. We inherit the setting in the previous section. Let $X_I \cong K/H_I$ be an R-space for a non-empty subset $I \subset I_{\text{reg}}$. We consider the group Γ^I generated by the elements γ^i $(i \in I)$ given in (3.3);

$$\Gamma^I = \langle \gamma^i : i \in I \rangle \subset \mathrm{Aut}(K).$$

Then Γ^I is a finite abelian subgroup of $\mathrm{Aut}(K)$ which is isomorphic to $(\mathbb{Z}_2)^k$, where k is the cardinality of I. We call a non-empty subset $I \subset I_{\text{reg}}$ *admissible* if (K, H_I, Γ^I) is a Γ^I-symmetric triple, that is if

$$F(\Gamma^I, K)_0 \subset H_I \subset F(\Gamma^I, K) \tag{4.1}$$

holds. Therefore, as we have seen in Section 2, if $I \subset I_{\text{reg}}$ is admissible, then the R-space X_I has Γ^I-symmetric structure, where $\Gamma^I \cong (\mathbb{Z}_2)^k$.

Note that $H_I \subset H_{\{i\}}$ holds for any $i \in I$. Thus for any $k \in H_I$ and any $i \in I$ we have

$$\gamma^i(k) = \exp(\pi\xi_i)k\exp(-\pi\xi_i) = \exp(\pi\xi_i)\exp\big(-\pi\mathrm{Ad}_G(k)\xi_i\big)k = k,$$

since $\mathrm{Ad}_G(k)\xi_i = \xi_i$. Hence the right hand side inclusion in (4.1), $H_I \subset F(\Gamma^I, K)$ holds. Therefore the left hand side inclusion in (4.1), $F(\Gamma^I, K)_0 \subset H_I$, holds if and only if

$$\mathfrak{h}_I = \mathfrak{f}(\Gamma^I, K),$$

where $\mathfrak{h}_I$ is the Lie algebra of H_I and $\mathfrak{f}(\Gamma^I, K)$ is the Lie algebra of $F(\Gamma^I, K)_0$. These two Lie algebras can be described in terms of the root system $\mathcal{R}$ of $P \cong G/K$ as follows:

$$\mathfrak{h}_I = \{X \in \mathfrak{k} \mid [X, \xi_I] = 0\} = \mathfrak{m} \oplus \sum_{\alpha \in \mathcal{R}_I^0} \mathfrak{k}_\alpha$$

where

$$\mathcal{R}_I^0 := \{\alpha \in \mathcal{R}^+ \mid \alpha(\xi_I) = 0\} = \{\alpha \in \mathcal{R}^+ \mid \alpha(\xi_i) = 0 \text{ for all } i \in I\}, \tag{4.2}$$

and

$$\mathfrak{f}(\Gamma^I, K) = \{X \in \mathfrak{k} \mid \mathrm{Ad}_G(g_i)X = X \text{ for all } i \in I\} = \mathfrak{m} \oplus \sum_{\alpha \in \mathcal{R}_I^{\mathrm{ev}}} \mathfrak{k}_\alpha$$

where

$$\mathcal{R}_I^{\mathrm{ev}} := \{\alpha \in \mathcal{R}^+ \mid \alpha(\xi_i) \text{ is even for all } i \in I\}.$$

Consequently we obtain the following characterization of admissible subsets $I \subset I_{\mathrm{reg}}$ in terms of $\mathcal{R}$.

THEOREM 4.1 ([**22**]). *A non-empty subset I of I_{reg} is admissible if and only if $\mathcal{R}_I^{\mathrm{ev}} = \mathcal{R}_I^0$, that is if and only if*

$$\left(\forall i \in I: \ c_i^\alpha \text{ even}\right) \Longrightarrow \left(\forall i \in I : c_i^\alpha = 0\right)$$

holds for all $\alpha = \sum_{j=1}^r c_j^\alpha \alpha_j \in \mathcal{R}^+$.

From Theorem 4.1 we have the following propositions.

PROPOSITION 4.2 ([**22**]). *Unions of admissible subsets of I_{reg} are admissible.*

PROPOSITION 4.3 ([**22**]). *If $\mathcal{R}$ is a reduced root system, that is, if $\mathbb{Z}\alpha \cap \mathcal{R} = \{-\alpha, \alpha\}$ for any $\alpha \in \mathcal{R}$, then I_{reg} is admissible. Thus, in this case, a principal orbit $X_{I_{\mathrm{reg}}} = \mathrm{Ad}_G(K)\xi_{I_{\mathrm{reg}}}$ has a $\Gamma^{I_{\mathrm{reg}}}$-symmetric structure, where $\Gamma^{I_{\mathrm{reg}}} \cong (\mathbb{Z}_2)^r$.*

PROPOSITION 4.4 ([**22**]). *If for all $i \in I$ the element ξ_i is extrinsically symmetric, that is ξ_I is a sum of extrinsically symmetric elements, then I is admissible.*

Applying Theorem 4.1 to the list of root systems of symmetric spaces of compact types, we can determine all admissible $I \subset I_{\mathrm{reg}}$ when $\mathcal{R}$ is irreducible (see [**22**, Section 4]).

EXAMPLE 4.5. We give a simple example of the natural Γ-symmetric structures on flag manifolds. These Γ-symmetric structures are also discussed in [**10**] and [**21**].

Let $G/K = \mathrm{SU}(r+1)/\mathrm{SO}(r+1)$ be the compact symmetric space of type AI, where the involution σ on G is the complex conjugation $\sigma(g) = \bar{g}$ for $g \in G$. Then we have the canonical decomposition $\mathfrak{g} = \mathfrak{k} \oplus \mathfrak{p}$ of the Lie algebra $\mathfrak{g} = \mathfrak{su}(r+1)$ of $G = \mathrm{SU}(r+1)$, where $\mathfrak{k} = \mathfrak{o}(r+1)$ is the Lie algebra of $K = \mathrm{SO}(r+1)$ and

$$\mathfrak{p} = \left\{\sqrt{-1}X \mid X \text{ is a real trace-free } (r+1) \times (r+1) \text{ symmetric matrix }\right\}.$$

As a maximal abelian subspace $\mathfrak{a}$ of $\mathfrak{p}$ we take the set of diagonal matrices in $\mathfrak{p}$:

$$\mathfrak{a} = \left\{H = \sqrt{-1}\mathrm{diag}(t_1, t_2, \ldots, t_{r+1}) \;\middle|\; t_j \in \mathbb{R}, \sum_{j=1}^{r+1} t_j = 0\right\}.$$

Then the root system of G/K with respect to $\mathfrak{a}$ is

$$\mathcal{R} = \{\pm(e_i - e_j) \mid 1 \leq i, j \leq r+1\},$$

where $e_i \in \mathfrak{a}^*$ is $e_i(H) = t_i$ for $H \in \mathfrak{a}$. We choose a fundamental system Σ of $\mathcal{R}$ as

$$\Sigma = \{\alpha_1 = e_1 - e_2, \dots, \alpha_r = e_r - e_{r+1}\},$$

and let $\{\xi_1, \dots, \xi_r\}$ be the basis of $\mathfrak{a}$ dual to Σ.

For a non-empty subset $I = \{i_1, \dots i_k\}$ of $I_{\mathrm{reg}} = \{1, 2, \dots, r\}$ where $1 \leq i_1 < i_2 < \cdots < i_k \leq r$, we set

$$\xi_I := \sum_{i \in I} \xi_i \in \mathfrak{a} \qquad \text{and} \qquad X_I := \mathrm{Ad}_G(K)\xi_I \subset \mathfrak{p}.$$

Then the isotropy subgroup of ξ_I in K is $H_I \cong \mathrm{S}(\mathrm{O}(i_1) \times \mathrm{O}(i_2 - i_1) \times \cdots \mathrm{O}(r+1-i_k))$, hence X_I can be identified with the flag manifold $F_I(\mathbb{R}^{r+1})$ as follows:

$$\begin{array}{ccccc} X_I = \mathrm{Ad}_G(K)\xi_I & \cong & K/H_I & \cong & F_I(\mathbb{R}^{r+1}) \\ \mathrm{Ad}_G(k)\xi_I & \longleftrightarrow & kH_I & \longleftrightarrow & k\mathcal{F}_o. \end{array}$$

Here $F_I(\mathbb{R}^{r+1})$ is the flag manifold of sequences of subspaces in $\mathbb{R}^{r+1}$ defined as

$$F_I(\mathbb{R}^{r+1}) = \left\{ (V_1, \dots, V_k) \;\middle|\; \begin{array}{l} V_j \text{ is an } i_j\text{-dimensional subspace of } \mathbb{R}^{r+1} \\ \{0\} \subset V_1 \subset V_2 \subset \cdots \subset V_k \subset \mathbb{R}^{r+1} \end{array} \right\},$$

and $\mathcal{F}_o = (\langle v_1, \dots, v_{i_1} \rangle_{\mathbb{R}}, \langle v_1, \dots, v_{i_2} \rangle_{\mathbb{R}}, \dots, \langle v_1, \dots, v_{i_k} \rangle_{\mathbb{R}})$ is the origin of $F_I(\mathbb{R}^{r+1})$, where $v_1, \dots, v_{r+1}$ is the standard basis of $\mathbb{R}^{r+1}$. Since the root system $\mathcal{R}$ is of type A_r, by Theorem 4.1, any non-empty subset I of I_{reg} is admissible, i.e., X_I admits the natural Γ^I-symmetric structure, where $\Gamma^I \cong (\mathbb{Z}_2)^k$.

When $I = \{i\}$, $X_I = \mathrm{Ad}_G(K)\xi_I$ is an isolated orbit type. Then $F_{\{i\}}(\mathbb{R}^{r+1})$ is the Grassmannian manifold of i-dimensional subspaces in $\mathbb{R}^{r+1}$, which is a symmetric R-space. In this case, the natural Γ^I-symmetric structure is described as the geodesic symmetry of the Grassmannian manifold $F_{\{i\}}(\mathbb{R}^{r+1})$, that is, for an i-dimensional subspace V of $\mathbb{R}^{r+1}$, the reflection s_V in $\mathbb{R}^{r+1}$ with respect to V gives the geodesic symmetry of $F_{\{i\}}(\mathbb{R}^{r+1})$ at V.

When $I = \{i_1, i_2\}$, the flag manifold $F_{\{i_1,i_2\}}(\mathbb{R}^{r+1})$ has fibrations over the Grassmannian manifolds $F_{\{i_1\}}(\mathbb{R}^{r+1})$ and $F_{\{i_2\}}(\mathbb{R}^{r+1})$. By these fibrations, geodesic symmetries s_{V_1} (resp. s_{V_2}) of the Grassmannian manifold $F_{\{i_1\}}(\mathbb{R}^{r+1})$ at V_1 (resp. $F_{\{i_2\}}(\mathbb{R}^{r+1})$ at V_2) can be lifted to $F_{\{i_1,i_2\}}(\mathbb{R}^{r+1})$. Then they provide the natural Γ^I-symmetric structure on $F_{\{i_1,i_2\}}(\mathbb{R}^{r+1})$, where $\Gamma^I \cong (\mathbb{Z}_2)^2$.

In the same way, geodesic symmetries of Grassmannian manifolds can be lifted to the flag manifold $F_I(\mathbb{R}^{r+1})$ for any non-empty subset $I \subset I_{\mathrm{reg}}$, and they give the natural Γ^I-symmetric structure on $F_I(\mathbb{R}^{r+1})$. As we mentioned in Proposition 4.4, if ξ_I is a sum of extrinsically symmetric elements ξ_i, then the natural Γ^I-symmetric structure on X_I is induced from geodesic symmetries of each $X_{\{i\}}$ $(i \in I)$.

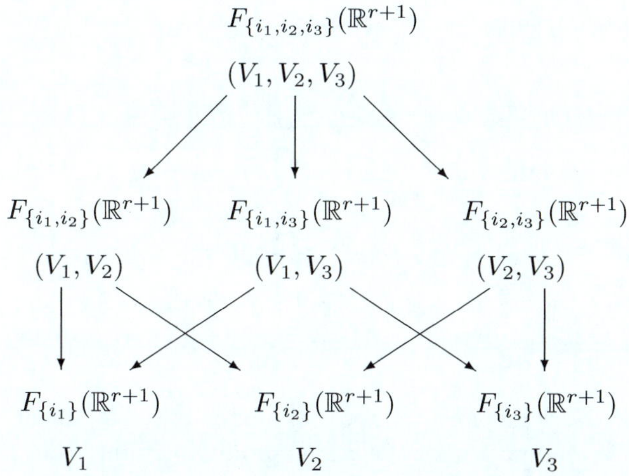

When $I = \{i_i, i_2, i_3\}$, $F_{\{i_1,i_2,i_3\}}(\mathbb{R}^{r+1})$ has the natural Γ^I-symmetric structure, where $\Gamma = \langle \gamma^{i_1}, \gamma^{i_2}, \gamma^{i_3} \rangle \cong (\mathbb{Z}_2)^3$. However this Γ^I-symmetric structure is not a minimal one, that is there exists a *proper* subgroup $\hat{\Gamma}$ of Γ^I which gives a $\hat{\Gamma}$-symmetric structure on $F_{\{i_1,i_2,i_3\}}(\mathbb{R}^{r+1})$. Indeed, take the subgroup

$$\hat{\Gamma} := \{e,\, \gamma^{i_1}\gamma^{i_3},\, \gamma^{i_2},\, \gamma^{i_1}\gamma^{i_2}\gamma^{i_3}\} = \langle \gamma^{i_1}\gamma^{i_3},\, \gamma^{i_2} \rangle \cong (\mathbb{Z}_2)^2$$

of $\Gamma^I \cong (\mathbb{Z}_2)^3$. Then $(K, H_I, \hat{\Gamma})$ is a $\hat{\Gamma}$-symmetric triple, hence $K/H_I \cong X_I$ has a $\hat{\Gamma}$-symmetric structure. See Example 10 in [**22**] for details.

5. Extrinsically Γ-symmetric spaces

As an extrinsic analog of symmetric spaces, Ferus [**9**] introduced the notion of *symmetric submanifolds* in a Euclidean space. They are also called *extrinsically symmetric spaces.* He classified all extrinsically symmetric spaces by proving that a compact extrinsically symmetric spaces is a standard embedding of a symmetric R-space. Here we extend the definition of extrinsically symmetric spaces to extrinsically Γ-symmetric spaces.

DEFINITION 5.1. Let Γ be a finite abelian group. A connected submanifold M of a Euclidean space $\mathbb{R}^n$ is called an *extrinsically* Γ*-symmetric space* if for each $x \in M$ there exists an injective homomorphism $\varphi_x : \Gamma \to \mathrm{Isom}(\mathbb{R}^n)$; $\gamma \mapsto \gamma_x$ which satisfies the following conditions:

(1) $\gamma_x(M) = M$ for all $x \in M$ and for all $\gamma \in \Gamma$,
(2) for all $\gamma \in \Gamma$, the map $\mu^\gamma : M \times M \to M$; $(x, y) \mapsto \gamma_x(y)$ is smooth, and $\mu = \{\mu^\gamma\}_{\gamma \in \Gamma}$ provides a Γ-symmetric structure on M,
(3) for each $x \in M$, the fixed point set

$$F(\Gamma_x, \mathbb{R}^n) := \{z \in \mathbb{R}^n \mid \gamma_x(z) = z \text{ for all } \gamma \in \Gamma\}$$

is the affine normal subspace $T_x^\perp M$ of M at x.

We consider an R-space which admits the natural Γ-symmetric structure given in the previous section. We now proceed as in [**22**, Section 3.4]. Let I be an admissible subset of I_{reg}. Then (K, H_I, Γ^I) is a Γ^I-symmetric triple and the homogeneous space K/H_I has the Γ^I-symmetric structure. Using the K-equivariant identification

$$\iota : K/H_I \to X_I := \mathrm{Ad}_G(K)\xi_I, \quad kH_I \mapsto \mathrm{Ad}_G(k)\xi_I,$$

the Γ^I-symmetric structure on K/H_I can be transferred to $X_I \subset \mathfrak{p}$ equivariantly by

$$\gamma_{\iota(x)}(\iota(y)) := \iota(\gamma_x(y))$$

for all $x, y \in K/H_I$ and for all $\gamma \in \Gamma^I$. By the definitions (3.2) and (3.3) we have

$$\gamma^i_{\xi_I}(\mathrm{Ad}_G(k)\xi_I) = \mathrm{Ad}_G(\gamma^i(k))\xi_I = \mathrm{Ad}_G(g_i k g_i^{-1})\xi_I = \mathrm{Ad}_G(g_i)\mathrm{Ad}_G(k)\xi_I$$

for $k \in K$ and $i \in I$. Therefore the map $\gamma^i_{\xi_I}$ extends to the linear endomorphism $\mathrm{Ad}_G(g_i)|_{\mathfrak{p}}$ of $\mathfrak{p}$. We endow $\mathfrak{p}$ with the scalar product coming from the Riemannian metric of P. This scalar product is actually the restriction of an $\mathrm{Ad}_G(G)$-invariant scalar product on $\mathfrak{g}$. We set

$$\Gamma^I_{\xi_I} := \{\gamma_{\xi_I} \mid \gamma \in \Gamma^I\} = \langle \gamma^i_{\xi_I} : i \in I \rangle \subset \mathrm{Diff}(X_I), \tag{5.1}$$

$$\Gamma^I_{\xi_I,\mathfrak{p}} := \langle \mathrm{Ad}_G(g_i)|_{\mathfrak{p}} : i \in I \rangle \subset \mathrm{End}(\mathfrak{p}). \tag{5.2}$$

Since I is admissible, by Theorem 4.1 we have

$$\begin{aligned} F(\Gamma^I_{\xi_I,\mathfrak{p}}, \mathfrak{p}) &= \{Y \in \mathfrak{p} \mid \mathrm{Ad}_G(g_i)(Y) = Y \text{ for all } i \in I\} \\ &= \mathfrak{a} \oplus \sum_{\alpha \in \mathcal{R}_I^{\mathrm{ev}}} \mathfrak{p}_\alpha = \mathfrak{a} \oplus \sum_{\alpha \in \mathcal{R}_I^0} \mathfrak{p}_\alpha = \mathfrak{a} \oplus \sum_{\substack{\alpha \in \mathcal{R}^+ \\ \alpha(\xi_I) = 0}} \mathfrak{p}_\alpha \\ &= \{Y \in \mathfrak{p} \mid [Y, \xi_I] = 0\}. \end{aligned} \tag{5.3}$$

Thus $F(\Gamma^I_{\xi_I,\mathfrak{p}}, \mathfrak{p})$ is the normal space of X_I at ξ_I in $\mathfrak{p}$. Therefore an R-spaces X_I with admissible I is an extrinsically Γ-symmetric space.

6. Maximal antipodal sets of Γ-symmetric R-spaces

In 1982, Chen and Nagano [**7**] introduced he notions of antipodal sets and the 2-number for Riemannian manifolds using closed geodesics: In a Riemannian manifold M, two points x and y of M are said to be antipodal if y is the midpoint of a closed geodesic in M emanating from x. A subset $\mathcal{A}$ of M is called an antipodal set if any two points in $\mathcal{A}$ are antipodal, and the supremum cardinality of antipodal sets in M is a geometric invariant of M, called 2-number. In particular, for a compact Riemannian symmetric space M, the definition of an antipodal set is equivalent to a subset $\mathcal{A}$ of M such that the geodesic symmetry s_x fixes every points of $\mathcal{A}$ for every $x \in \mathcal{A}$. Antipodal sets and 2-numbers of compact symmetric spaces were studied in Chen and Nagano [**8**]. In recent years, maximal antipodal sets of compact symmetric spaces and compact Lie groups are investigated in detail by Tanaka and Tasaki ([**28**], [**29**], [**30**]). The paper [**6**] by Chen is a comprehensive survey on 2-numbers and their applications.

The definition of antipodal sets of compact symmetric spaces naturally extends to Γ-symmetric spaces.

DEFINITION 6.1. Let Γ be a finite abelian group, and let $\mu = \{\mu^\gamma\}_{\gamma\in\Gamma}$ be a Γ-symmetric structure on a C^∞-manifold M. A subset $\mathcal{A}$ of the Γ-symmetric space (M, μ) is called *antipodal* if

$$\gamma_x(y) = y \quad \text{for all } x, y \in \mathcal{A} \text{ and for all } \gamma \in \Gamma,$$

equivalently if

$$y \in F(\Gamma_x, M) \quad \text{for all } x, y \in \mathcal{A}.$$

An antipodal set $\mathcal{A}$ of (M, μ) is called *maximal* if $\mathcal{A}$ is not a proper subset of another antipodal set of M. The supremum of the cardinalities of antipodal sets of (M, μ) is called the *antipodal number* denoted by $\#_\Gamma M$. An antipodal set $\mathcal{A}$ of (M, μ) is called *great* if the cardinality of $\mathcal{A}$ is equal to $\#_\Gamma M$.

When M is a compact symmetric space, i.e. $\Gamma = \mathbb{Z}_2$, the antipodal number is originally called the *two-number* by Chen and Nagano [**8**], and denoted by $\#_2 M$.

We shall describe maximal antipodal sets of R-spaces X_I with respect to the natural Γ^I-symmetric structures given in Section 4, when $I \subset I_{\text{reg}}$ is admissible. The fixed point set $F(\Gamma^I_{\xi_I,\mathfrak{p}}, \mathfrak{p})$ is obtained in (5.3). Thus, since the Γ^I-symmetric structure on X_I is K-equivariant, we can verify that $y \in F(\Gamma^I_x, X_I)$ holds for two points $x, y \in X_I$ if and only if $[y, x] = 0$, that is y commutes with x as vectors of the Lie algebra $\mathfrak{g}$.

Let $\mathcal{A}$ be a maximal antipodal set of $X_I \subset \mathfrak{p}$. The linear subspace $\text{span}_\mathbb{R}(\mathcal{A})$ of $\mathfrak{p}$ spanned by $\mathcal{A}$ is abelian. Hence there exists a maximal abelian subspace $\mathfrak{a}'$ of $\mathfrak{p}$ containing $\text{span}_\mathbb{R}(\mathcal{A})$. Then $X_I \cap \mathfrak{a}'$ is an antipodal set of X_I containing $\mathcal{A}$, hence we have $\mathcal{A} = X_I \cap \mathfrak{a}'$, since $\mathcal{A}$ is a maximal antipodal set of X_I. The intersection of an orbit of the s-representation of P and a maximal abelian subspace $\mathfrak{a}' \subset \mathfrak{p}$ is the orbit of the Weyl group $W(P, \mathfrak{a}')$ of P with respect to $\mathfrak{a}'$ (c.f. [**4**, eq. (2.12), p. 1021]). Consequently we have the following description of the structure of maximal antipodal sets of X_I.

THEOREM 6.2 ([**22**]). *Every maximal antipodal set of X_I is of the form $X_I \cap \mathfrak{a}'$ for some maximal abelian subspace $\mathfrak{a}'$ in $\mathfrak{p}$ and therefore an orbit of the Weyl group $W(P, \mathfrak{a}')$ of P with respect to $\mathfrak{a}'$. Moreover, any two maximal antipodal sets of X_I are conjugate by an element of K.*

Theorem 6.2 is a generalization of the result on maximal antipodal sets of symmetric R-spaces by Tanaka and Tasaki [**28**]. Theorem 6.2 shows that the maximal antipodal set of X_I with respect to the natural Γ^I-symmetric structure is unique up to conjugation. In contrast, it is known that some compact symmetric spaces, which are not symmetric R-spaces, may have different kinds of maximal antipodal sets. For example, oriented real Grassmannian manifolds $\widetilde{G}_k(\mathbb{R}^n)$ have several maximal antipodal sets ([**31**]).

As in Theorem 6.2, a maximal antipodal set $\mathcal{A}$ of X_I is the orbit of a Weyl group of P. From the results of Berndt, Console and Fino [**2**] and Sánchez [**23**], we have the following corollary concerning a topological meaning of the antipodal number $\#_{\Gamma^I} X_I$ of X_I, which is the cardinality $|\mathcal{A}|$ of $\mathcal{A}$. That is a generalization of a result of Takeuchi [**26**] for symmetric R-spaces.

COROLLARY 6.3 ([**22**]). *For any maximal antipodal set $\mathcal{A}$ of X_I,*

$$\#_{\Gamma^I} X_I = |\mathcal{A}| = \dim H_*(X_I, \mathbb{Z}_2)$$

holds, that is the antipodal number $\#_{\Gamma'} X_I$ *is the sum of the* $\mathbb{Z}_2$*-Betti numbers of* X_I.

Terauchi [**32**] studied maximal antipodal sets and the antipodal number of a flag manifold with respect to the Γ-symmetric structure explained in Example 4.5.

References

[1] Yuri Bahturin and Michel Goze, *$Z_2 \times Z_2$-symmetric spaces*, Pacific J. Math. **236** (2008), no. 1, 1–21, DOI 10.2140/pjm.2008.236.1. MR2398984

[2] Jürgen Berndt, Sergio Console, and Anna Fino, *On index number and topology of flag manifolds*, Differential Geom. Appl. **15** (2001), no. 1, 81–90, DOI 10.1016/S0926-2245(01)00050-X. MR1845178

[3] Jürgen Berndt, Sergio Console, and Carlos Olmos, *Submanifolds and holonomy*, Chapman & Hall/CRC Research Notes in Mathematics, vol. 434, Chapman & Hall/CRC, Boca Raton, FL, 2003, DOI 10.1201/9780203499153. MR1990032

[4] Raoul Bott and Hans Samelson, *Applications of the theory of Morse to symmetric spaces*, Amer. J. Math. **80** (1958), 964–1029, DOI 10.2307/2372843. MR105694

[5] Francis E. Burstall and John H. Rawnsley, *Twistor theory for Riemannian symmetric spaces*, Lecture Notes in Mathematics, vol. 1424, Springer-Verlag, Berlin, 1990. With applications to harmonic maps of Riemann surfaces, DOI 10.1007/BFb0095561. MR1059054

[6] Bang-Yen Chen, *Two-numbers and their applications—a survey*, Bull. Belg. Math. Soc. Simon Stevin **25** (2018), no. 4, 565–596. MR3896273

[7] Bang-yen Chen and Tadashi Nagano, *Un invariant géométrique riemannien* (French, with English summary), C. R. Acad. Sci. Paris Sér. I Math. **295** (1982), no. 5, 389–391. MR684733

[8] Bang-Yen Chen and Tadashi Nagano, *A Riemannian geometric invariant and its applications to a problem of Borel and Serre*, Trans. Amer. Math. Soc. **308** (1988), no. 1, 273–297, DOI 10.2307/2000963. MR946443

[9] Dirk Ferus, *Symmetric submanifolds of Euclidean space*, Math. Ann. **247** (1980), no. 1, 81–93, DOI 10.1007/BF01359868. MR565140

[10] Michel Goze and Elisabeth Remm, *Riemannian Γ-symmetric spaces*, Differential geometry, World Sci. Publ., Hackensack, NJ, 2009, pp. 195–206, DOI 10.1142/9789814261173_0019. MR2523505

[11] Sigurdur Helgason, *Differential geometry, Lie groups, and symmetric spaces*, Pure and Applied Mathematics, vol. 80, Academic Press, Inc. [Harcourt Brace Jovanovich, Publishers], New York-London, 1978. MR514561

[12] David Joyce, *A classifying invariant of knots, the knot quandle*, J. Pure Appl. Algebra **23** (1982), no. 1, 37–65, DOI 10.1016/0022-4049(82)90077-9. MR638121

[13] S. Kobayashi, T. Nagano, On filtered Lie algebras and geometric structures I, II, J. Math. Mech. **13** (1964), 875–907, **14** (1965) 513–521.

[14] Andreas Kollross, *Exceptional $Z_2 \times Z_2$-symmetric spaces*, Pacific J. Math. **242** (2009), no. 1, 113–130, DOI 10.2140/pjm.2009.242.113. MR2525505

[15] Oldřich Kowalski, *Generalized symmetric spaces*, Lecture Notes in Mathematics, vol. 805, Springer-Verlag, Berlin-New York, 1980. MR579184

[16] A. J. Ledger, *Espaces de Riemann symétriques généralisés* (French), C. R. Acad. Sci. Paris Sér. A-B **264** (1967), A947–A948. MR221435

[17] A. J. Ledger and M. Obata, *Affine and Riemannian s-manifolds*, J. Differential Geometry **2** (1968), 451–459. MR244893

[18] O. Loos, Symmetric spaces I, II, W. A. Benjamin, New York 1969.

[19] Robert Lutz, *Sur la géométrie des espaces Γ-symétriques* (French, with English summary), C. R. Acad. Sci. Paris Sér. I Math. **293** (1981), no. 1, 55–58. MR633562

[20] Tadashi Nagano, *Transformation groups on compact symmetric spaces*, Trans. Amer. Math. Soc. **118** (1965), 428–453, DOI 10.2307/1993971. MR182937

[21] Paola Piu and Elisabeth Remm, *Riemannian symmetries in flag manifolds*, Arch. Math. (Brno) **48** (2012), no. 5, 387–398, DOI 10.5817/AM2012-5-387. MR3007620

[22] Peter Quast and Takashi Sakai, *Natural Γ-symmetric structures on R-spaces* (English, with English and French summaries), J. Math. Pures Appl. (9) **141** (2020), 371–383, DOI 10.1016/j.matpur.2020.02.010. MR4134460

[23] Cristián U. Sánchez, *The index number of an R-space: an extension of a result of M. Takeuchi's*, Proc. Amer. Math. Soc. **125** (1997), no. 3, 893–900, DOI 10.1090/S0002-9939-97-03517-X. MR1343722

[24] Mituhisa Takasaki, *Abstraction of symmetric transformations* (Japanese), Tôhoku Math. J. **49** (1943), 145–207. MR21002

[25] Masaru Takeuchi, *Cell decompositions and Morse equalities on certain symmetric spaces*, J. Fac. Sci. Univ. Tokyo Sect. I **12** (1965), 81–192 (1965). MR216517

[26] Masaru Takeuchi, *Two-number of symmetric R-spaces*, Nagoya Math. J. **115** (1989), 43–46, DOI 10.1017/S0027763000001513. MR1018081

[27] Masaru Takeuchi and Shoshichi Kobayashi, *Minimal imbeddings of R-spaces*, J. Differential Geometry **2** (1968), 203–215. MR239007

[28] Makiko Sumi Tanaka and Hiroyuki Tasaki, *Antipodal sets of symmetric R-spaces*, Osaka J. Math. **50** (2013), no. 1, 161–169. MR3080635

[29] Makiko Sumi Tanaka and Hiroyuki Tasaki, *Maximal antipodal subgroups of some compact classical Lie groups*, J. Lie Theory **27** (2017), no. 3, 801–829. MR3611097

[30] Makiko Sumi Tanaka and Hiroyuki Tasaki, *Maximal antipodal sets of compact classical symmetric spaces and their cardinalities I*, Differential Geom. Appl. **73** (2020), 101682, 32, DOI 10.1016/j.difgeo.2020.101682. MR4149631

[31] Hiroyuki Tasaki, *Antipodal sets in oriented real Grassmann manifolds*, Internat. J. Math. **24** (2013), no. 8, 1350061, 28, DOI 10.1142/S0129167X13500614. MR3103877

[32] Y. Terauchi, Γ 対称空間の対蹠集合 (Japanese), Master thesis, Tokyo Metropolitan University, 2018. http://hdl.handle.net/10748/00010398

Institut für Mathematik, Universität Augsburg, 86135 Augsburg, Germany
Email address: peter.quast@math.uni-augsburg.de

Department of Mathematical Sciences, Tokyo Metropolitan University, 1-1 Minami-Osawa, Hachioji-shi, Tokyo 192-0397, Japan
Email address: sakai-t@tmu.ac.jp

Contemporary Mathematics
Volume **777**, 2022
https://doi.org/10.1090/conm/777/15626

On the first eigenvalue of the p-Laplacian on Riemannian manifolds

Shoo Seto

Abstract. We survey results on the first (nontrivial) eigenvalue of the p-Laplace operator for both the Dirichlet and Neumann/closed condition on Riemannian manifolds. We also discuss an extension of the p-Laplace operator to act on differential forms. Some potential future directions of work are also given.

Contents

1. Introduction

The Laplace operator, denoted by Δ, is a second order elliptic linear operator which has been extensively studied. Of particular interest are its eigenvalues and eigenfunctions on Riemannian manifolds. The eigenvalues (or spectrum) contains much information regarding the geometry of the underlying space and has found applications in many areas of mathematics, physics, statistics, etc. By the spectral theory of compact self-adjoint operators, the (Dirichlet) eigenvalues of the Laplacian are real, positive, and discrete. The Laplace eigenvalue equation has an extremal characterization given as the Euler-Lagrange equation of a normalized energy functional. Let (M^n, g) be an n-dimensional compact Riemannian manifold with boundary ∂M. The normalized Dirichlet energy functional is given by

$$\mathcal{R}[f] := \frac{\int_M |\nabla f|^2}{\int_M |f|^2}$$

for functions $f \in C_0^\infty(M)$. We can also consider this as the Rayleigh quotient of the Laplacian and we get the corresponding Dirichlet eigenvalue equation

$$\begin{cases} \Delta f = -\lambda f & \text{on } M \\ f \equiv 0 & \text{on } \partial M. \end{cases}$$

In particular, the first eigenvalue has the variational characterization

$$\lambda_1 = \inf\{\mathcal{R}[f] \mid f \in C_0^\infty(M)\},$$

and for higher eigenvalues,

$$\lambda_k = \inf\left\{\mathcal{R}[f] \mid f \in C_0^\infty(M), \int_M f f_i = 0, i = 1, \dots, k-1\right\}$$

where f_i are eigenfunctions for λ_i.

Here the discussion has been regarding the Dirichlet eigenvalue. If we prescribe the boundary condition that the outward derivative of the function is zero at the boundary or if the domain is closed (no boundary), we get the Neumann eigenvalues, the first nonzero eigenvalue having the characterization

$$\mu_1 = \inf\left\{\mathcal{R}[f] \mid f \in C^\infty(M), \int_M f = 0\right\}$$

and for higher eigenvalues

$$\mu_k = \inf\left\{\mathcal{R}[f] \mid f \in C^\infty(M), \int_M f f_i = 0, i = 0, 1, \dots, k-1\right\}$$

where f_i are eigenfunctions for μ_i. Note that $\mu_0 = 0$ with constant functions being their eigenfunctions.

We now consider the following generalization. For $p > 1$, define the L^p-Dirichlet energy by

$$\mathcal{R}_p[f] := \frac{\int_M |\nabla f|^p}{\int_M |f|^p}.$$

By calculus of variations, we get the corresponding p-Laplace eigenvalue equation

$$\Delta_p f = -\lambda |f|^{p-2} f. \tag{1}$$

where we define the p-Laplace operator by

$$\Delta_p f := \operatorname{div}(|\nabla f|^{p-2} \nabla f).$$

When f is sufficiently regular,

$$\Delta_p f = |\nabla f|^{p-2} \Delta f + (p-2)|\nabla f|^{p-4} \operatorname{Hess} f(\nabla f, \nabla f). \tag{2}$$

However, the regularity of solutions to (1) is very different when $p = 2$ and $p \neq 2$. When $p = 2$, by elliptic regularity, the solutions are smooth. However when $p \neq 2$, we only have $C^{1,\alpha}$. In fact, when $p > 2$ and on compact manifolds, the nontrivial solutions of (1) are never C^2. If f is C^2, then using (1) and (2),

$$\Delta_p f = |\nabla f|^{p-2} \Delta f + (p-2)|\nabla f|^{p-4} \operatorname{Hess} f(\nabla f, \nabla f) = -\lambda |f|^{p-2} f.$$

but at the maximum and minimum point, $|\nabla f| = 0$, which implies that $f = 0$ everywhere. Consequently, we will be dealing with functions in the Sobolev space $W^{1,p}(M)$ for $p \geq 1$ defined by

$$W^{1,p}(M) = \{f \in L^p(M) \mid |\nabla f| \in L^p(M)\}.$$

The first Dirichlet eigenvalue for the p-Laplacian is given by

$$\lambda_{1,p} = \inf\left\{\mathcal{R}_p[f] \mid f \in W_0^{1,p}(M) \backslash \{0\}\right\}$$

for f which are compactly supported and in the Sobolev space $W^{1,p}$ This is a generalization of the usual Laplace eigenvalue equation ($p = 2$) and was introduced by E. Lieb in [**5**] and independently by F. de Thélin in [**12**].

Similarly, the first Neumann/closed eigenvalue for the p-Laplacian is given by

$$\mu_{1,p} = \inf\left\{\mathcal{R}_p[f] \mid f \in W^{1,p}(M)\backslash\{0\}, \int_M |f|^{p-2}f = 0\right\}.$$

For higher eigenvalues, when $p = 2$ and on compact manifolds, by the spectral theory of compact self-adjoint operators, the eigenvalues are discrete. However the situation drastically changes when $p \neq 2$. While the first eigenvalue is isolated, it is an open problem on whether the higher eigenvalues have finite multiplicity. Furthermore, it is an open problem if the variational characterization give all the possible eigenvalues of the p-Laplace eigenvalue equation. However, in the case of one-dimension, the eigenvalues are all explicitly known.

Computing the eigenvalues explicitly (even for $p = 2$) is impossible except for certain special cases, often involving some form of symmetry. Instead, great efforts are given in obtaining upper and lower bounds as well as the asymptotics of the eigenvalues. These lead to investigations regarding isospectral rigidity properties of the domain. One such example is the case of the sphere, where knowing just the first (non-trivial) eigenvalue leads to a rigidity result (c.f. Theorem 2.2). Obtaining lower bounds for the first eigenvalue of the Laplacian have been a topic of interest and we present some lower bounds for the first eigenvalue of the p-Laplacian which generalize the estimates for the case $p = 2$.

2. Estimates for the first eigenvalue of the p-Laplacian

2.1. Closed manifold case.

2.1.1. *Upper bounds.* By the variational characterization, an upper bound for the first eigenvalue can be obtained by simply computing the Rayleigh quotient $\mathcal{R}_p[f]$ with any function f satisfying the boundary condition(s). For optimal upper bounds, we first introduce a local estimate, the p-Laplacian generalization of Cheng's estimate proved by Matei [**6**] for $p \geq 2$.

THEOREM 2.1. *Let M be an n-dimensional complete Riemannian manifold with* $\mathrm{Ric} \geq (n-1)K$, $K \in \mathbb{R}$. *Then for any $x_0 \in M$, $r_0 \in (0, D)$, $D = \mathrm{diam}(M)$, and $p \geq 2$,*

$$\lambda_{1,p}(B(x_0, r_0)) \leq \lambda_{1,p}(V_n(K, r_0))$$

where $B(x_0, r_0)$ is the geodesic ball of radius r_0 centered at x_0 and $V_n(K, r_0)$ is a geodesic ball of radius r_0 on the n-dimensional simply connected space form with constant sectional curvature K. Equality holds if and only if $B(x_0, r_0)$ is isometric to $V_n(K, r_0)$.

The proof essentially follows from a p-Laplace comparison argument, similar to the $p = 2$ case. This can be extended to the following global estimate.

COROLLARY 2.1. *Let M be an n-dimensional closed Riemannian manifold with* $\mathrm{Ric} \geq (n-1)K > 0$. *Then for any $p \geq 2$,*

$$\mu_{1,p}(M) \leq \lambda_{1,p}(V_n(K, \tfrac{D}{2})).$$

2.1.2. *Lower bounds.* On closed (compact, empty boundary) manifolds, there are two optimal lower bounds. First for positive Ricci curvature lower bound we have

THEOREM 2.2 (Lichnerowicz-Obata). *Let M^n be an n-dimensional closed Riemannian manifold with* $\mathrm{Ric} \geq (n-1)K > 0$. *Then*

$$\mu_1 \geq nK = \mu_1(S^n_K)$$

where S^n_K is the n-dimensional space of constant curvature K.

For non-negative Ricci curvature lower, we have

THEOREM 2.3 (Zhong-Yang). *Suppose* $\mathrm{Ric} \geq 0$ *and* $\mathrm{diam}(M) = D$. *Then*

$$\mu_1 \geq \frac{\pi^2}{D^2}$$

with equality attained on S^1.

The above theorems have been generalized to the p-Laplacian setting. Matei [**6**] showed that the first eigenvalue of the p-Laplacian on closed manifolds with $\mathrm{Ric} \geq (n-1)K$ is bounded below by the first eigenvalue of the p-Laplacian on spheres with radius $\frac{1}{K^2}$.

THEOREM 2.4. *Let M^n be an n-dimensional closed Riemannian manifold with* $\mathrm{Ric} \geq (n-1)K > 0$. *Then for any $p > 1$,*

$$\mu_{1,p} \geq \mu_{1,p}(S^n_K)$$

with equality if and only if M^n is isometric to S^n_K.

Matei also provided a quantitative lower bound however is non-optimal and does not recover the Lichnerowicz-Obata estimate when $p = 2$. In [**10**], the following estimate is given:

THEOREM 2.5. *Let (M^n, g) be a closed Riemannian manifold. For $q > \frac{n}{2}$, $p \geq 2$ and $K > 0$, we have*

$$\mu_{1,p}^{\frac{2}{p}} \geq \frac{\sqrt{n}(p-2)+n}{\sqrt{n}(p-2)+n-1}\frac{(n-1)K}{p-1}. \tag{3}$$

REMARK 2.1. In [**10**], this is proved in a more general context of integral Ricci curvature bound. The bound is most likely not the optimal lower bound when $p \neq 2$ as the eigenvalues of the p-Laplacian for the sphere are not known however does recover the Lichnerowicz-Obata estimate when $p = 2$.

An important tool that used in obtaining the lower bounds is the p-Bochner formula

LEMMA 2.1 (p-Bochner formula).

$$\begin{aligned}\frac{1}{p}\Delta(|\nabla f|^p) =&(p-2)|\nabla f|^{p-2}|\nabla|\nabla f||^2\\ &+ |\nabla f|^{p-2}\{|\operatorname{Hess} f|^2 + \langle \nabla f, \nabla \Delta f\rangle + \mathrm{Ric}(\nabla f, \nabla f)\}.\end{aligned}$$

Integrating the above equation gives us the p-Reilly formula

LEMMA 2.2 (p-Reilly formula).

$$\begin{aligned}(4)\qquad &\int_{\partial M} |\nabla f|^{p-2}\{(\Delta f)\nabla_n f - (\Delta_{\partial M} f)\nabla_n f - H(\nabla_n f)^2 \\ &- \mathrm{II}(\nabla_{\partial M} f, \nabla_{\partial M} f) + \langle(\nabla_n f), \nabla f\rangle_{\partial M}\} \\ &= (p-2)\int_M |\nabla f|^{p-2}|\nabla|\nabla f||^2 - \int_M (\Delta f)(\Delta_p f) \\ &+ \int_M |\nabla f|^{p-2}(|\operatorname{Hess} f|^2 + \operatorname{Ric}(\nabla f, \nabla f)).\end{aligned}$$

REMARK 2.2. In [**1**], we used the above formula along with the Kähler condition to improve the lower bound of the first eigenvalue on Kähler manifolds.

PROOF OF THEOREM 2.5. On closed manifolds, the boundary terms of (4) vanish so that

$$\begin{aligned}0 =&(p-2)\int_M |\nabla f|^{p-2}|\nabla|\nabla f||^2 - \int_M (\Delta f)(\Delta_p f) \\ &+ \int_M |\nabla f|^{p-2}(|\operatorname{Hess} f|^2 + \operatorname{Ric}(\nabla f, \nabla f)).\end{aligned}$$

Let μ be the first eigenvalue and f the first eigenfunction. Then applying (1), integrating by parts and Holder's inequality gives.

$$\int_M (\Delta f)(\Delta_p f) \leq (\mu)^{\frac{2}{p}}(p-1)\int_M |\nabla f|^p.$$

To control the Hessian term, we use the Cauchy-Schwarz inequalities

$$|\operatorname{Hess}(\nabla f, \nabla f)|^2 \leq |\nabla f|^4 |\operatorname{Hess} f|^2, \quad |\Delta f|^2 \leq n|\operatorname{Hess} f|^2.$$

and from the definition of the p-Laplacian

$$\Delta_p f - |\nabla f|^{p-2}\Delta f = (p-2)|\nabla f|^{p-4}\operatorname{Hess} f(\nabla f, \nabla f).$$

Combining the above, we get

$$\int_M |\nabla f|^{p-2}|\operatorname{Hess} f|^2 \geq \frac{1}{n+\sqrt{n}(p-2)}\int_M \Delta f \Delta_p f.$$

Using the Ricci lower bound, we get

$$\int_M |\nabla f|^{p-2}\operatorname{Ric}(\nabla f, \nabla f) \geq (n-1)K\int_M |\nabla f|^p.$$

Since $|\nabla|\nabla f||^2$ is non-negative, we simply drop it. (In [**10**] this term is used in controlling the integral curvature term). Combining the inequalities gives the estimate. □

When the manifold is Kähler, an improvement can be made in the lower bound,

THEOREM 2.6. *Let (M, J, g) be an $n = 2m$ (real) dimensional closed Kähler manifold. Assume the underlying (real) Ricci curvature satisfies* $\operatorname{Ric} \geq (n-1)K$ *for $K > 0$. Then for $p \geq 2$,*

$$(5)\qquad \mu_{1,p}^{\frac{2}{p}} \geq \left(1+\frac{2}{p}\right)\frac{(n-1)K}{p-1}.$$

REMARK 2.3. The improvement in the Kähler setting for the case $p = 2$ was also shown by Lichnerowicz and if equality is achieved in that case, there is a nontrivial holomorphic vector field on M.

To briefly explain the Kähler case, on complex manifolds, we may use the complex structure J to decompose the Hessian into the sum of a J-symmetric bilinear form and a J-skew-symmetric bilinear form:

$$\operatorname{Hess} f = H_1 f + H_2 f$$

where

$$\begin{aligned} H_1 f(X, Y) &= \frac{1}{2}(\operatorname{Hess} f(X, Y) + \operatorname{Hess} f(JX, JY)) \\ H_2 f(X, Y) &= \frac{1}{2}(\operatorname{Hess} f(X, Y) - \operatorname{Hess} f(JX, JY)). \end{aligned}$$

The $(1, 1)$-form associated to $H^1 f$ by the complex structure J is $i\partial\overline{\partial} f$ and $H^2 f$ is called the complex Hessian.

Under this decomposition we have

$$\begin{aligned} 2\|H_1 f\|^2 &= \|\operatorname{Hess} f\|^2 + \langle \operatorname{Hess} f, J^* \operatorname{Hess} f\rangle \\ 2\|H_2 f\|^2 &= \|\operatorname{Hess} f\|^2 - \langle \operatorname{Hess} f, J^* \operatorname{Hess} f\rangle. \end{aligned}$$

Focusing on the term $\langle \operatorname{Hess} f, J^* \operatorname{Hess} f\rangle$, note that this is term which involves four derivatives. After some integration by parts and Ricci commutation, we can bring out a term $\operatorname{Ric}(\nabla f, \nabla f)$, which also involves four derivatives, so that we can use the additional curvature term to give an improvement on the lower bound estimate. We use the decomposition for H_2 since the sign is in the correct direction. See a result by [**14**] for application of the decomposition using H_1 where the Ricci term is canceled instead.

For the Ricci non-negative case, the following theorem by Valtorta [**13**] gives a sharp lower bound

THEOREM 2.7. *Let M be a compact Riemannian manifold with nonnegative Ricci curvature and diameter D, and possibly with convex boundary. Then for any $p \in (1, \infty)$,*

$$\frac{\mu_{1,p}}{p-1} \geq \left(\frac{\pi_p}{D}\right)^p$$

where equality holds only if M is a one-dimensional manifold. Here $\pi_p := \frac{2\pi}{p\sin(\pi/p)}$.

REMARK 2.4. Valtorta's proof uses a gradient estimate technique which compares to the one-dimensional model.

Furthermore, a general lower bound which includes the negative setting is given by a one-dimensional comparison theorem is given by Li and Wang [**4**] in the more general setting of Bakry-Émery manifolds

THEOREM 2.8. *Let (M^n, g, f) be a compact Bakry-Émery manifold with diameter D and $\operatorname{Ric} + \nabla^2 f \geq \kappa g$ for $\kappa \in \mathbb{R}$. Then for $1 < p \leq \infty$,*

$$\mu_{1,p} \geq \tilde{\mu}_{1,p}(\kappa, D)$$

where $\tilde{\mu}_{1,p}(\kappa, D)$ *is the first nonzero Neumann eigenvalue of the one-dimensional problem*

$$(p-1)|\varphi'|^{p-2}\varphi'' - \kappa t|\varphi'|^{p-2}\varphi' = -\tilde{\mu}|\varphi|^{p-2}\varphi$$

on $[-D/2, D/2]$.

The method of proof for the non-positively curved case involves a maximum principle argument. Since the p-Laplacian for $p \neq 2$ is non-linear, we must consider the linearization of Δ_p at a function v given by

$$\begin{aligned} L_v(f) :=& |\nabla v|^{p-2}\Delta f + (p-2)|\nabla v|^{p-4}\operatorname{Hess} f(\nabla v, \nabla v) + (p-2)\Delta_p v\frac{\langle \nabla v, \nabla f\rangle}{|\nabla v|^2} \\ &+ 2(p-2)|\nabla v|^{p-4}\operatorname{Hess} v(\nabla v, \nabla f - \frac{\nabla v}{|\nabla v|}\langle \frac{\nabla v}{|\nabla v|}, \nabla f\rangle). \end{aligned}$$

At points where $\nabla v \neq 0$, the linearized operator is elliptic. We define the second order part to be

$$L_v^{II}(f) := |\nabla v|^{p-2}\Delta f + (p-2)|\nabla v|^{p-4}\operatorname{Hess} f(\nabla v, \nabla v).$$

We then have the corresponding linearized p-Bochner formula

LEMMA 2.3.

$$\begin{aligned} \frac{1}{p}L_f^{II}(|\nabla f|^p) = |\nabla f|^{2p-4}\Bigg(&|\nabla f|^{2-p}\left\{\langle \nabla \Delta_p f, \nabla f\rangle - (p-2)\frac{\operatorname{Hess} f(\nabla f, \nabla f)}{|\nabla f|^2}\Delta_p f\right\} \\ &+|\operatorname{Hess} f|^2 + p(p-2)\left(\frac{\operatorname{Hess} f(\nabla f, \nabla f)}{|\nabla f|^2}\right)^2 + \operatorname{Ric}(\nabla f, \nabla f)\Bigg). \end{aligned}$$

It would be interesting to prove an Zhong-Yang type eigenvalue estimate with integral curvature bounds. The case for $p = 2$ was done in [**8**] and the essential tool was to use an auxiliary function J to control the Ricci curvature term. When applying the maximum principle to an appropriately chosen function, we will need the following.

PROPOSITION 2.1. *Let* $f, g, h \in C^\infty$. *The product rule for* L_f^{II} *is given by*

$$\begin{aligned} L_f^{II}(gh) =& hL_f^{II}[g] + gL_f^{II}[h] \\ &+ 2|\nabla f|^{p-2}\langle \nabla g, \nabla h\rangle + 2(p-2)|\nabla f|^{p-4}\langle \nabla h, \nabla f\rangle\langle \nabla g, \nabla f\rangle. \end{aligned}$$

The auxiliary function will satisfy an elliptic equation involving the linearized p-Laplacian and obtaining estimates to its solution is also of its own independent interest. A similar method could yield a p-Laplace generalization of the Li-Yau inequality for the p-heat equation under integral curvature condition.

2.2. With boundary. When the boundary of the manifold is nonempty, we must impose boundary conditions on our eigenvalue problem. The two most common boundary conditions are

$$\begin{cases} f \equiv 0 & \text{on } \partial M \text{ (Dirichlet)} \\ \nabla_n f \equiv 0 & \text{on } \partial M \text{ (Neumann)}. \end{cases}$$

It is easy to see when $p = 2$ that the first eigenfunction for the Neumann boundary are nonzero constant functions with $\mu_1 = 0$, however in the Dirichlet case, constant functions cannot be the eigenfunction with eigenvalue 0 since the boundary condition will imply that $\Delta f = 0$ only if $f \equiv 0$, which we do not consider an eigenfunction.

Reilly considered the Lichnerowicz-Obata Theorem on compact manifolds with boundary and showed that the lower bound estimate for the first nonzero eigenvalue still hold with the additional assumption that the mean curvature with respect to the outward normal is non-negative. Furthermore he proved the equality case characterizes a hemisphere in S^n_K. By applying the first eigenfunctions to (4), we can obtain the same lower bound, i.e.,

THEOREM 2.9. *Let Ω be a convex domain. Then the same lower bound of* (3) *still hold under Dirichlet and Neumann boundary condition.*

REMARK 2.5. In the Kähler case, it was proved in [**3**] for $p = 2$ that the lower bound given in (5) holds under a convexity assumption and the Dirichlet lower bound. Furthermore, if equality is achieved, then the boundary is totally geodesic. The lower bound estimate also holds when $p > 2$ under the Dirichlet condition. The problem is still open under the Neumann boundary condition. See [**1**] for details.

3. Extension to differential forms

3.1. Closed manifold case. We now consider the Laplacian extended to act on differential forms ω and is called the Hodge-Laplacian:

$$\Delta\omega = dd^*\omega + d^*d\omega$$

where d is the exterior derivative and d^* is its L^2-adjoint. The associated functional is

$$\mathcal{F}[\alpha] = \int_M \|d\alpha\|^2 + \|d^*\alpha\|^2.$$

In [**9**], we give an L^p generalization of the function so that

$$\mathcal{F}_p[\alpha] := \int_M \|d\alpha\|^p + \|d^*\alpha\|^p$$

and taking the variation leads to defining the p-Hodge Laplacian

$$\Delta_p\alpha := d^*(\|d\alpha\|^{p-2}d\alpha) + d(\|d^*\alpha\|^{p-2}d^*\alpha).$$

Note that for k-form α, $\mathcal{F}_p[\alpha] = 0$ if and only if $\alpha \in \mathcal{H}^k(M)$ the space of harmonic k-forms. Consider the space

$$A_k := \left\{\alpha \in \mathcal{W}^{1,p}(\Omega^k(M)) \mid \int_M \|\alpha\|^p = \mathrm{vol}(M), \int_M \|\alpha\|^{p-2}\langle\alpha,\omega\rangle = 0, \omega \in \mathcal{H}^k(M)\right\}.$$

Then the first eigenvalue is given variationally by

$$\mu_1 = \inf\{\mathcal{F}[\alpha] \mid \alpha \in A_k\}.$$

The corresponding Lichnerowicz type estimate is given by the following.

THEOREM 3.1 ([**9**]). *Let M^n be a closed Riemannian manifold with eigenvalues of the curvature operator bounded below by $H \in \mathbb{R}$ and $p \geq 2$. Then*

$$\mu_1 \geq \left(\frac{k(n-k)}{2^{\frac{2}{p}-1}(C + \frac{(p-2)}{2})}H\right)^{\frac{p}{2}},$$

where $C = \max\left\{\frac{k}{k+1}, \frac{n-k}{n-k+1}\right\}$.

REMARK 3.1. For $p = 2$ and $1 \leq k \leq \frac{n}{2}$ the above recovers an estimate due to Gallot-Meyer [**2**]

When working on differential forms, instead of the Ricci tensor, we need to work with the Weitzenböck tensor. Let $p \in M$ and let $\{E_i\}_{i=1}^n$ be an orthonormal frame at p. For $\alpha \in \Omega^k M)$, define

$$W_k(\alpha)(X_1, X_k) := \sum_{i,j} R(E_j, X_i)\alpha(X_1, \dots, E_j, \dots X_k)$$

where R is the Riemann curvature tensor and $X_i \in \mathfrak{X}(M)$. If the eigenvalues of the curvature operator are bounded below by $H \in \mathbb{R}$, then we can show that

$$(W_k(\alpha), \alpha) \geq k(n-k)H\|\alpha\|^2.$$

The generalization of the Bochner formula to forms is the Bochner-Weitzenböck formula

$$\frac{1}{2}\Delta\|\alpha\|^2 = (\Delta\alpha, \alpha) - \|\nabla\alpha\|^2 - (W_k(\alpha), \alpha),$$

here $\Delta := dd^* + d^*d$ is the Hodge-Laplacian. On exact 1-forms $\alpha = df$, since $\nabla df = \operatorname{Hess} f$, we can control $\|\nabla\alpha\|^2$ by the usual Cauchy-Schwarz inequality. For k-forms, we have

LEMMA 3.1. *Let $\alpha \in \Omega^k(M)$, $1 \leq k \leq n-1$. Then*

$$\|\nabla\alpha\|^2 \geq \frac{1}{k+1}\|d\alpha\|^2 + \frac{1}{n-k+1}\|d^*\alpha\|^2.$$

A proof of this inequality can be found in various sources, c.f. [**9**] using twistor operator decomposition, however here we show a straightforward proof given in [**11**]

PROOF. Denote $a_{i_0,i_1,\dots,i_k} = \nabla_{i_0}\alpha_{i_1,\dots,i_k}$. Let $i_0, i_1, \dots, i_k$ be indices such that no two are the same. Using normal coordinates, we have

$$\begin{aligned} k!\|\nabla\alpha\|^2 &= \sum_{i_1,\dots,i_k}\left(\sum_{j\neq i_1,\dots,i_k}(a_{j,i_1,\dots,i_k})^2 + \sum_{s=1}^{k}(a_{i_s,i_1,\dots,i_k})^2\right) \\ &= \frac{1}{k+1}\sum_{i_0,\dots,i_k}\sum_{s=0}^{k}(a_{i_s,i_0,\dots,\hat{i}_s,\dots,i_k})^2 + k\sum_{i_2,\dots i_k}\sum_{j\neq i_2,\dots,i_k}(a_{j,j,i_2,\dots,i_k})^2, \end{aligned}$$

where in the first line, we split the sum over terms where the first index is not in the indices $i_1, \dots, i_k$ and term which are included. In the second line, for the first term we include j in the set of indices however there are $k+1$ many ways to do so and the second term we match the first and second indices. Using the Cauchy-Schwarz inequality, we have

$$\sum_{s=0}^{k}(a_{i_s,i_0,\dots,\hat{i}_s,\dots i_k})^2 \geq \frac{1}{k+1}\left(\sum_{s=0}^{k}(-1)^s a_{i_s,i_0,\dots,\hat{i}_s,\dots,i_k}\right)^2 = k!\|d\alpha\|^2$$

and

$$\sum_{j\neq i_2,\dots,i_k}(a_{j,j,i_2,\dots,i_k})^2 \geq \frac{1}{n-k+1}\left(-\sum_j a_{j,j,i_2,\dots,i_k}\right)^2 = \frac{(k-1)!}{(n-k+1)}\|d^*\alpha\|^2$$

and the result follows. □

3.2. With boundary. In this section we briefly discuss the situation of a compact manifold M with nonempty smooth boundary ∂M. Let n denote the unit outer normal vector and let $I : \partial M \to M$ be the inclusion. Then $I^*\alpha$ is the restriction of a form to the boundary. Then d and its adjoint d^* are related with an additional boundary term given by

$$\int_M \langle d\alpha, \beta\rangle = \int_M \langle \alpha, d^*\beta\rangle + \int_{\partial M} \langle I^*(\alpha), \iota_n\beta\rangle, \quad \alpha \in \Omega^k(M), \beta \in \Omega^{k+1}(M).$$

and the corresponding Green's formula for the p-Laplacian is

$$\begin{aligned}(\Delta_p\alpha, \beta) = &\int_M \|d\alpha\|^{p-2}\langle d\alpha, d\beta\rangle + \int_M \|d^*\alpha\|^{p-2}\langle d^*\alpha, d^*\beta\rangle \\ &- \int_{\partial M} \langle \iota_n(\|d\alpha\|^{p-2}d\alpha), I^*(\beta)\rangle + \int_{\partial M} \langle \|d^*\alpha\|^{p-2} I^*(d^*\alpha), \iota_n\beta\rangle.\end{aligned}$$

The two most common boundary conditions for the classical Laplacian eigenvalue problem are the Dirichlet and Neumann boundary condition. For the Hodge-Laplacian, the analogous boundary conditions are the absolute boundary condition

$$\begin{cases} \iota_n\alpha = 0 \\ \iota_n d\alpha = 0, \quad \text{on } \partial M \end{cases}$$

and the relative boundary condition

$$\begin{cases} I^*(\alpha) = 0 \\ I^*(d^*\alpha) = 0, \quad \text{on } \partial M. \end{cases}$$

The essential feature of the boundary condition is that if α satisfies either of the boundary conditions, then $\Delta_p\alpha = 0$ implies $d\alpha = 0$ and $d^*\alpha = 0$. The new boundary terms that will be introduced when integrating the Bochner-Weitzenböck formula are

$$\begin{aligned}&\int_M \langle d(\|\alpha\|^{p-2}) \wedge \alpha, d\alpha\rangle - \int_M \langle \iota_{\nabla\|\alpha\|^{p-2}}\alpha, d^*\alpha\rangle + \int_M \|\alpha\|^{p-2}\|d\alpha\|^2 + \int_M \|\alpha\|^{p-2}\|d^*\alpha\|^2 \\ &- \int_{\partial M} \|\alpha\|^{p-2}\langle I^*(\alpha), \iota_n(d\alpha)\rangle + \int_{\partial M} \|\alpha\|^{p-2}\langle I^*(d^*\alpha), \iota_n(\alpha)\rangle \\ &= \int_M \left((p-2)\|\alpha\|^{p-2}|\nabla\|\alpha\||^2 + \|\alpha\|^{p-2}\|\nabla\alpha\|^2 + \|\alpha\|^{p-2}(W_k(\alpha), \alpha)\right).\end{aligned}$$

Since the boundary terms will vanish under either of the boundary conditions, we get the same estimate for the boundary value problem as well. It would be interesting to see what the Reilly formula, for instance a generalization of Theorem 3 in [**7**] would be in this context. However due to the asymmetry of the weight function in the p-Laplacian, it is not immediate what the appropriate Bochner-Weitzenböck type formula would be for Δ_p.

References

[1] C. Blacker and S. Seto, *First eigenvalue of the p-Laplacian on Kähler manifolds*, Proc. Amer. Math. Soc. **147** (2019), no. 5, 2197–2206, DOI 10.1090/proc/14395. MR3937693

[2] S. Gallot and D. Meyer, *Sur la première valeur propre du p-spectre pour les variétés à opérateur de courbure positif* (French), C. R. Acad. Sci. Paris Sér. A-B **276** (1973), A1619–A1621. MR322735

[3] V. Guedj, B. Kolev, and N. Yeganefar, *A Lichnerowicz estimate for the first eigenvalue of convex domains in Kähler manifolds*, Anal. PDE **6** (2013), no. 5, 1001–1012, DOI 10.2140/apde.2013.6.1001. MR3125547

[4] X. Li and K. Wang, *Sharp lower bound for the first eigenvalue of the Weighted p-Laplacian II*, to appear in Mathematical Research letters (2020).

[5] E. H. Lieb, *On the lowest eigenvalue of the Laplacian for the intersection of two domains*, Invent. Math. **74** (1983), no. 3, 441–448, DOI 10.1007/BF01394245. MR724014

[6] A.-M. Matei, *First eigenvalue for the p-Laplace operator*, Nonlinear Anal. **39** (2000), no. 8, Ser. A: Theory Methods, 1051–1068, DOI 10.1016/S0362-546X(98)00266-1. MR1735181

[7] S. Raulot and A. Savo, *A Reilly formula and eigenvalue estimates for differential forms*, J. Geom. Anal. **21** (2011), no. 3, 620–640, DOI 10.1007/s12220-010-9161-0. MR2810846

[8] X. Ramos Olivé, S. Seto, G. Wei, and Q. S. Zhang, *Zhong-Yang type eigenvalue estimate with integral curvature condition*, Math. Z. **296** (2020), no. 1-2, 595–613, DOI 10.1007/s00209-019-02448-w. MR4140755

[9] S. Seto, *The first nonzero eigenvalue of the p-Laplacian on differential forms*, Pacific J. Math. **309** (2020), no. 1, 213–222, DOI 10.2140/pjm.2020.309.213. MR4202009

[10] S. Seto and G. Wei, *First eigenvalue of the p-Laplacian under integral curvature condition*, Nonlinear Anal. **163** (2017), 60–70, DOI 10.1016/j.na.2017.07.007. MR3695968

[11] S.-i. Tachibana, *On conformal Killing tensor in a Riemannian space*, Tohoku Math. J. (2) **21** (1969), 56–64, DOI 10.2748/tmj/1178243034. MR242078

[12] F. de Thélin, *Quelques résultats d'existence et de non-existence pour une EDP elliptique non linéaire* (French, with English summary), C. R. Acad. Sci. Paris Sér. I Math. **299** (1984), no. 18, 911–914. MR774666

[13] D. Valtorta, *Sharp estimate on the first eigenvalue of the p-Laplacian*, Nonlinear Anal. **75** (2012), no. 13, 4974–4994, DOI 10.1016/j.na.2012.04.012. MR2927560

[14] X. Wang, *An integral formula in Kähler geometry with applications*, Commun. Contemp. Math. **19** (2017), no. 5, 1650063, 12, DOI 10.1142/S0219199716500632. MR3670795

Department of Mathematics, California State University, Fullerton, California 92831

Email address: shoseto@fullerton.edu

Contemporary Mathematics
Volume **777**, 2022
https://doi.org/10.1090/conm/777/15632

Polars of disconnected compact Lie groups

Makiko Sumi Tanaka and Hiroyuki Tasaki

Dedicated to the memory of Professor Tadashi Nagano

ABSTRACT. A compact Lie group is a Riemannian symmetric space with respect to a biinvariant Riemannian metric. We give an explicit description of the polars of a compact Lie group. We show that if a connected component of a compact Lie group contains a polar, it generates a subgroup isomorphic to a certain semidirect product. We determine the polars of some disconnected compact Lie groups.

1. Introduction

A compact Lie group has a biinvariant Riemannian metric, with respect to which it is a Riemannian symmetric space. Chen-Nagano [**CN1**] introduced the notion of a polar and investigated polars of connected Riemannian symmetric spaces. In [**CN2**] they also introduced the notion of an antipodal set and studied the maxima of the cardinalities of antipodal sets, called the 2-numbers. In [**TT3**] the authors classified maximal antipodal sets of some classical compact Riemmanian symmetric spaces M by using a realization of M as a polar of a connected compact Lie group and a classification of maximal antipodal subgroups of compact classical Lie groups obtained in [**TT2**]. In order to continue the classification of maximal antipodal sets for some other classical compact Riemmanian symmetric spaces M, we need a realization of M as a polar of a disconnected compact Lie group. Therefore we study general theory of polars of disconnected compact Lie groups in this paper.

The present paper is organized as follows. In Section 2 we give an explicit description of the polars as well as the poles of compact Lie groups. In Section 3 we show that if a connected component of a disconnected compact Lie group contains a polar, the union of the identity component and the connected component is a subgroup. Since the subgroup is equipped the structure of a semidirect product, we treat semidirect products in Section 4. We give an explicit description of the fixed point set of the symmetry at the identity element of a semideirec product $G \rtimes \langle\sigma\rangle$ of a connected compact Lie group G and the subgroup of the automorphism

2020 *Mathematics Subject Classification.* Primary 53C35; Secondary 53C40.

Key words and phrases. Symmetric space, compact Lie group, polar.

The first author was partly supported by the Grant-in-Aid for Science Research (C) 2019 (No. 19K03478), Japan Society for the Promotion of Science. The second author was partly supported by the Grant-in-Aid for Science Research (C) 2018 (No. 18K03268), Japan Society for the Promotion of Science.

group of G generated by an involutive automorphism σ of G. In Section 5 we study polars of the orthogonal groups, the unitary groups and the symplectic groups as well as polars of semidirect products defined by each of them. In particular, we give realizations, as a polar of a semidirect product, of compact Riemannian symmetric spaces $U(n)/O(n)$, $SU(n)/SO(n)$, $U(2n)/Sp(n)$, and $SU(2n)/Sp(n)$ each of which is not realized as a polar of a connected compact Lie group.

2. Compact Lie groups

Let G be a compact Lie group. There exists a biinvariant Riemannian metric on G, with respect to which G is a Riemannian symmetric space. For each point $x \in G$ the symmetry at x, denoted by s_x, satisfies $s_x(y) = xy^{-1}x$ $(y \in G)$. The symmetry s_x is uniquely determined on the connected component of G which contains x since s_x is an isometry fixing x but it is not uniquely determined on the other connected components. However, since $s_x(y)$ is expressed by operations and the inverse of each y of the connected component, it can be naturally defined on G even if G is not connected.

Let e denote the identity element of G. A *polar* of G (with respect to e) is defined as a connected component of the fixed point set $F(s_e, G)$ of s_e in G. A polar which consists of a single point is called a *pole* of e. We also call the point a pole of e. Since $s_e(y) = y$ if and only if $y^2 = e$, we have $F(s_e, G) = \{y \in G \mid y^2 = e\}$. Thus a polar of G is a connected component of the set of involutive elements of G. We denote by $Z(G)$ the center of G and we set

$$Z_2(G) = Z(G) \cap F(s_e, G).$$

LEMMA 2.1. *$Z_2(G)$ is a finite abelian subgroup of G and $Z_2(G)$ is isomorphic to a product of some copies of $\mathbb{Z}_2$.*

PROOF. Since $Z_2(G)$ is a subset of $Z(G)$, any two elements of $Z_2(G)$ commute. Thus for $z_1, z_2 \in Z_2(G)$ we have $(z_1z_2)^2 = z_1^2z_2^2 = e$ and hence $z_1z_2 \in Z_2(G)$. If $z \in Z_2(G)$, $z^{-1} = z \in Z_2(G)$ since $z^2 = e$. Therefore $Z_2(G)$ is a subgroup of $Z(G)$, which is an abelian subgroup. Since $Z(G)$ and $F(s_e, G)$ are closed in G, $Z_2(G)$ is closed and hence $Z_2(G)$ is a closed Lie subgroup of G. Let $\mathfrak{g}$ be the Lie algebra of G. There exists an open neighborhood U of $0 \in \mathfrak{g}$ on which the exponential map from $\mathfrak{g}$ to G gives a diffeomorphism. We take $V = \frac{1}{2}U$ and let $X \in V$ be an element of the Lie algebra of $Z_2(G)$. Then $(\exp X)^2 = \exp 2X = e$ which implies $X = 0$ since $2X \in U$. Therefore $\dim Z_2(G) = 0$ and hence $Z_2(G)$ is a finite abelian subgroup since G is compact. By the fundamental theorem of finite abelian groups $Z_2(G)$ is isomorphic to a product of some copies of $\mathbb{Z}_2$. □

We denote by G_0 the identity component of G. Let $Z_G(G_0)$ denote the centralizer of G_0 in G and we set $\tilde{Z}_2(G) = Z_G(G_0) \cap F(s_e, G)$. Then $Z_2(G) \subset \tilde{Z}_2(G)$.

PROPOSITION 2.2. *The set of poles of G coincides with $\tilde{Z}_2(G)$. Any polar of G is described as $G^+(x) := \{gxg^{-1} \mid g \in G_0\}$ for a point x belonging to it.*

Before we prove the proposition we state the following lemma.

LEMMA 2.3 (cf. Theorem 5.44(ii) in [**HM**]). *Let H be a Lie group and we denote by $\mathfrak{h}$ the Lie algebra of H. Let H_0 be the identity component of H. Then the centralizer of H_0 in H coincides with the kernel of the adjoint representation* $\mathrm{Ad} : H \to GL(\mathfrak{h})$.

PROOF OF PROPOSITION 2.2. Let U be an open neighborhood of $0 \in \mathfrak{g}$ on which the exponential map from $\mathfrak{g}$ to G gives a diffeomorphism. For $x \in F(s_e, G)$, $x \exp U$ is an open neighborhood of x. We take U sufficiently small so that $U = -U$ and $\mathrm{Ad}(x)U = U$ hold. For $X \in U$, $x \exp X \in F(s_e, G)$ if and only if $e = (x \exp X)^2 = x \exp X x \exp X = x \exp X x^{-1} \exp X = \exp \mathrm{Ad}(x) X \exp X$, which is equivalent to $\exp \mathrm{Ad}(x)X = \exp(-X)$, and moreover, it is equivalent to $\mathrm{Ad}(x)X = -X$. Therefore the tangent space at x of the connected component of $F(s_e, G)$ coincides with $dL_x\{X \in \mathfrak{g} \mid \mathrm{Ad}(x)X = -X\}$, where L_x denotes the left translation by x and dL_x denotes its differential. Since $\mathrm{Ad}(x)^2$ is the identity map on $\mathfrak{g}$ by $x^2 = e$, each eigenvalue of $\mathrm{Ad}(x)$ is ± 1. Thus x is isolated in $F(s_e, G)$ if and only if 1 is the unique eigenvalue of $\mathrm{Ad}(x)$, that is, $\mathrm{Ad}(x)$ is the identity map of $\mathfrak{g}$, which is equivalent to $x \in Z_G(G_0)$ by Lemma 2.3. Therefore the set of isolated points of $F(s_e, G)$ coincides with $\tilde{Z}_2(G) = Z_G(G_0) \cap F(s_e, G)$.

Let M be a polar of G and x be an element of M. We obtain

$$M = \{x \exp X \mid X \in \mathfrak{g}, \mathrm{Ad}(x)X = -X\}$$

by the argument above. If $x \exp X \in M$, we have $x \exp X = x \exp \frac{1}{2}X \exp \frac{1}{2}X = \exp \frac{1}{2}\mathrm{Ad}(x)X x \exp \frac{1}{2}X = \exp(-\frac{1}{2}X) x \exp \frac{1}{2}X$. Thus $x \exp X$ is an element of $G^+(x)$ and hence $M \subset G^+(x)$. On the other hand, since $G^+(x)$ is a connected subset of $F(s_e, G)$, we obtain $M = G^+(x)$. □

By Proposition 2.2 we obtain the following corollaries.

COROLLARY 2.4. *There exit $x_1, \dots, x_k \in F(s_e, G)$ such that $G^+(x_i) \cap G^+(x_j) = \emptyset$ when $i \neq j$ and*

$$F(s_e, G) = \tilde{Z}_2(G) \cup \bigcup_{i=1}^{k} G^+(x_i).$$

In particular, $\tilde{Z}_2(G)$ is the set of all poles and $G^+(x_i)\,(i = 1, \dots, k)$ are all polars of positive dimension.

COROLLARY 2.5. *For $x \in F(s_e, G)$, the tangent space $T_xG^+(x)$ of $G^+(x)$ at x is given by*

$$T_xG^+(x) = dL_x\{X \in \mathfrak{g} \mid \mathrm{Ad}(x)X = -X\}.$$

In particular, $\mathrm{Ad}(x)$ is the identity map if and only if x is a pole.

The *meridian* corresponding to $G^+(x)$, denoted by $G^-(x)$, is the connected component of $F(s_x \circ s_e, G)$ through x.

COROLLARY 2.6. *For $x \in F(s_e, G)$, the meridian $G^-(x)$ corresponding to $G^+(x)$ coincides with $xZ_G(x)_0$, where $Z_G(x)_0$ denotes the identity component of the centralizer $Z_G(x)$ of x in G.*

PROOF. Since the differential of $s_x \circ s_e$ at x is -1 times the identity on $T_xG^+(x)$ and the identity on the orthogonal complement $T_x^\perp G^+(x)$ to $T_xG^+(x)$ in T_xG. We obtain

$$T_xG^-(x) = T_x^\perp G^+(x) = dL_x\{X \in \mathfrak{g} \mid \mathrm{Ad}(x)X = X\}$$

by Corollary 2.5 and $\mathrm{Ad}(x)^2 = \mathrm{id}_{\mathfrak{g}}$. Thus $T_xG^-(x)$ coincides with $T_xxZ_G(x)_0$. Since $G^-(x)$ and $xZ_G(x)$ are totally geodesic submanifolds, we obtain $G^-(x) = xZ_G(x)_0$ by the connectivity. □

Some of the assertions of the following proposition are stated in the case where G is connected in [**CN2**].

PROPOSITION 2.7. *Let G be a compact Lie group and let G_0 be the identity component of G. Then the following hold for $g \in G$.*

(1) *If $s_g = s_e$, g is a pole of e.*
(2) *If g is a pole of e, $s_g = s_e$ holds on G_0.*
(3) *If $g \in Z(G)$ and g is a pole of e, $s_g = s_e$ holds on G.*

PROOF. (1) Since g is an isolated fixed point of s_g and $s_g = s_e$, we have g is an isolated fixed point of s_e and hence g is a pole of e. (2) By Corollary 2.4 we have $g \in \tilde{Z}_2(G)$. For each $x \in G_0$ we have $s_g(x) = gx^{-1}g = g^2x^{-1} = x^{-1} = s_e(x)$ since $x^{-1} \in G_0$. Thus $s_g = s_e$ on G_0. (3) Since $g \in Z(G)$, for each $x \in G$ we obtain $s_g(x) = s_e(x)$ as mentioned above. Thus $s_g = s_e$ on G. □

By Corollary 2.4 we obtain the following in the case where a compact Lie group G is connected.

THEOREM 2.8. *Let G be a connected compact Lie group and denote by e the identity element of G. There exist $x_1, \ldots, x_k \in F(s_e, G)$ such that $G^+(x_i) \cap G^+(x_j) = \emptyset$ when $i \neq j$ and*

$$F(s_e, G) = Z_2(G) \cup \bigcup_{i=1}^{k} G^+(x_i).$$

The right-hand side is the decomposition into a disjoint union of the polars of G and $Z_2(G)$ is the set of poles of G.

We refer to [**N**] for explicit description of the polars of each connected compact Lie group.

EXAMPLE 2.9. Let $O(n)$ and $SO(n)$ denote the orthogonal group and the special orthogonal group of degree n respectively. We denote by 1_n the identity matrix of size n. The centers of $O(n)$ and $SO(n)$, and the centralizer of $SO(n)$ in $O(n)$ are the following:

$$Z(O(n)) = \{\pm 1_n\},$$

$$Z_{O(n)}(SO(n)) = \begin{cases} SO(2) & (n = 2), \\ \{\pm 1_n\} & (n \neq 2), \end{cases}$$

$$Z(SO(n)) = \begin{cases} SO(2) & (n = 2), \\ \{\pm 1_n\} & (n : \text{even}, n \neq 2), \\ \{1_n\} & (n : \text{odd}). \end{cases}$$

Hence we get

$$Z_2(O(n)) = \tilde{Z}_2(O(n)) = \{\pm 1_n\},$$

$$Z_2(SO(n)) = \begin{cases} \{\pm 1_n\} & (n : \text{even}), \\ \{1_n\} & (n : \text{odd}). \end{cases}$$

Therefore the poles of $O(n)$ and $SO(n)$ $(n : \text{even})$ are $\pm 1_n$ and the pole of $SO(n)$ $(n : \text{odd})$ is 1_n by Corollary 2.4.

EXAMPLE 2.10. Let $U(n)$ and $SU(n)$ denote the unitary group and the special unitary group of degree n respectively. We define $U^{pm}(n), U^+(n), U^-(n)$ as follows:

$$\begin{aligned} U^{pm}(n) &= \{g \in U(n) \mid \det g \in \{\pm 1\}\}, \\ U^+(n) &= \{g \in U(n) \mid \det g = 1\} = SU(n), \\ U^-(n) &= \{g \in U(n) \mid \det g = -1\}. \end{aligned}$$

Then $U^{pm}(n)$ and $U^+(n)$ are Lie subgroups of $U(n)$ and $U^{pm}(n) = U^+(n) \cup U^-(n)$ is the decomposition into a disjoint union of the connected components. We get

$$\begin{aligned} Z(U^{pm}(n)) &= \{z1_n \mid z^{2n} = 1\} \cong \mathbb{Z}_{2n}, \\ Z_{U^{pm}(n)}(U^+(n)) &= \{z1_n \mid z^{2n} = 1\} \cong \mathbb{Z}_{2n}, \\ Z(U^+(n)) &= \{z1_n \mid z^n = 1\} \cong \mathbb{Z}_n. \end{aligned}$$

Hence we obtain

$$\begin{aligned} Z_2(U^{pm}(n)) &= \tilde{Z}_2(U^{pm}(n)) = \{\pm 1_n\}, \\ Z_2(U^+(n)) &= \begin{cases} \{\pm 1_n\} & (n : \text{even}), \\ \{1_n\} & (n : \text{odd}). \end{cases} \end{aligned}$$

Therefore the poles of $U^{pm}(n)$ and $SU(n)\,(n : \text{even})$ are $\pm 1_n$ and the pole of $SU(n)\,(n : \text{odd})$ is 1_n by Corollary 2.4.

3. Disconnected compact Lie groups

In this section we investigate polars of disconnected compact Lie groups.

Let G be a disconnected compact Lie group and let

$$G = \bigcup_{\lambda \in \Lambda} G_\lambda$$

be the decomposition into a disjoint union of the connected components of G. We assume $0 \in \Lambda$ and let G_0 be the identity component of G.

We introduce two actions which we will use in the following. For an element x of a Lie group G, we define $I_x : G \to G$ by $I_x(g) = xgx^{-1}$ $(g \in G)$. I_x preserves the identity component G_0 and defines an automorphism of G_0. For an automorphism σ of G, we define $\rho_\sigma(g) : G \to G$ by $\rho_\sigma(g)(h) = gh\sigma(g)^{-1}$ $(h \in G)$ for $g \in G$. We call the action ρ_σ the twisted conjugate action by σ, which is called σ-action in [**I**].

In order to determine the polars of G, it is enough to determine each connected component of $G_\lambda \cap F(s_e, G)$ for $\lambda \in \Lambda$ which satisfies $G_\lambda \cap F(s_e, G) \neq \emptyset$. If $G_\lambda \cap F(s_e, G) \neq \emptyset$, we take a point $x_\lambda \in G_\lambda \cap F(s_e, G)$. Then $G_\lambda = G_0 x_\lambda$ holds. Let T_λ be a maximal torus of $F(I_{x_\lambda}, G_0)$. The following lemma is useful for disconnected compact Lie groups and we used it in [**TT1**].

The following Lemma 3.1 and its proof depend on personal communication with Osamu Ikawa.

LEMMA 3.1.

$$G_\lambda = \bigcup_{g \in G_0} g(x_\lambda T_\lambda)g^{-1}.$$

PROOF. We denote I_{x_λ} by σ. Let θ_1 and θ_2 be involutive automorphisms of $G_0 \times G_0$ defined by $\theta_1(g,h) = (\sigma(h), \sigma(g))$ and $\theta_2(g,h) = (h,g)$ for $g, h \in G_0$. Then $\theta_1 \circ \theta_2 = \theta_2 \circ \theta_1$. We have $F(\theta_1, G_0 \times G_0) = \{(g, \sigma(g)) \mid g \in G_0\}$, which we denote by F_{θ_1}, and $F(\theta_2, G_0 \times G_0) = \{(g,g) \mid g \in G_0\}$, which we denote by ΔG_0. Then $(G_0 \times G_0, F_{\theta_1})$ and $(G_0 \times G_0, \Delta G_0)$ are symmetric pairs and the latter defines a compact symmetric space G_0. Under the correspondence $G_0 \times G_0/\Delta G_0 \ni (x,y)\Delta G_0 \mapsto xy^{-1} \in G_0$, the Hermann action defined by $F_{\theta_1} \cong G_0$ coincides with the twisted conjugate action ρ_σ. By a property of Hermann actions (cf. [**I**] Equation (1.2)) $G_0 = \bigcup_{g \in G_0} \rho_\sigma(g)(T_\lambda)$ holds. We have $\rho_\sigma(g)(T_\lambda) = gT_\lambda\sigma(g)^{-1} = gT_\lambda I_{x_\lambda}(g)^{-1} = gT_\lambda x_\lambda g^{-1}x_\lambda = gx_\lambda x_\lambda T_\lambda x_\lambda g^{-1}x_\lambda = gx_\lambda T_\lambda g^{-1}x_\lambda$ since $x_\lambda^2 = e$ and $T_\lambda \subset F(I_{x_\lambda}, G_0)$. Hence we get $G_0 x_\lambda = \bigcup_{g \in G_0} g(x_\lambda T_\lambda)g^{-1}$. □

By this lemma we obtain

$$\begin{aligned} G_\lambda \cap F(s_e, G) &= \bigcup_{g \in G_0} g(x_\lambda T_\lambda)g^{-1} \cap F(s_e, G) = \bigcup_{g \in G_0} g(x_\lambda T_\lambda \cap F(s_e, G))g^{-1} \\ &= \bigcup_{g \in G_0} g\{x \in x_\lambda T_\lambda \mid x^2 = e\}g^{-1}. \end{aligned}$$

Therefore in order to determine $G_\lambda \cap F(s_e, G)$, it is enough to determine $\{x \in x_\lambda T_\lambda \mid x^2 = e\}$ and conjugacy classes of each element of the set.

PROPOSITION 3.2. *As an element of the quotient group G/G_0, the order of G_λ is two if $G_\lambda \cap F(s_e, G) \neq \emptyset$. Moreover, if $G_\lambda \cap F(s_e, G) \neq \emptyset$, $G_0 \cup G_\lambda$ is a subgroup of G.*

PROOF. Assume $G_\lambda \cap F(s_e, G) \neq \emptyset$ and let $x_\lambda \in G_\lambda \cap F(s_e, G)$. Since $G_\lambda = G_0 x_\lambda$, we have $G_\lambda G_\lambda = G_0 x_\lambda G_0 x_\lambda = G_0 G_0 = G_0$. Thus as an element of the quotient group G/G_0, the order of G_λ is two. The latter part is clear. □

In order to investigate the structure of $G_0 \cup G_\lambda$ we use the notion of semidirect product, which we will treat in the next section.

4. Semidirect products

In this section we review the definition and some properties of the semidirect product. After that we describe the structure of $G_0 \cup G_\lambda$ mentioned in Proposition 3.2 using the semidirect product.

Let G be a Lie group and let H be a discrete topological group. We denote by $\mathrm{Aut}(G)$ the automorphism group of G. Let $\sigma : H \to \mathrm{Aut}(G)$ be a group homomorphim. We define a group operation on the direct product $G \times H$ by

$$(g_1, h_1)(g_2, h_2) = (g_1\sigma(h_1)(g_2), h_1h_2) \quad (g_1, g_2 \in G, h_1, h_2 \in H).$$

It turns out that $G \times H$ is a Lie group with respect to the group operation. This Lie group is called a *semidirect product* of G and H with respect to σ and denoted by $G \rtimes_\sigma H$. We write an element of $G \rtimes_\sigma H$ determined by $(g,h) \in G \times H$ as $(g,h)_\sigma$. When we denote the identity elements of G and H by e_G and e_H respectively, the identity element of $G \rtimes_\sigma H$ is $(e_G, e_H)_\sigma$. The inverse element of $(g,h)_\sigma$ $(g \in G, h \in H)$ is given by $(g,h)_\sigma^{-1} = (\sigma(h^{-1})(g^{-1}), h^{-1})_\sigma$. When H is a subgroup of $\mathrm{Aut}(G)$ and $\sigma : H \to \mathrm{Aut}(G)$ is the inclusion map, we denote $G \rtimes_\sigma H$ (resp. $(g,h)_\sigma$) by $G \rtimes H$ (resp. (g,h)) simply.

Let G be a disconnected compact Lie group and let

$$G = \bigcup_{\lambda \in \Lambda} G_\lambda$$

be the decomposition into a disjoint union of the connected components of G. We assume $0 \in \Lambda$ and let G_0 be the identity component of G. If $G_\lambda \cap F(s_e, G) \neq \emptyset$, $G_0 \cup G_\lambda$ is a subgroup of G by Proposition 3.2. Moreover, we have the following.

PROPOSITION 4.1. *If $G_\lambda \cap F(s_e, G) \neq \emptyset$, $G_0 \cup G_\lambda$ is isomorphic to $G_0 \rtimes \langle I_{x_\lambda} \rangle$ as a Lie group for any $x_\lambda \in G_\lambda \cap F(s_e, G)$, where $\langle I_{x_\lambda} \rangle$ denotes the subgroup of* $\mathrm{Aut}(G_0)$ *generated by I_{x_λ}.*

PROOF. We take $x_\lambda \in G_\lambda \cap F(s_e, G)$. We define a map $\psi : G_0 \rtimes \langle I_{x_\lambda} \rangle \to G_0 \cup G_\lambda$ by $\psi(g, e) = g$, $\psi(g, I_{x_\lambda}) = g x_\lambda$ for $g \in G_0$. Then we can show that ψ is a Lie group isomorphism. □

By this proposition the determination of the polars of G is reduced to the determination of the polars of $G_0 \rtimes \langle I_{x_\lambda} \rangle$ for $x_\lambda \in G_\lambda \cap F(s_e, G) \neq \emptyset$.

Since the map $G \to G \rtimes_\sigma H$ defined by $g \mapsto (g, e_H)_\sigma$ is an injective homomorphism, the image $(G, e_H)_\sigma$ of the map is isomorphic to G and it is a Lie subgroup of $G \rtimes_\sigma H$. Moreover, $(G, e_H)_\sigma$ is a normal subgroup of $G \rtimes_\sigma H$. If G is connected, $(G, e_H)_\sigma$ is the identity component of $G \rtimes_\sigma H$.

THEOREM 4.2. *Let G be a connected Lie group and let $\mathbb{Z}_2 = \{\epsilon_0, \epsilon_1\}$ be the cyclic group of order 2, where ϵ_0 is the identity element. Let $\sigma, \sigma' : \mathbb{Z}_2 \to \mathrm{Aut}(G)$ be homomorphisms. We set $\sigma_1 = \sigma(\epsilon_1)$ and $\sigma'_1 = \sigma'(\epsilon_1)$. Then the following two conditions are equivalent.*

(1) *$G \rtimes_\sigma \mathbb{Z}_2$ and $G \rtimes_{\sigma'} \mathbb{Z}_2$ are isomorphic as Lie groups.*
(2) *There exist an automorphism φ of G and an element x in G such that*

$$\sigma'_1(x) = x^{-1}, \quad \varphi \circ \sigma_1 \circ \varphi^{-1} = I_x \circ \sigma'_1.$$

PROOF. Let e denote the identity element of G. We show (1) implies (2). Let $\Phi : G \rtimes_\sigma \mathbb{Z}_2 \to G \rtimes_{\sigma'} \mathbb{Z}_2$ be a Lie group isomorphism. Since Φ maps the identity component of $G \rtimes_\sigma \mathbb{Z}_2$ to that of $G \rtimes_{\sigma'} \mathbb{Z}_2$, we have $\Phi((G, \epsilon_0)_\sigma) = (G, \epsilon_0)_{\sigma'}$. Hence we get $\Phi((G, \epsilon_1)_\sigma) = (G, \epsilon_1)_{\sigma'}$. Therefore there is an element $x \in G$ such that $\Phi((e, \epsilon_1)_\sigma) = (x, \epsilon_1)_{\sigma'}$. Since on the one hand we have $\Phi((e, \epsilon_1)_\sigma (e, \epsilon_1)_\sigma) = \Phi((e, \epsilon_0)_\sigma) = (e, \epsilon_0)_{\sigma'}$, and the other hand we have

$$\Phi((e, \epsilon_1)_\sigma (e, \epsilon_1)_\sigma) = \Phi((e, \epsilon_1)_\sigma) \Phi((e, \epsilon_1)_\sigma) = (x, \epsilon_1)_{\sigma'} (x, \epsilon_1)_{\sigma'} = (x \sigma'_1(x), \epsilon_0)_{\sigma'},$$

we get $x \sigma'_1(x) = e$, that is, $\sigma'_1(x) = x^{-1}$.

The restriction of Φ to $(G, \epsilon_0)_\sigma$ induces an automorphism of G, which we denote by $\varphi : G \to G$. For $g \in G$, on the one hand we have

$$\begin{aligned} \Phi((e, \epsilon_1)_\sigma) \Phi((g, \epsilon_0)_\sigma) &= (x, \epsilon_1)_{\sigma'} (\varphi(g), \epsilon_0)_{\sigma'} = (x \sigma'(\epsilon_1)(\varphi(g)), \epsilon_1 \epsilon_0)_{\sigma'} \\ &= (x \sigma'_1(\varphi(g)), \epsilon_1)_{\sigma'}, \end{aligned}$$

and on the other hand we have

$$\begin{aligned} \Phi((e, \epsilon_1)_\sigma) \Phi((g, \epsilon_0)_\sigma) &= \Phi((e, \epsilon_1)_\sigma (g, \epsilon_0)_\sigma) = \Phi((\sigma_1(g), \epsilon_1)_\sigma) \\ &= \Phi((\sigma_1(g), \epsilon_0)_\sigma (e, \epsilon_1)_\sigma) = \Phi((\sigma_1(g), \epsilon_0)_\sigma) \Phi((e, \epsilon_1)_\sigma) \\ &= (\varphi(\sigma_1(g)), \epsilon_0)_{\sigma'} (x, \epsilon_1)_{\sigma'} = (\varphi(\sigma_1(g)) x, \epsilon_1)_{\sigma'}. \end{aligned}$$

Thus we get $x \sigma'_1(\varphi(g)) = \varphi(\sigma_1(g)) x$ for any $g \in G$, which implies $\varphi(\sigma_1(g)) = I_x(\sigma'_1(\varphi(g)))$. Therefore we obtain $\varphi \circ \sigma_1 \circ \varphi^{-1} = I_x \circ \sigma'_1$.

We show (2) implies (1). Using $\varphi \in \mathrm{Aut}(G)$ and $x \in G$ in (2), we define a map $\Phi : G \rtimes_\sigma \mathbb{Z}_2 \to G \rtimes_{\sigma'} \mathbb{Z}_2$ by $\Phi((g, \epsilon_0)_\sigma) = (\varphi(g), \epsilon_0)_{\sigma'}$ and $\Phi((g, \epsilon_1)_\sigma) = (\varphi(g) x, \epsilon_1)_{\sigma'}$

for $g \in G$. Then Φ is a bijection. By direct calculation we can show Φ is a homomorphism. Thus Φ is an isomorphism and $G \rtimes_\sigma \mathbb{Z}_2$ is isomrphic to $G \rtimes_{\sigma'} \mathbb{Z}_2$. $\square$

By Theorem 4.2 we obtain the following.

COROLLARY 4.3. *Let G be a connected Lie group and let $\sigma : \mathbb{Z}_2 \to \mathrm{Aut}(G)$ be a homomorphism, where $\mathbb{Z}_2 = \{\epsilon_0, \epsilon_1\}$ is the cyclic group of order 2 with the identity element ϵ_0. We set $\sigma_1 = \sigma(\epsilon_1)$. Then the following two conditions are equivalent.*

(1) *$G \rtimes_\sigma \mathbb{Z}_2$ is isomorphic to $G \times \mathbb{Z}_2$ as a Lie group.*
(2) *There exists an element x in G such that $x = x^{-1}$ and $\sigma_1 = I_x$.*

PROOF. We define a homomorphism $\tau : \mathbb{Z}_2 \to \mathrm{Aut}(G)$ by $\tau(\epsilon_0) = \tau(\epsilon_1) = 1_G$, the identity map of G. Then $G \rtimes_\tau \mathbb{Z}_2 = G \times \mathbb{Z}_2$ holds. Applying Theorem 4.2 to σ and τ, we obtain that (1) holds if and only if there exists $\varphi \in \mathrm{Aut}(G)$ and $y \in G$ such that $\sigma_1(y) = y^{-1}$ and $I_y \circ \sigma_1 = \varphi \circ 1_G \circ \varphi^{-1} = 1_G$. It turns out that this condition is equivalent to (2). $\square$

We have the following.

THEOREM 4.4. *Let G be a connected Lie group and let τ be a nontrivial involutive automorphism of G. We set $\hat{G} = G \rtimes \langle\tau\rangle$. We denote by e the identity element of $\langle\tau\rangle$. Then the center $Z(\hat{G})$ of $\hat{G}$ and the centralizer $Z_{\hat{G}}(G, e)$ of the identity component (G, e) of $\hat{G}$ in $\hat{G}$ are given as follows.*

$$Z(\hat{G}) = \{(z, e) \mid z \in Z(G) \cap F(\tau, G)\} \cup \{(z, \tau) \mid \tau = I_{z^{-1}}\},$$
$$Z_{\hat{G}}(G, e) = \{(z, e) \mid z \in Z(G)\} \cup \{(z, \tau) \mid \tau = I_{z^{-1}}\}.$$

PROOF. Since $\hat{G} = (G, e) \cup (G, \tau)$ is the decomposition into a disjoint union of the connected components of $\hat{G}$, we have $Z(\hat{G}) = ((G, e) \cap Z(\hat{G})) \cup ((G, \tau) \cap Z(\hat{G}))$. If we take $(z, e) \in (G, e) \cap Z(\hat{G})$, we have $(g, e)(z, e) = (z, e)(g, e)$ and $(g, \tau)(z, e) = (z, e)(g, \tau)$ for every $g \in G$. The former equation implies $z \in Z(G)$. By the latter equation we have $(e, \tau)(z, e) = (z, e)(e, \tau)$ when $g = e$, that is, $(\tau(z), \tau) = (z, \tau)$. Hence $z \in F(\tau, G)$. Therefore $z \in Z(G) \cap F(\tau, G)$. Conversely, if $z \in Z(G) \cap F(\tau, G)$, we have

$$(g, e)(z, e) = (gz, e) = (zg, e) = (z, e)(g, e),$$
$$(g, \tau)(z, e) = (g\tau(z), \tau) = (gz, \tau) = (zg, \tau) = (z, e)(g, \tau).$$

Thus $(z, e) \in Z(\hat{G})$. Therefore we get $(G, e) \cap Z(\hat{G}) = \{(z, e) \mid z \in Z(G) \cap F(\tau, G)\}$. If we take $(z, \tau) \in (G, \tau) \cap Z(\hat{G})$, we have $(g, e)(z, \tau) = (z, \tau)(g, e)$ and $(g, \tau)(z, \tau) = (z, \tau)(g, \tau)$ for every $g \in G$. The former equation implies $(gz, \tau) = (z\tau(g), \tau)$ and hence $gz = z\tau(g)$. Therefore we have $\tau(g) = I_{z^{-1}}(g)$ for every $g \in G$, which implies $\tau = I_{z^{-1}}$. The latter equation implies $(g\tau(z), e) = (z\tau(g), e)$, which holds when $\tau = I_{z^{-1}}$. Conversely, if $z \in G$ satisfies $\tau = I_{z^{-1}}$, we have

$$(g, e)(z, \tau) = (gz, \tau) = (zz^{-1}gz, \tau) = (zI_{z^{-1}}(g), \tau) = (z\tau(g), \tau) = (z, \tau)(g, e),$$
$$\begin{aligned}(g, \tau)(z, \tau) &= (g\tau(z), e) = (gI_{z^{-1}}(z), e) = (gz, e) = (zz^{-1}gz, e) = (zI_{z^{-1}}(g), e) \\ &= (z\tau(g), e) = (z, \tau)(g, \tau).\end{aligned}$$

Therefore we get $(G, \tau) \cap Z(\hat{G}) = \{(z, \tau) \mid \tau = I_{z^{-1}}\}$.

As for $Z_{\hat{G}}(G, e)$, we have $Z_{\hat{G}}(G, e) = ((G, e) \cap Z_{\hat{G}}(G, e)) \cup ((G, \tau) \cap Z_{\hat{G}}(G, e))$. We get $(G, e) \cap Z_{\hat{G}}(G, e) = (Z(G), e)$. By the argument above we obtain $(G, \tau) \cap Z_{\hat{G}}(G, e) = \{(z, \tau) \mid \tau = I_{z^{-1}}\}$. $\square$

COROLLARY 4.5. *Under the assumption of Theorem* 4.4,

(1) *if* τ *is an inner automorphism* I_{z_0} $(z_0 \in G)$, *we have*

$$Z(\hat{G}) = (Z(G) \cap F(\tau, G), e) \cup \{(z, \tau) \mid z \in Z(G)z_0^{-1}\},$$
$$Z_{\hat{G}}(G, e) = (Z(G), e) \cup \{(z, \tau) \mid z \in Z(G)z_0^{-1}\}.$$

(2) *If* τ *is an outer automorphism, we have*

$$Z(\hat{G}) = (Z(G) \cap F(\tau, G), e),$$
$$Z_{\hat{G}}(G, e) = (Z(G), e).$$

PROOF. Since (2) is an immediate consequent of Theorem 4.4, we show (1). For $z \in G$ we have $I_{z^{-1}} = \tau = I_{z_0}$ if and only if $I_{zz_0} = 1_G$. Since $I_{zz_0} = 1_G$ is equivalent to $zz_0 \in Z(G)$, we obtain (1). □

COROLLARY 4.6. *When* τ *is an outer automorphism of* G, *there exists no pole of* $\hat{G} = G \rtimes \langle\tau\rangle$ *contained in* (G, τ).

PROOF. By Corollary 2.4 the set of poles of $\hat{G}$ is $\tilde{Z}_2(\hat{G}) = Z_{\hat{G}}(G, e) \cap F(s_{\hat{e}}, \hat{G})$, where $\hat{e}$ denotes the identity element of $\hat{G}$. Since $\tilde{Z}_2(\hat{G})$ is contained in (G, e) by Corollary 4.5, we obtain the conclusion. □

THEOREM 4.7. *Let* G *be a connected compact Lie group and let* σ *be an involutive automorphism of* G. *Then we have*

$$F(s_{\hat{e}}, G \rtimes \langle\sigma\rangle) = (F(s_{e_G}, G), e) \cup (F(s_{e_G} \circ \sigma, G), \sigma),$$

where we denote by $\hat{e}$ *and* e_G *the identity elements of* $G \rtimes \langle\sigma\rangle$ *and* G *respectively. In particular, each connected component of* $(F(s_{e_G} \circ \sigma, G), \sigma)$ *is a polar of* $G \rtimes \langle\sigma\rangle$. *The connected component of* $(F(s_{e_G} \circ \sigma, G), \sigma)$ *containing* (e_G, σ) *coincides with* $(\rho_\sigma(G) \cdot e_G, \sigma)$, *where* $\rho_\sigma(G) \cdot e_G$ *is the orbit of the twisted conjugate action of* $\rho_\sigma(G)$ *through* e_G. $\rho_\sigma(G) \cdot e_G$ *is a symmetric space defined by a symmetric pair* $(G, F(\sigma, G))$, *which is realized by the imbedding*

$$G/F(\sigma, G) \to G\ ;\ gF(\sigma, G) \mapsto g\sigma(g)^{-1}.$$

PROOF. We denote $F(s_{e_G} \circ \sigma, G)$ by M. Since we have the decomposition $G \rtimes \langle\sigma\rangle = (G, e) \cup (G, \sigma)$ into a disjoint union of the connected components and $s_{\hat{e}}$ preserves each connected component, we get $F(s_{\hat{e}}, G \rtimes \langle\sigma\rangle) = F(s_{\hat{e}}, (G, e)) \cup F(s_{\hat{e}}, (G, \sigma))$. It is easy to see $F(s_{\hat{e}}, (G, e)) = (F(s_{e_G}, G), e)$. For $g \in G$, $s_{\hat{e}}(g, \sigma) = (g, \sigma)$ if and only if $\hat{e} = (g, \sigma)^2 = (g\sigma(g), e)$, that is, $\sigma(g) = g^{-1}$. Hence we obtain $F(s_{\hat{e}}, (G, \sigma)) = (M, \sigma)$. Each connected component of (M, σ) is a polar of $G \rtimes \langle\sigma\rangle$, which is a conjugate orbit of (G, e) by Proposition 2.2. For any $g \in G$ and $x \in M$, we have

$$(g, e)(x, \sigma)(g, e)^{-1} = (g, e)(x, \sigma)(g^{-1}, e) = (gx\sigma(g^{-1}), \sigma) = (\rho_\sigma(g)(x), \sigma).$$

Thus the conjugate action of (G, e) on (G, σ) induces the twisted conjugate action of G by σ on M. Therefore for any $(x, \sigma) \in (M, \sigma)$, the connected component of (M, σ) containing (x, σ) coincides with $(\rho_\sigma(G) \cdot x, \sigma)$. In particular $\rho_\sigma(G) \cdot e_G$ is a connected component of $F(s_{e_G} \circ \sigma, G)$, which is a symmetric space defined by a symmetric pair $(G, F(\sigma, G))$. This is realized by the imbedding

$$G/F(\sigma, G) \to G\ ;\ gF(\sigma, G) \mapsto g\sigma(g)^{-1}.$$

□

When σ in Theorem 4.7 is the identity map of G, we have $G \rtimes \langle\sigma\rangle = G \times \mathbb{Z}_2$ and $F(s_{\hat{e}}, G \times \mathbb{Z}_2) = F(s_{e_G}, G) \times \mathbb{Z}_2$.

5. Classical compact Lie groups

In this section we determine the polars of some compact Lie groups by using results in the previous sections. In the following we denote a diagonal matrix

$$\begin{bmatrix} \epsilon_1 & & \\ & \ddots & \\ & & \epsilon_n \end{bmatrix}$$

by $\mathrm{diag}(\epsilon_1, \dots, \epsilon_n)$ and we set $x_i = \mathrm{diag}(\underbrace{-1, \dots, -1}_{i}, \underbrace{1, \dots, 1}_{n-i})$ for $0 \le i \le n$. Moreover, we write $N \cong G/K$ when a symmetric space N is a symmetric space defined by a symmetric pair (G, K).

5.1. Orthogonal groups.

Example 5.1. $G = O(n)$ is a disconnected compact Lie group with the identity component $G_0 = SO(n)$. We simply write the identity element 1_n of $O(n)$ as e. When we set $G_1 = G_0 x_1$, we have $G = G_0 \cup G_1$, which is the decomposition into a disjoint union of the connected components of G. Since each eigenvalue of an element x of $F(s_e, O(n)) = \{x \in O(n) \mid x^2 = 1_n\}$ is ± 1, there exist $g \in SO(n)$ and i $(0 \le i \le n)$ such that $g x_i g^{-1} = x$. Hence we obtain

$$F(s_e, O(n)) = \bigcup_{i:\text{even}} G^+(x_i) \cup \bigcup_{i:\text{odd}} G^+(x_i),$$

where $G^+(x_i) = \{g x_i g^{-1} \mid g \in SO(n)\}$. We have $\bigcup_{i:\text{even}} G^+(x_i) \subset G_0$ and $\bigcup_{i:\text{odd}} G^+(x_i) \subset G_1$. The above equation gives all polars of $O(n)$ as the following: $G^+(x_0) = \{1_n\}, G^+(x_n) = \{-1_n\}$, and for $1 < i < n$, $G^+(x_i) \cong SO(n)/S(O(i) \times O(n-i))$ which is the real Grassmann manifold. We have

$$F(s_e, SO(n)) = \bigcup_{i:\text{even}} G^+(x_i),$$

which gives all polars of $SO(n)$: $G^+(x_0) = \{1_n\}, G^+(x_n) = \{-1_n\}$ when n is even, and for even i $(1 < i < n)$, $G^+(x_i) \cong SO(n)/S(O(i) \times O(n-i))$ which is the real Grassmann manifold.

Since $x_i \in G_1$ when i is odd, $O(n)$ is isomorphic to $SO(n) \rtimes \langle I_{x_i}\rangle$ as a Lie group for any odd i by Proposition 4.1. For

$$\varphi_i = \begin{bmatrix} 0 & 1_{n-i} \\ 1_i & 0 \end{bmatrix} \in O(n),$$

we have $I_{\varphi_i} \circ I_{x_{n-i}} \circ I_{\varphi_i}^{-1} = I_{\varphi_i x_{n-i} {}^t\varphi_i} = I_{-x_i} = I_{x_i}$. Hence $SO(n) \rtimes \langle I_{x_{n-i}}\rangle$ is isomorphic to $SO(n) \rtimes \langle I_{x_i}\rangle$ by Theorem 4.2. When both n and i are odd, we have $x_{n-i} \in SO(n)$ and hence $SO(n) \rtimes \langle I_{x_i}\rangle$ is isomorphic to $SO(n) \times \mathbb{Z}_2$ by Corollary 4.3. Therefore $O(n)$ is isomorphic to $SO(n) \times \mathbb{Z}_2$ when n is odd. As for the case where n is even, it turns out that $O(n)$ is not isomorphic to $SO(n) \times \mathbb{Z}_2$. It is because the center of $SO(n) \times \mathbb{Z}_2$ is:

$$Z(SO(n) \times \mathbb{Z}_2) = Z(SO(n)) \times Z(\mathbb{Z}_2) = \begin{cases} SO(2) \times \mathbb{Z}_2 & (n = 2), \\ \{\pm 1_n\} \times \mathbb{Z}_2 & (n : \text{even}, n \neq 2), \end{cases}$$

on the other hand, $Z(O(n)) = \{\pm 1_n\}$ for any n.

EXAMPLE 5.2. We set

$$J'_n = \begin{bmatrix} J'_1 & & \\ & \ddots & \\ & & J'_1 \end{bmatrix} \in SO(2n), \quad J'_1 = \begin{bmatrix} 0 & 1 \\ -1 & 0 \end{bmatrix} \in SO(2).$$

Then $I_{J'_n}$ is an involutive automorphism of $SO(2n)$. Set $\hat{G} = SO(2n) \rtimes \langle I_{J'_n} \rangle$ and we denote by $\hat{e}$ (resp. e, e') the identity element of $\hat{G}$ (resp. $SO(2n), \langle I_{J'_n} \rangle$). By Theorem 4.7 we obtain

$$F(s_{\hat{e}}, \hat{G}) = (F(s_e, SO(2n)), e') \cup (M, I_{J'_n}),$$

where $M = \{g \in SO(2n) \mid I_{J'_n}(g) = g^{-1}\}$.

We get $M = J'_n\{g \in SO(2n) \mid g^2 = -1_{2n}\}$. In order to prove this it is sufficient to show that when we write an element $g \in SO(2n)$ as $g = J'_n h$, $g \in M$ if and only if $h^2 = -1_{2n}$. We can prove this by a straight calculation. We have

$$\{g \in SO(2n) \mid g^2 = -1_{2n}\} = DIII(n) \cup I_{x_1}(DIII(n)),$$

where $DIII(n) \cong SO(2n)/U(n)$ (cf. [**TT3**] Section 6). In particular M consists of two connected components $J'_n DIII(n)$ and $J'_n I_{x_1}(DIII(n))$. Therefore a compact Riemannian symmetric space $SO(2n)/U(n)$ is realized as a polar of $SO(2n) \rtimes \langle I_{J'_n} \rangle$. It is known that $SO(2n)/U(n)$ is not realized as a polar of any connected compact Lie group when n is even with $n \geq 4$ (cf. [**CN2**] Appendix).

5.2. Unitary groups.

EXAMPLE 5.3. $U(n)$ and $SU(n)$ are connected compact Lie groups. We simply write the identity element 1_n of $U(n)$ as e. Since each eigenvalue of an element x of $F(s_e, U(n)) = \{x \in U(n) \mid x^2 = 1_n\}$ is ± 1, there exist $g \in U(n)$ and i $(0 \leq i \leq n)$ such that $gx_ig^{-1} = x$. Hence we obtain

$$F(s_e, U(n)) = \bigcup_{i=0}^{n} G^+(x_i),$$

where $G^+(x_i) = \{gx_ig^{-1} \mid g \in U(n)\}$. The above equation gives all polars of $U(n)$ as the following: $G^+(x_0) = \{1_n\}, G^+(x_n) = \{-1_n\}$, and for $1 < i < n$, $G^+(x_i) \cong U(n)/(U(i) \times U(n-i))$, which is the complex Grassmann manifold. We have

$$F(s_e, SU(n)) = \bigcup_{i:\text{even}} G^+(x_i),$$

where $G^+(x_i) = \{gx_ig^{-1} \mid g \in SU(n)\} = \{gx_ig^{-1} \mid g \in U(n)\}$. The above equation gives all polars of $SU(n)$ as the following: $G^+(x_0) = \{1_n\}$, $G^+(x_n) = \{-1_n\}$ when n is even, and for an even $i\,(1 < i < n)$, $G^+(x_i) \cong U(n)/(U(i) \times U(n-i))$, which is the complex Grassmann manifold.

We use the notation $U^{pm}(n), U^+(n), U^-(n)$ defined in Example 2.10. $U^{pm}(n) = U^+(n) \cup U^-(n)$ is the decomposition into a disjoint union of the connected components of $U^{pm}(n)$, where $U^+(n) = SU(n)$ is the identity component of $U^{pm}(n)$. If i is even, $x_i \in U^+(n)$ and if i is odd, $x_i \in U^-(n)$. Since s_e preserves $U^+(n)$ as well as $U^-(n)$, we have $F(s_e, U^{pm}(n)) = F(s_e, U^+(n)) \cup F(s_e, U^-(n))$. $F(s_e, U^+(n)) = F(s_e, SU(n))$ is obtained in the above. We have

$$F(s_e, U^-(n)) = \bigcup_{i:\text{odd}} G^+(x_i),$$

where $G^+(x_i) = \{g x_i g^{-1} \mid g \in SU(n)\} = \{g x_i g^{-1} \mid g \in U(n)\}$. The above equation gives all polars of $U(n)^{pm}$ contained in $U^-(n)$ as follows: $G^+(x_n) = \{-1_n\}$ when n is odd, and for an odd $i\,(1 < i < n)$, $G^+(x_i) \cong U(n)/(U(i) \times U(n-i))$, which is the complex Grassmann manifold.

$U^{pm}(n)$ is isomorphic to $SU(n) \rtimes \langle I_{x_i} \rangle$ as a Lie group for any odd i by Proposition 4.1. By the similar argument in Example 5.1, $SU(n) \rtimes \langle I_{x_{n-i}} \rangle$ is isomorphic to $SU(n) \rtimes \langle I_{x_i} \rangle$. When both n and i are odd, we have $x_{n-i} \in SU(n)$ and hence $U^{pm}(n)$ is isomorphic to $SU(n) \times \mathbb{Z}_2$ by Corollary 4.3.

EXAMPLE 5.4. Let σ_I be an involutive automorphism of $U(n)$ defined by $\sigma_I(g) = \bar{g}$ $(g \in U(n))$, where $\bar{g}$ is the complex conjugation of g. σ_I preserves $SU(n)$ and defines an involutive automorphism of $SU(n)$. We consider $U(n) \rtimes \langle \sigma_I \rangle$ and its subgroup $SU(n) \rtimes \langle \sigma_I \rangle$. We denote by $\hat{e}$ (resp. e, e'), the identity element of $U(n) \rtimes \langle \sigma_I \rangle$ (resp. $U(n), \langle \sigma_I \rangle$). By Theorem 4.7 we have

$$F(s_{\hat{e}}, U(n) \rtimes \langle \sigma_I \rangle) = (F(s_e, U(n)), e') \cup (M, \sigma_I),$$

where $M = \{g \in U(n) \mid \sigma_I(g) = g^{-1}\}$. Since we studied $F(s_e, U(n))$ in Example 5.3, it is enough to study (M, σ_I). In order to do this we take a maximal torus T in $F(\sigma_I, U(n)) = O(n)$ as

$$T = \left\{ \begin{bmatrix} R(\theta_1) & & & \\ & \ddots & & \\ & & R(\theta_k) & \\ & & & (1) \end{bmatrix} \middle| \theta_1, \dots, \theta_k \in \mathbb{R} \right\},$$

where

$$R(\theta) = \begin{bmatrix} \cos\theta & -\sin\theta \\ \sin\theta & \cos\theta \end{bmatrix}, \quad k = \left\lfloor \frac{n}{2} \right\rfloor,$$

and (1) in the definition of T denotes 1 when n is odd and nothing when n is even. We write $(U(n), e')$ as $U(n)$ and $(U(n), \sigma_I)$ as $U(n)\sigma_I$ in the following. By the argument in the proof of Theorem 4.7, the conjugate action of $U(n)$ on $M\sigma_I$ induces the twisted conjugate action by σ_I on M. Using Lemma 3.1 we get

$$M\sigma_I = \{\rho_{\sigma_I}(g)(t) \mid t \in T,\, t^2 = e,\, g \in U(n)\}\sigma_I = \{g t\, {}^t g \mid t \in T,\; t^2 = e,\, g \in U(n)\}\sigma_I.$$

When $t \in T$ satisfies $t^2 = e$,

$$t = \begin{bmatrix} \epsilon_1 & & \\ & \ddots & \\ & & \epsilon_n \end{bmatrix}, \quad \epsilon_i = \pm 1_2 \ (1 \le i \le n).$$

Since

$$\begin{bmatrix} \mathbf{i} & 0 \\ 0 & \mathbf{i} \end{bmatrix} \begin{bmatrix} -1 & 0 \\ 0 & -1 \end{bmatrix} \begin{bmatrix} \mathbf{i} & 0 \\ 0 & \mathbf{i} \end{bmatrix} = \begin{bmatrix} 1 & 0 \\ 0 & 1 \end{bmatrix},$$

there exists $g \in U(n)$ such that $g 1_n {}^t g = t$ for any $t \in T$ satisfying $t^2 = e$. Thus we get $M\sigma_I = \{g 1_n {}^t g \mid g \in U(n)\}\sigma_I$. Hence $M\sigma_I$ is connected and

$M\sigma_I \cong U(n)/O(n)$. Therefore a compact Riemannian symmetric space $U(n)/O(n)$ is realized as a polar of $U(n) \rtimes \langle\sigma_I\rangle$.

As for $SU(n) \rtimes \langle\sigma_I\rangle$, we have

$$F(s_{\hat{e}}, SU(n) \rtimes \langle\sigma_I\rangle) = (F(s_e, SU(n)), e') \cup (M', \sigma_I),$$

where $M' = \{g \in SU(n) \mid \sigma_I(g) = g^{-1}\}$ by Theorem 4.7. Since we studied $F(s_e, SU(n))$ in Example 5.3, it is enough to study (M', σ_I). By the argument above we get

$$\begin{aligned} M'\sigma_I &= (SU(n) \cap M)\sigma_I = (SU(n) \cap \{g 1_n {}^t g \mid g \in U(n)\})\sigma_I \\ &= \{g 1_n {}^t g \mid g \in U^{pm}(n)\}\sigma_I, \end{aligned}$$

because $\det(g 1_n {}^t g) = (\det g)^2$. If $g \in U^-(n)$, we have $g x_1 \in SU(n)$ and we get $(g x_1) 1_n {}^t(g x_1) = g 1_n {}^t g$. Thus we obtain $M'\sigma_I = \{g 1_n {}^t g \mid g \in SU(n)\}\sigma_I$. Hence (M', σ_I) is connected and $(M', \sigma_I) \cong SU(n)/SO(n)$. Therefore a compact Riemannian symmetric space $SU(n)/SO(n)$ is realized as a polar of $SU(n) \rtimes \langle\sigma_I\rangle$.

It is known that $U(n)/O(n)$ and $SU(n)/SO(n)$ are not realized as polars of any connected compact Lie group when $n \geq 3$ (cf. [**CN2**] Appendix).

EXAMPLE 5.5. Let σ_{II} be an involutive automorphism of $U(2n)$ defined by $\sigma_{II}(g) = J_n \bar{g} J_n^{-1}$ $(g \in U(2n))$, where $J_n = \begin{bmatrix} 0 & -1_n \\ 1_n & 0 \end{bmatrix} \in SO(2n)$. σ_{II} preserves $SU(2n)$ and defines an involutive automorphism of $SU(2n)$. We consider $U(2n) \rtimes \langle\sigma_{II}\rangle$ and its subgroup $SU(2n) \rtimes \langle\sigma_{II}\rangle$. We denote by $\hat{e}$ (resp. e, e') the identity element of $U(2n) \rtimes \langle\sigma_{II}\rangle$ (resp. $U(2n), \langle\sigma_{II}\rangle$).

First, we determine the polars of $U(2n) \rtimes \langle\sigma_{II}\rangle$. By Theorem 4.7 we have $F(s_{\hat{e}}, U(2n) \rtimes \langle\sigma_{II}\rangle) = (F(s_e, U(2n)), e') \cup (M, \sigma_{II})$, where $M = \{g \in U(2n) \mid \sigma_{II}(g) = g^{-1}\}$. Since we studied $F(s_e, U(2n))$ in Example 5.3, it is enough to study (M, σ_{II}). In order to do this we take a maximal torus T in $F(\sigma_{II}, U(2n))$, which is isomorphic to $Sp(n)$. Since the Lie algebra of $F(\sigma_{II}, U(2n))$ is

$$\left\{ \begin{bmatrix} X & -\bar{Y} \\ Y & \bar{X} \end{bmatrix} \middle| \begin{array}{l} X : \text{a skew-Hermitian matrix of size } n, \\ Y : \text{a complex symmetric matrix of size } n \end{array} \right\},$$

we obtain

$$\mathfrak{t} = \left\{ \begin{bmatrix} t & 0 \\ 0 & -t \end{bmatrix} \middle| t = \operatorname{diag}(\sqrt{-1}t_1, \ldots, \sqrt{-1}t_n),\ t_1, \ldots, t_n \in \mathbb{R} \right\}$$

is a maximal abelian subalgebra of the Lie algebra of $F(\sigma_{II}, U(2n))$. Then

$$T = \exp \mathfrak{t} = \left\{ \begin{bmatrix} z & 0 \\ 0 & \bar{z} \end{bmatrix} \middle| z = \operatorname{diag}(z_1, \ldots, z_n),\ z_1, \ldots, z_n \in U(1) \right\}$$

is a maximal torus of $F(\sigma_{II}, U(2n))$.

We write $(U(2n), e')$ as $U(2n)$ and $(U(2n), \sigma_{II})$ as $U(2n)\sigma_{II}$ in the following. By the argument in the proof of Theorem 4.7, the conjugate action of $U(2n)$ on $M\sigma_{II}$ induces the twisted conjugate action by σ_{II} on M. Using Lemma 3.1 we get

$$\begin{aligned} M\sigma_{II} &= \{\rho_{\sigma_{II}}(g)(t) \mid t \in T,\ t^2 = e,\ g \in U(2n)\}\sigma_{II} \\ &= \{g t J_n {}^t g J_n^{-1} \mid t \in T,\ t^2 = e,\ g \in U(2n)\}\sigma_{II}. \end{aligned}$$

For $t \in T$, we have $t^2 = e$ if and only if

$$t = \begin{bmatrix} \epsilon & 0 \\ 0 & \epsilon \end{bmatrix}, \quad \epsilon = \operatorname{diag}(\epsilon_1, \ldots, \epsilon_n), \quad \epsilon_1 = \cdots = \epsilon_n = \pm 1.$$

For $g_1, g_2 \in U(n)$,

$$\begin{bmatrix} g_1 & 0 \\ 0 & g_2 \end{bmatrix} \begin{bmatrix} \epsilon & 0 \\ 0 & \epsilon \end{bmatrix} \begin{bmatrix} 0 & -1_n \\ 1_n & 0 \end{bmatrix} \begin{bmatrix} {}^t g_1 & 0 \\ 0 & {}^t g_2 \end{bmatrix} \begin{bmatrix} 0 & 1_n \\ -1_n & 0 \end{bmatrix} = \begin{bmatrix} g_1 \epsilon\, {}^t g_2 & 0 \\ 0 & g_2 \epsilon\, {}^t g_1 \end{bmatrix}.$$

Hence if we set $g_1 = g_2 = \mathrm{diag}(\sqrt{-1}, 1, \ldots, 1) \in U(n)$, we get $g_1 \epsilon\, {}^t g_2 = g_2 \epsilon\, {}^t g_1 = \mathrm{diag}(-\epsilon_1, \epsilon_2, \ldots, \epsilon_n)$. Therefore for any $t \in T$ satisfying $t^2 = e$, there exists $g \in U(2n)$ such that $\rho_{\sigma_{II}}(g)(t) = 1_{2n}$, which concludes $M\sigma_{II}$ is connected. The isotropy subgroup is $\{g \in U(2n) \mid \rho_{\sigma_{II}}(g)(1_{2n}) = 1_{2n}\} = F(\sigma_{II}, U(2n))$, which is isomorphic to $Sp(n)$. Thus we get $M\sigma_{II} \cong U(2n)/Sp(n)$. Therefore a compact Riemannian symmetric space $U(2n)/Sp(n)$ is realized as a polar of $U(2n) \rtimes \langle \sigma_{II} \rangle$.

Next, we determine the polars of $SU(2n) \rtimes \langle \sigma_{II} \rangle$. By Theorem 4.7 we have $F(s_{\hat{e}}, SU(2n) \rtimes \langle \sigma_{II} \rangle) = (F(s_e, SU(2n)), e') \cup (M', \sigma_{II})$, where $M' = \{g \in SU(2n) \mid \sigma_{II}(g) = g^{-1}\}$. Since we studied $F(s_e, SU(2n))$ in Example 5.3, it is enough to study (M', σ_{II}). By the argument above we get

$$M'\sigma_{II} = (SU(2n) \cap \{\rho_{\sigma_{II}}(g)(t) \mid t \in T,\, t^2 = e,\, g \in U(2n)\})\sigma_{II}.$$

Since we have $\det \rho_{\sigma_{II}}(g)(e) = \det(g J_n\, {}^t g J_n^{-1}) = (\det g)^2$ for $g \in U(2n)$, we get

$$\begin{aligned} &SU(2n) \cap \{\rho_{\sigma_{II}}(g)(t) \mid t \in T,\, t^2 = e,\, g \in U(2n)\} = \{\rho_{\sigma_{II}}(g)(e) \mid g \in U^{pm}(2n)\} \\ &= \{\rho_{\sigma_{II}}(g)(e) \mid g \in U^+(2n)\} \cup \{\rho_{\sigma_{II}}(g)(e) \mid g \in U^-(2n)\}. \end{aligned}$$

Since $U^+(2n) = SU(2n)$ and $U^-(2n) = g_0 U^+(2n)$ for any $g_0 \in U^-(2n)$, $U^+(2n)$ and $U^-(2n)$ are connected. Hence $\{\rho_{\sigma_{II}}(g)(e) \mid g \in U^+(2n)\}$ and $\{\rho_{\sigma_{II}}(g)(e) \mid g \in U^-(2n)\}$ are connected. In order to show $\{\rho_{\sigma_{II}}(g)(e) \mid g \in U^+(2n)\} \cap \{\rho_{\sigma_{II}}(g)(e) \mid g \in U^-(2n)\} = \emptyset$, we assume $\rho_{\sigma_{II}}(g_1)(e) = \rho_{\sigma_{II}}(g_1)(e)$ for $g_1 \in U^+(n), g_2 \in U^-(n)$. Then we get $g_2^{-1} g_1 \in F(\sigma_{II}, U(2n))$. Thus we get $\det(g_2^{-1} g_1) = 1$. Therefore $\det g_1 = \det g_2$, which is a contradiction. Hence

$$M'\sigma_{II} = \{\rho_{\sigma_{II}}(g)(e) \mid g \in U^+(2n)\}\sigma_{II} \cup \{\rho_{\sigma_{II}}(g)(e) \mid g \in U^-(2n)\}\sigma_{II}$$

is the decomposition into a disjoint union of the connected components. Each connected component of $M'\sigma_{II}$ is a symmetric space defined by a symmetric pair $(SU(2n), Sp(n))$. Therefore a compact Riemannian symmetric space $SU(2n)/Sp(n)$ is realized as a polar of $SU(2n) \rtimes \langle \sigma_{II} \rangle$.

It is known that $U(2n)/Sp(n)$ and $SU(2n)/Sp(n)$ are not realized as a polar of any connected compact Lie group when $n \geq 2$ (cf. [**CN2**] Appendix).

5.3. Symplectic groups.

EXAMPLE 5.6. Let $Sp(n) = \{g \in M_n(\mathbb{H}) \mid g\, {}^t\bar{g} = 1_n\}$ be the group of symplectic matrices of size n, where $M_n(\mathbb{H})$ denotes the set of quaternion square matrices of size n and $\bar{g}$ denotes the conjugation of g in the meaning of the quaternions. $Sp(n)$ is a connected compact Lie group. We simply write the identity element 1_n of $Sp(n)$ as e. Since $T = \{\mathrm{diag}(e^{i\theta_1}, \ldots, e^{i\theta_n}) \mid \theta_1, \ldots, \theta_n \in \mathbb{R}\}$ is a maximal torus of $Sp(n)$, we have

$$F(s_e, Sp(n)) \cap T = \{\mathrm{diag}(\epsilon_1, \ldots, \epsilon_n) \mid \epsilon_1, \ldots, \epsilon_n = \pm 1\}.$$

By

$$\begin{bmatrix} 0 & 1 \\ 1 & 0 \end{bmatrix} \begin{bmatrix} 1 & 0 \\ 0 & -1 \end{bmatrix} \begin{bmatrix} 0 & 1 \\ 1 & 0 \end{bmatrix}^{-1} = \begin{bmatrix} -1 & 0 \\ 0 & 1 \end{bmatrix},$$

it follows that each element of $F(s_e, Sp(n)) \cap T$ is conjugate to x_i for some i $(0 \leq i \leq n)$. Therefore we obtain

$$F(s_e, Sp(n)) = \bigcup_{i=0}^{n} G^+(x_i),$$

where $G^+(x_i) = \{gx_ig^{-1} \mid g \in Sp(n)\}$. The above equation gives all polars of $Sp(n)$ as the following: $G^+(x_0) = \{1_n\}, G^+(x_n) = \{-1_n\}$, and for $1 < i < n$, $G^+(x_i) \cong Sp(n)/(Sp(i) \times Sp(n-i))$, which is the quaternion Grassmann manifold.

Let $\{1, \mathbf{i}, \mathbf{j}, \mathbf{k}\}$ be the standard basis of the quaternions. $I_{\mathbf{i}1_n}$ is an involutive automorphism of $Sp(n)$. We denote $I_{\mathbf{i}1_n}$ by $I_{\mathbf{i}}$ simply. Set $\hat{G} = Sp(n) \rtimes \langle I_{\mathbf{i}} \rangle$ and we denote by $\hat{e}$ (resp. e') the identity element of $\hat{G}$ (resp. $\langle I_{\mathbf{i}} \rangle$). By Theorem 4.7 we obtain

$$F(s_{\hat{e}}, \hat{G}) = (F(s_e, Sp(n)), e') \cup (M, I_{\mathbf{i}}),$$

where $M = \{g \in Sp(n) \mid I_{\mathbf{i}}(g) = g^{-1}\}$. We can show $M = \mathbf{i}\{g \in Sp(n) \mid g^2 = -1_n\}$ similarly to Example 5.2. Therefore we get $M = \mathbf{i}CI(n)$, where $CI(n) = \{g \in Sp(n) \mid g^2 = -1_n\} \cong Sp(n)/U(n)$ (cf. [**TT3**] Section 5). In particular M is connected. Therefore a compact Riemannian symmetric space $Sp(n)/U(n)$ is realized as a polar of $Sp(n) \rtimes \langle I_{\mathbf{i}} \rangle$. It is known that $Sp(n)/U(n)$ is not realized as a polar of any connected compact Lie group when $n \geq 2$ (cf. [**CN2**] Appendix).

References

[CN1] Bang-yen Chen and Tadashi Nagano, *Totally geodesic submanifolds of symmetric spaces. II*, Duke Math. J. **45** (1978), no. 2, 405–425. MR487902

[CN2] Bang-Yen Chen and Tadashi Nagano, *A Riemannian geometric invariant and its applications to a problem of Borel and Serre*, Trans. Amer. Math. Soc. **308** (1988), no. 1, 273–297, DOI 10.2307/2000963. MR946443

[HM] Karl H. Hofmann and Sidney A. Morris, *The structure of compact groups: A primer for the student—a handbook for the expert*, De Gruyter Studies in Mathematics, vol. 25, De Gruyter, Berlin, 2013. Third edition, revised and augmented, DOI 10.1515/9783110296792. MR3114697

[I] Osamu Ikawa, *σ-actions and symmetric triads*, Tohoku Math. J. (2) **70** (2018), no. 4, 547–565, DOI 10.2748/tmj/1546570825. MR3896137

[N] Tadashi Nagano, *The involutions of compact symmetric spaces*, Tokyo J. Math. **11** (1988), no. 1, 57–79, DOI 10.3836/tjm/1270134261. MR947946

[TT1] Makiko Sumi Tanaka and Hiroyuki Tasaki, *Maximal antipodal subgroups of the automorphism groups of compact Lie algebras*, Hermitian-Grassmannian submanifolds, Springer Proc. Math. Stat., vol. 203, Springer, Singapore, 2017, pp. 39–47, DOI 10.1007/978-981-10-5556-0_4. MR3710831

[TT2] Makiko Sumi Tanaka and Hiroyuki Tasaki, *Maximal antipodal subgroups of some compact classical Lie groups*, J. Lie Theory **27** (2017), no. 3, 801–829. MR3611097

[TT3] Makiko Sumi Tanaka and Hiroyuki Tasaki, *Maximal antipodal sets of compact classical symmetric spaces and their cardinalities I*, Differential Geom. Appl. **73** (2020), 101682, 32, DOI 10.1016/j.difgeo.2020.101682. MR4149631

Department of Mathematics, Faculty of Science and Technology, Tokyo University of Science, Noda, Chiba, 278-8510 Japan

Email address: tanaka_makiko@ma.noda.tus.ac.jp

Department of Mathematics, Faculty of Pure and Applied Sciences, University of Tsukuba, Tsukuba, Ibaraki, 305-8571 Japan

Email address: tasaki@math.tsukuba.ac.jp

SELECTED PUBLISHED TITLES IN THIS SERIES

777 **Bang-Yen Chen, Nicholas D. Brubaker, Takashi Sakai, Bogdan D. Suceavă, Makiko Sumi Tanaka, Hiroshi Tamaru, and Mihaela B. Vajiac, Editors,** Differential Geometry and Global Analysis, 2022

776 **Aaron Wootton, S. Allen Broughton, and Jennifer Paulhus, Editors,** Automorphisms of Riemann Surfaces, Subgroups of Mapping Class Groups and Related Topics, 2022

775 **Fernando Galaz-García, Cecilia González-Tokman, and Juan Carlos Pardo Millán, Editors,** Mexican Mathematicians in the World, 2021

774 **Randall J. Swift, Alan Krinik, Jennifer M. Switkes, and Jason H. Park, Editors,** Stochastic Processes and Functional Analysis, 2021

773 **Nicholas R. Baeth, Thiago H. Freitas, Graham J. Leuschke, and Victor H. Jorge Pérez, Editors,** Commutative Algebra, 2021

772 **Anatoly M. Vershik, Victor M. Buchstaber, and Andrey V. Malyutin, Editors,** Topology, Geometry, and Dynamics, 2021

771 **Nicolás Andruskiewitsch, Gongxiang Liu, Susan Montgomery, and Yinhuo Zhang, Editors,** Hopf Algebras, Tensor Categories and Related Topics, 2021

770 **Stéphane Ballet, Gaetan Bisson, and Irene Bouw, Editors,** Arithmetic, Geometry, Cryptography and Coding Theory, 2021

769 **Kiyoshi Igusa, Alex Martsinkovsky, and Gordana Todorov, Editors,** Representations of Algebras, Geometry and Physics, 2021

768 **Dražen Adamović, Andrej Dujella, Antun Milas, and Pavle Pandžić, Editors,** Lie Groups, Number Theory, and Vertex Algebras, 2021

767 **Moshe Jarden and Tony Shaska, Editors,** Abelian Varieties and Number Theory, 2021

766 **Paola Comparin, Eduardo Esteves, Herbert Lange, Sebastián Reyes-Carocca, and Rubí E. Rodríguez, Editors,** Geometry at the Frontier, 2021

765 **Michael Aschbacher,** Quaternion Fusion Packets, 2021

764 **Gabriel Cunningham, Mark Mixer, and Egon Schulte, Editors,** Polytopes and Discrete Geometry, 2021

763 **Tyler J. Jarvis and Nathan Priddis, Editors,** Singularities, Mirror Symmetry, and the Gauged Linear Sigma Model, 2021

762 **Atsushi Ichino and Kartik Prasanna,** Periods of Quaternionic Shimura Varieties. I., 2021

761 **Ibrahim Assem, Christof Geiß, and Sonia Trepode, Editors,** Advances in Representation Theory of Algebras, 2021

760 **Olivier Collin, Stefan Friedl, Cameron Gordon, Stephan Tillmann, and Liam Watson, Editors,** Characters in Low-Dimensional Topology, 2020

759 **Omayra Ortega, Emille Davie Lawrence, and Edray Herber Goins, Editors,** The Golden Anniversary Celebration of the National Association of Mathematicians, 2020

758 **Jan Šťovíček and Jan Trlifaj, Editors,** Representation Theory and Beyond, 2020

757 **Kaïs Ammari and Stéphane Gerbi, Editors,** Identification and Control: Some New Challenges, 2020

756 **Joeri Van der Veken, Alfonso Carriazo, Ivko Dimitrić, Yun Myung Oh, Bogdan D. Suceavă, and Luc Vrancken, Editors,** Geometry of Submanifolds, 2020

755 **Marion Scheepers and Ondřej Zindulka, Editors,** Centenary of the Borel Conjecture, 2020

754 **Susanne C. Brenner, Igor Shparlinski, Chi-Wang Shu, and Daniel B. Szyld, Editors,** 75 Years of Mathematics of Computation, 2020